MIL SPEC RADIO GEAR

KOREAN WAR TO PRESENT DAY

By Mark Francis, KIØPF

CQ Communications, Inc.

Library of Congress Control Number: 2005930867
ISBN 0-943016-33-9

Layout and design: Elizabeth Ryan
Cover Photo: Mark Francis, KIØPF

Published by CQ Communications, Inc.
25 Newbridge Road
Hicksville, NY 11801 USA

Printed in the United States of America.

To Mom and Jack,

All my friends in the Seven-Meg Gang

and

Everyone who loves the smell of
moisture and fungus proofing

CONTENTS

Forward—A Personal Note

I had my first encounter with military radio surplus in 1976 around the 4th of July. I was fresh out of college with a degree in Computer Science and didn't know a megahertz from a microfarad. I was a software geek fer cryin' out loud.

But all that was about to change.

I met John K9WT at his place during some sort of a holiday get-together. My big contribution to the festivities was a lighter-fluid-powered tin can cannon, which I demonstrated several times by launching tennis balls 200 feet in the air before the bottom of the first can blew out.

John, an EE, gathered I was interested in gadgets and suggested I take a look at his radio setup. What I found in his room was completely unlike anything I'd ever seen. He had what I later learned was an operational RT-68 VHF transceiver and a PP-112 power supply set up on a big desk. Numerous other pieces of gear were strewn about, including one massive green box with what looked like 3000 knobs sticking out of it. In fact, *everything* seemed to be the same shade of dark green, and *everything* seemed to have 3000 knobs.

John explained the big green box on the floor was an HF transmitter and offered to show me what it looked like inside. The allen screws were already loose because he'd been working on it, and it took both of us to drag it out of the case.

I was flabbergasted.

This thing was five or six cubic feet of solid gears, cams, doodads and doohickeys. (Presumably there had to be some electronics in there *somewhere*). The few previous encounters I'd had with amateur radio stuff left me with the impression that ham gear was generally small, gray, and civilized and sat in carefully carpentered wooden hutches in closets or on desks in school radio clubs. So what in tarnation was *this* beast?

If you guessed a T-195 AM transmitter, you get two 4X150D attaboys and an autotune pat-on-the-back! A Collins design, the T-195 was probably the most outrageous electromechanical contrivance since the ASR-33 teletype. (And the ASR-33 didn't produce RF….at least not on purpose).

I was fascinated. And over a quarter century later, I'm *still* fascinated. There's something really cool about the smell of moisture and fungus proofing inside an RT-70 that's been sealed up for forty years, or the satisfaction of getting fifty vacuum tubes worth of Technical Materiel Corporation transmitter up and running, or demonstrating the water-tightness of a PRC-74 by heaving it into a lake. (No kidding).

Collecting coins or stamps or Hummel plates is fine. But once you acquire them, they go in an album or a cabinet, and just sit there. I want my stuff to do *something*. Once you get them going, surplus radios fill that bill. Sometimes "doing something" is going up in a puff of smoke and the smell of ozone, but even that has its own special charm.

Surplus "collecting" isn't really the right term. It's really surplus "using." The hobby demands active participation before, during, and after the actual acquisition of the equipment. Last I looked, you could buy an MS-63 Saint-Gaudens $20 gold piece for around five hundred bucks, but once you have it in the folder, you're done. It just sits there. But you can buy a nice R-390A receiver for about the same amount (or one that needs some work for a lot less) and listen to stations all over the world. Or integrate it into a transceiver setup. You can't use a coin or a stamp or a Hummel plate. But I've been using my surplus collection for the last twenty-five years!

Collecting rare (but static) objects like stamps and coins demands research, patience, and persistence. Collecting surplus adds three extra dimensions—*restoring, use, and maintenance.*

Restoring: Equipment often doesn't work right when acquired, or works on major functions but requires alignment. Either situation requires further effort in the form of knowledge-gathering and action by the collector.

Use: This is the period of trouble-free operation between failures.

Maintenance: Like any other piece of electromechanical gear, once the equipment is working, it won't work forever. Guaranteed. Sooner or later, you *will* have to troubleshoot and repair your gear because there is no equivalent to the Yaesu repair service in L.A. where you can send your busted RT-68!

Surplus offers these extra dimensions, BUT you must expect to put out the additional effort to make that operating period as long as possible. Restoring and maintenance are integral parts of the surplus hobby. If you aren't prepared to spend a significant amount of time in those areas, (or otherwise a huge amount of money) the hobby may not be the best for you.

If you like gadgets, appreciate beautifully built equipment, enjoy learning new things all the time, and aren't afraid to get right in there with a soldering iron, read on! Whether you're just starting out and have yet to acquire your first military radio, or you have a whole basement full of rack 'n' stacked "heavy metal," my goal is to enhance your participation in this truly strange and wonderful hobby.

Mark Francis, KI0PF

Introduction

The goal of this book is to offer the relative newcomer (who may've gotten hooked the same way I did: with one look at a T-195) with some basic tools and useful information. Many of the tips herein aren't in any technical manuals (TMs). They've been gathered by my friends and me during more than 25 years of messing with surplus. The emphasis will be on the practical—how to figure out what your interests are, how to choose equipment that meets those interests, and how to recognize that equipment when you run across it. Once you've purchased a radio or ancillary equipment covered here, you'll find specs, operation info, and other helpful tips. An additional goal is to provide useful details and data to experienced collectors.

To use this book, you can go directly to a section in the Data Sheets area covering the particular equipment in which you're interested. Or you can check the table of contents for the part of surplus collecting that interests you.

For the most part, suggested prices are not given because there are usually big differences depending on where you find the equipment. On-line auction prices are usually quite a bit higher than swap meet prices. Just about any time the author has seen suggested price lists on-line, they've made about as much sense as an elevator in an outhouse. The right price is the price you want to pay at the time.

So let's get started!

Why Mess with Military Radios?

The author gets this question all the time from ham friends who often have no surplus experience. The query really comes in two parts, with the second half unsaid. "Why mess with military radios...*when you have all that great commercial stuff to choose from*?"

It's true that modern commercial gear offers virtually every bell and whistle anyone could possibly want. Yaesus, Icoms, Kenwoods—(the author's surplus group calls them Yay-com-woods)—can be pulled out of the box and put on the air in just a few minutes. Why put up with the hassles of adapting military sets? Hmmm, where to start...

Quality. The United States has the most effective, best-equipped fighting force on the planet—across virtually every category, including communications. The comm gear was and is amazingly well built with mil-spec devices and is often highly ruggedized.

As a comparison, the author thinks Commander linear amplifiers are among the best-built commercial units available. He's studied them up close at the Dayton Hamvention on several occasions and has always gone away duly impressed (but still in possession of his savings). The construction and design of his Technical Materiel Corporation PAL-350 linear is an order of magnitude better still, even though it was built in 1962.

The physical construction of military mobile and packset gear borders on the amazing. Gasketed covers and O-rings on control shafts provide waterproof seals. The author really did pitch his PRC-74 off a dock into a lake as a demo. It stayed under five feet of water for about thirty seconds until he dove in to retrieve it. No moisture was evident inside the case. Try this stunt with an FT-817 and you can kiss a $600 radio g'bye.

Obviously, this sort of thing is going a little overboard (literally). But there is a point to be made here. No one is going to take their FT-817 out into inclement weather for fear of getting it wet. The author hauls his mil gear out in the muck all the time and has never had a problem. Well, maybe the one time on a particularly cold and humid night when he had to chip rime ice off his whip antenna so the autotuner would function!

Cost. Using the Commander HF amplifiers as an example again, an HF-2500 1500 watt unit will cost you about $1.60 a watt. It's beautifully built, all right, but the author spent about $0.50 per watt for his TMC PAL-350 linear (two 4CX250Bs, about 400W PEP). And the amp has low level stages that require only 100mw of drive for rated output, thus allowing compatibility with numerous mil surplus 100mw exciters.

It's true that the cost of surplus has gone up over the last few years due to demand, but it's still the best buy around if you're willing to be patient and search for the right deal. In fact, much of the older equipment—Korean War into the 60s—is still a great bargain for those who are just starting out and want to work with quality equipment. You just have to be willing to put up with "hollow state" devices, the surplus hound's genial term for vacuum tubes.

Often, the latest surplus available is still in use by the U.S. military and can get quite expensive. But you can still generally find it at a fraction of the original cost. One example is the PRC-138, a very fancy 1.6 - 60 MHz HF packset with three built-in modems, ALE, a wide-range automatic antenna tuner, and a mess of other features. The original cost of these things is probably what you'd pay for a good used car, yet the author knows of sets in private hands.

Whether you're interested in older equipment or the newest state-of-the-art goodies, you have to do your research, be patient, and wait for your deal. This book will offer help on where and how to search.

Versatility. In some categories, especially HF manpacks, mil surplus offers distinct advantages. In 2000, Bonnie Crystal (KQ6XA) started a Yahoo group called <www.HFpack.com>, dedicated to furthering the use of manpack HF communication. She coined the term *pedestrian mobile*, which means one carries the entire station—transceiver, power source, antenna, and accessories—on his or her person. For more, see the "HFpack Phenomenon" section.

While the Yaesu FT-817 is often used pedestrian mobile, it's a five watt radio, and HFpackers found out quickly that a little extra power definitely helps. Also, the radio has no internal antenna tuner, so one is limited to resonant antennas. The solution is to build a small outboard amplifier to boost the power up into the 20-30 watt region. Of course this necessitates switching to a larger capacity battery. Outboard autotuners solve the non-resonant antenna problem when one wants to tune into a whip, etc. This gets rapidly awkward. All this stuff—outboard amplifier, battery, and tuner—has to get lashed up somehow and carried around. And the result is very likely not weatherproof. The solution, as we will find later, may well be a military packset.

One often hears military radios are somehow not suitable for ham use. This conception may have started in the post-WWII days due to experiences with gear like the ART-13, which is a known TV jammer until some simple shielding measures are taken. Another example was the BC-223/AX late-thirties vintage AM transmitter which, if the supply voltage to the local oscillator isn't carefully regulated, produces almost as much FM as AM! The author believes that up until the late 1950s, ham gear was probably better than most military gear in the TVI and spurious emission categories. That may have been simply because ham gear manufacturers and amateurs themselves had to take two important considerations into account: Amateurs lived in increasingly close proximity to one another, so standards concerning spurs and out of band emissions became tighter. Then television came along, and spur and harmonic levels suddenly took on a whole new importance.

An ART-13 flying at 25,000 feet in a bomber over WWII Germany wasn't going to cause any problems, TVI or spur-wise. But postwar military comm equipment started to run into the same challenges as ham gear—an explosion of electronic devices of all types in close proximity. The author believes that from the 1960s on, military gear was on par with ham gear (and often better) in the unwanted emissions category. Eventually, the military had numerous high and low power radios, fire control systems, battlefield-deployed computers, and all sorts of other RF-susceptible systems. They *had* to design their comm gear better.

There are certainly instances in which the older gear is fairly clean but, by design, produces a wanted emission unsuitable for general ham use. An example is most of the 50s-vintage GRC gear like the RT-68. It produces wideband FM at 8-10 kHz deviation typically but can balloon out to as much as 20 kHz if you really yell into the mike. Obviously, this would drive an amateur repeater bananas! A simple solution is outlined in the RT-70 section.

Tradition and History. The author has a PRC-74B with the following written on the bottom of the case:

75th Rangers An Khe

The author did some research and discovered the 75th Rangers were, in fact, based at An Khe and did LRRPs (long range patrols) in the 1969 time frame (as well as other times). 75th Rangers was active in-country from 1967 to 1971. During that time, the PRC-74B was standard-A for exactly the type of ops pursued by special forces. The author's personal favorite 74B was definitely a Vietnam bring-back, very well used, and looks it. His example was almost certainly used in special ops with, say, a GRA-71 burst coder to send back intel via HF because the recon patrol was too far out for VHF. It's been completely restored to as-new functionality, but all the dings and scratches have been left alone in deference to its heritage.

Tradition and history are part of the fun of military surplus. While some history will be offered in this book as part of a particular equipment's description, detailed historical data has already been handled excellently by Tasker and others. See the Appendices for further references.

Learning Opportunity. The author still has his original copy of TM11-290, "Receiver-Transmitter RT-70/GRC," that came with his first surplus set. It is a very significant manual, because he learned the basics of FM communication from it. Reading military technical manuals ("TMs") from the '50s and '60s is like taking a course in RF communications electronics! These manuals are all generally set up the same way:

Introduction: A basic description of the equipment, its purpose, what other equipment it is used with, and technical specifications.

Description: A basic description at the organizational level. More detail, possibly with the radio out of the case showing major assemblies.

Operation: Sometimes the manual will include operation with theory and repair; sometimes there is a separate operator's manual.

Detailed Theory: Theory of operation is given at a high level, then module by module, or stage by stage with extensive use of block diagrams, embedded schematics, and illustrations.

Troubleshooting: Detailed suggestions for troubleshooting based on symptoms. There are often extensive voltage/resistance charts.

Repairs: Direction regarding disassembly to replace modules or components. It's always well to read this section carefully because some gear can be damaged by disassembling things out of order!

Alignment: Gives the test equipment necessary to align the equipment, then gives a detailed alignment procedure. This procedure is either followed to return the equipment to specs after repair or, as the author has often done, used as a method of troubleshooting defective equipment.

Final Tests or "Depot Testing": Tests performed to determine the equipment is performing to specs.

Table of Parts: TMs often have exhaustive lists of

parts used in the equipment, which helps in tracking them down or identifying substitutes.

Schematics: Detailed schematic diagrams.

This section order is generally the same across all TMs; eventually they become as familiar as old friends.

Satisfaction. This is perhaps the most difficult aspect of surplus collecting to pin down. Maybe it's because for surplus enthusiasts, there are so many possible rewarding aspects. For the author, one of those has been building systems. More basically, racking and stacking or haywiring.

When you buy a Yay-Com-Wood, you get a very sophisticated piece of gear with a tremendous amount of functionality. It's also basically a black box with a line cord hanging out the back and a few audio connections to figure out. Operation is accomplished with the controls provided and, while there are often a lot of them and a lot of functionality to figure out, that functionality is finite. There comes a time when one has found the correct audio, ALC, DSP, bandwidth, and other settings, and the amateur experience is reduced to energizing the black box and talking. It's little wonder that hams sometimes get tired of their gear. Or quit altogether out of boredom. It seems with surplus, there's always another lash-up to try out.

These are the reasons the author messes with military communications gear!

The Surplus Enthusiast Strategy

Everyone has their own method for finding surplus. The tips given here have been accumulated by the author over a 25+ year period. There may be something you can adopt into your own methods.

FIGURE OUT what part of the hobby interests you.

This may seem obvious, but there have been too many examples of people buying haphazardly and ending up disappointed because they didn't concentrate their efforts on equipment they could really use. In today's environment of high surplus prices and attenuated supply, it's even more important to know what you want. Otherwise you may find yourself with a deflated wallet, and little to show for it!

What interests you? Base station high-power HF ops? HF packset ops? VHF FM ops, either handheld or packset? Start by deciding what mode and power level you want to use.

LEARN which equipment does what you want.

Scour the various sources—the Internet, hamfests, your surplus friends, and this book—for the various sets that might fill the bill. The Fair Radio Sales catalog is a great source for experienced surplus enthusiasts as well as beginners. The author's very first surplus set was an RT-70 bought from Fair. The goal was to communicate on 51MHz over a distance of about 11 miles as the crow flies. With a haywired quarter-wave vertical made out of an SO-239 bulkhead mount connector suspended from a broomstick stuck in the chimney (no kidding!) the little radio performed the task admirably.

The author still has this RT-70 and homebuilt power supply. As far as he has been able to determine, the Champion sticker does not enhance the radio's performance. It was exactly the right radio for the job, and the technical manual provided a great short course in FM transceiver theory. See the RT-70 section for more.

FIND equipment that will meet your wants.

The "Surplus is Where You Find It" section gives several places to look. The watchword here is persistence. Your efforts will be rewarded, but only if you stick with it, because there are a mess of other people out there all going after a limited pool of equipment.

Extensive research on your part will help ensure that what you finally find will be what you really want (and can use). The PRC-47 is an interesting radio, and useful for SSB work, but it's an utter failure at listening to shortwave broadcast stations because of its

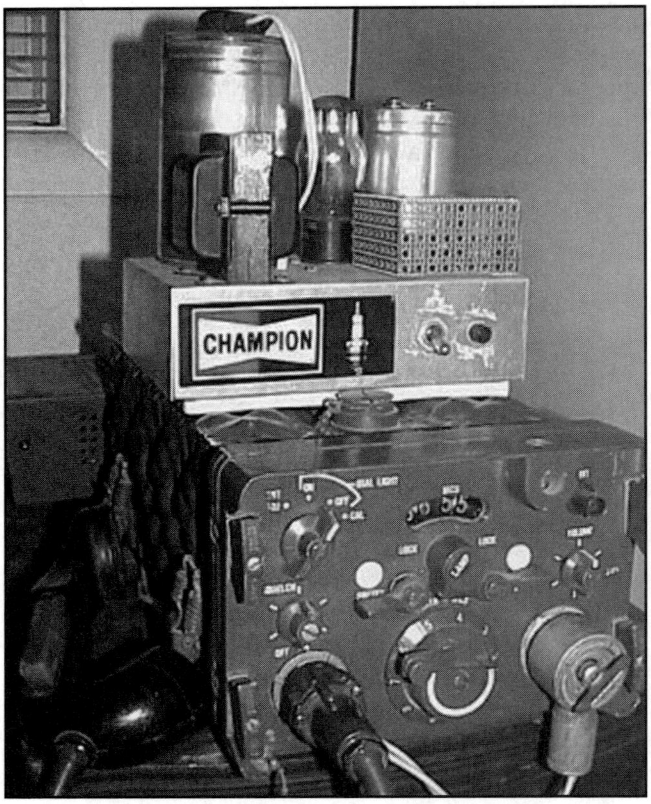

Surplus radio numero uno for KIOPF

audio-derived AGC. It also has a notoriously fragile DC-DC converter in the power supply. You'll find these sorts of things in the Equipment Data section of this book. Your research will really pay off when it comes to recognizing useful surplus. If there's just one thing the author wishes he'd done before he started acquiring equipment, it was find out more about what was out there!

Networking can be a better source of equipment than actually attending hamfests or searching the Internet. If you have friends who are into surplus, try them first.

POWER IT UP and use it correctly.

This book will be a great help in this area. The Equipment Data section includes complete hookup information and basic operating instructions for the covered equipment directly from personal experience. The author knows of at least one instance in which the happy purchaser of a PRC-104 took his unit home and promptly hooked power up backwards, destroying the power supply.

Information truly is King in the surplus hobby. Gather it any time you can!

In the author's opinion—and it ain't humble in this case—one place you don't need to look for information on military surplus gear is the ARRL. The author's group likes to call the ARRL the *Asian Radio Retailing League*. Their advertisers are commercial ham equipment manufacturers, the vast majority of which are out of the Pacific Rim. The ARRL's vested interests are in the Pacific Rim, not with mil surplus.

The Fair Radio Sales catalog is a great snapshot of what's available from one year to the next, as well as an idea on retail prices. Getting on their mailing list is a must—see the Surplus Dealers section in the Appendix.

There' s a common misconception that's floated around the surplus hobby for...well...forever. It's what the author likes to call the "must-have-in-pristine-condition" fallacy. Military comm gear, in keeping with its purpose, has been built to a higher standard of ruggedness than commercial gear, often *far* higher. Radios like PRC-10s, RT-524s, R-392s, RT-68s, and all the other mobile or packset examples are made to be splashed, bounced, cooked by the sun, and generally abused at levels that would turn typical ham gear into mush. A military set that's received a moderate amount of use in the field will look a little rough around the edges. The author's group has found repeatedly that sets showing some use are less likely to have serious problems, and more likely to operate straight off than the ones that appear new-out-of-the-box.

The PRC-74B the author has modified for LSB, 17m and 160m ops, and TX clarify is the grungiest one he owns—the dang thing looks like it's been run over by a truck and probably got dragged through every rice paddy in Southeast Asia. But it works great and was the closest to fully operational initially of all the 74s he's worked with.

When it comes to mil gear, it's not really what it looks like on the outside that counts; it's what it looks like behind those gaskets on the inside. Unless there's obvious evidence of serious abuse like handles broken off, holes punched in the outer case (demilling), broken connectors and things like that, there's no reason why you shouldn't consider a beat-up looking mil set. Plus, you can often get the unit at a discount due to its condition.

The concept here: *Grunge is Good!*

Working a Swap Meet

You've done your research, you know what you want (well, sort of...a true surplus fanatic never knows exactly what he wants, but you gotta start *somewhere*), you've got a wad of hard-earned cash and you're ready to hit the hamfests. Everyone develops their favorite hamfest strategy. This one has worked out well for the author in the past.

If there's a "setup day" or "setup period" wangle a way to get in for it. Beg, borrow, or (hopefully not) steal a way in, because this is when many of the real deals are made. The Dayton Hamvention bigwigs promulgate the rule "no selling on setup day." This is total horsepuckey and no one pays attention to it anyway. It's worth buying a dealer pass just to get into the Hara Arena flea market on Thursday. Period.

Another way: help a seller friend haul his stuff in, set up, and repack at the end of the 'fest. This will necessarily get you in during setup so you can peruse everyones' goodies as soon as they come in.

Once people are set up, do a fast pass through the entire 'fest, scanning for green- and gray-ness. The idea is not to let anything float away to the barbarian masses until you've gotten a chance to take a look at it. A lot of the most desirable reasonably-priced equipment will go very rapidly. Nowadays there are electronic auction sellers combing even the smaller hamfests in search of gear they can peddle, so you have to be on your toes and *ready to buy*.

KI0PF's Prime Directive of Surplus: When you see an item you really want, it's in your desired price range, and you have the funds...don't hesitate. In this market, it'll also be in someone else's price range and the next time you come looking, it'll be *gone*.

If you're at the 'fest with a friend, stay in radio contact. Preferably on an off-the-beaten-track VHF/UHF simplex freek. Two sets of eyes are better at spotting the goodies.

After the first couple of quick passes, slow down and pay more attention to what's on the tables. Sellers with any surplus at all are worth talking to. You never know what they might have stashed in their basement. The author has gotten quite a few "hits" this way. You'll also meet people while you're helping your friend sell at his own table.

Often you'll see an item you really want, with a price that's just too high even after some preliminary dickering. Decide on your maximum price as quickly as possible and have it ready. Don't show much interest after the preliminary query, but stake out the seller. See what kind of interest he's getting on the item. That will tell you whether you can forget about him until later, or you have to dicker immediately. A bolt-from-the-blue buyer can always appear and wreck your strategy.

About three-quarters of the way through the 'fest, which should just about be when you've seen everything several times and it's almost, but not quite, time for sellers to start packing up, make another slow pass through and look under the tables in the grungy card-

Survival Kit

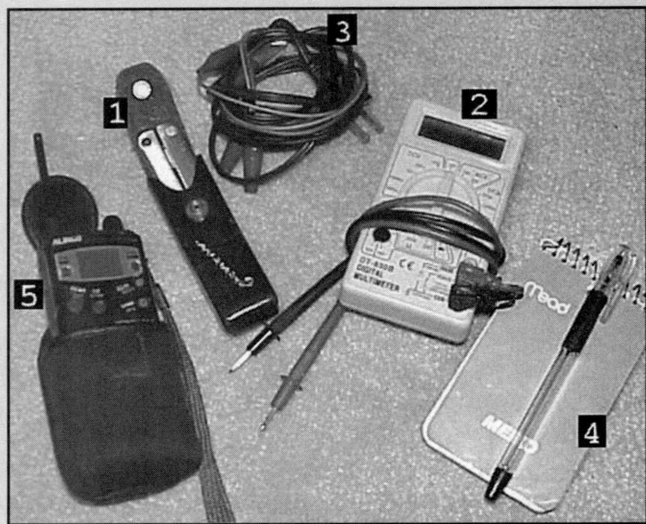

1) A Leatherman© tool or equivalent. You absolutely have to be able to remove covers, loosen screws and/or nuts, etc. to get inside equipment for physical inspections.

2) Cheap multimeter for simple resistance and voltage measurements. You can find digital multimeters for $10. At long 'fests like Dayton, you may have the ability to try the equipment overnight. Even rudimentary test gear can go a long way.

3) A suicide cord with removable clip leads for further haywiring. Simple line cords can be surprisingly tough to find at swap meets. The author, being an incurable cynic, thinks sellers hardly ever show up with line cords for obvious reasons. If there is 115VAC available anywhere, you may be able to test your find. If the seller won't let you test it, that's another factor to consider, hmmm?

4) A small notebook and pen. To remember where you saw that goodie, and record the various personal contacts you make. Contacts have become one of the most important aspects of hamfests. This is an absolute necessity at Dayton because of the sheer enormity of the flea market and the timespan of the event -- up to three and a half days, counting Thursday.

5) Radio for short-range comm, if you have a compadre with you. The author's Alinco DJ-S11T is shown. Check your batteries before the 'fest!

board boxes full of who-knows-what. Here's where you'll find the audio accessories, cables, and other stuff for your radios. Many sellers don't know what to do with military audio accessories due to the strange connectors and end up pitching 'em into "miscellaneous" boxes.

Each swap meet has its own rhythm. When you sense it starting to fold up, revisit the sellers with items that still interest you and make your BAFO (best and final offer). Success often comes at this point for those willing to hang on until pack-up time!

Tip: There are a few things you should always take to a swap meet. The author calls it his Survival Kit.

Above all, never lose sight of the Prime Directive of Surplus! By the way, there's a Secondary Directive (and even a Corollary)…

KI0PF's Secondary Directive of Surplus: If a seller tells you he has a manual for a piece of surplus gear he's selling, but he "forgot to bring it" or "he has it at home and will send it to you" or any other such horsepuckey, run away like a scalded ape.

KI0PF's Corollary to the Secondary Directive of Surplus: Or, if you end up buying the equipment anyway, assume you will never in a million billion gazillion years ever see such manual.

Surplus is Where You Find It

Now *that* is a Yogi Berra-ism if there ever was one!

Seriously, there are a lot of surplus sources out there, and they aren't all necessarily associated with a certain on-line auction house. The heyday of surplus has clearly ended, and may not return for a very long while. There were times in the '80s at the TRW swap meet in L.A. when the author spent $75 and hauled home a whole trunk full of great finds. Those days are kaput, unfortunately. There are several reasons:

1) Much of the U.S. military is no longer releasing comm gear as scrap at government auctions and DRMOs. This used to be a primary way for dealers to get the stuff in the first place. Instead, a large amount of perfectly good materiel is being physically destroyed. Several years ago, a friend witnessed first-hand the destruction of a large number of H-250 handsets. The military scattered them on the ground and drove over 'em with an M-60 tank. This was shortly before the M-60 tank was *itself* scrapped. He also learned of perfectly good PRC-70s being smashed by E-zeros with sledgehammers and axes. The author believes the refuse was eventually buried. The PRC-70 was a $16,000+ radio.

More recently, the fed has become much more efficient in its destructive rampage. Instead of wasting time and manpower to smash the sets, there are now reports of the equipment being simply fed into enormous metal crushers that turn five-figure pieces of comm gear into shavings in a blink of an eye. The most recent example of this idiocy was an eyewitness account of PRC-104 suites being pulled out of vehicles for an upgrade to the newer PRC-138 sets. The PRC-104s and RF amplifiers were all marked with DRMO destruct tags and thrown unceremoniously into truck beds in a heap. The witness estimated there were 30-50 complete sets. Each cost us taxpayers upwards of five figures.

Either the PRC-70 or PRC-104 will sell for around $1500 give-or-take in the civilian surplus enthusiast market. Yet apparently the fed finds it easier to destroy them than auction them through DRMO. The author finds all of this supremely infuriating. After all, *we as taxpayers paid for this equipment. As long as it isn't dangerous, and has no strategic value to U.S. enemies, why shouldn't we as hobbyists/taxpayers have access to it?*

2) All of a sudden, military surplus has become popular. Back in the 1980s, nobody wanted the stuff. The author bought TMC gear for practically nothing, and WWII/Korean War stuff for less than that. People more or less threw away vehicular mounting bases as useless items. Now Fair is asking $65 for MT-298 bases (see the "GRC-8.5" section) and $360 (!!) for an AM-2060 (see AM-2060 section). One possible explanation may be the military vehicle community grabbing up bases for use in vehicle restoration. Vehicle restoration is, of course, completely different than the government destruction deal, and the author says more power (so to speak) to the vehicle enthusiasts. It's great to see the equipment being used instead of being run through what amounts to a garbage disposal on steroids.

3) Ebay brought the advent of electronic auctions, making it possible to advertise a piece of gear to the entire planet. This means every surplus type worldwide who's plugged into the net can see the auction. If there's even one person who is willing to pay an outrageous price, the guys who used to find the stuff at swap meets are out of luck. Period. Ebay is a double-edged sword, a definite love/hate relationship. If you're a seller, you love it; if you're a buyer, well... you know. The author has done a little of both—and he knows. He's found that relatively modern (post-50s) surplus gear in claimed working condition on Ebay inevitably sells for prices well outside anything he'd ever be willing to pay, so he stopped looking there as a buyer long ago.

One bright side, however. Gear advertised as inoperative sells for more reasonable prices. The conclusion here would seem to be that there are a lot of people out there who want fancy gear but can't work on it. It begs the question, "Well, okay, what do these people do when their fancy stuff breaks?" So Ebay still has possibilities for those willing to tinker.

4) There's a basic shortage of the "old guard" surplus. By about the mid-1980s the enormous supply of WWII and Korean War period surplus finally started to dry up. No more NIB command sets. No more cheap ART-13s. No more $15 PRC-10s.

A combination of all these factors has produced a pretty sparse surplus environment. But don't despair, there's still hope. Places where you'll find surplus:

Swap Meets and Hamfests

Even with Ebay shoving mil radio prices through the roof and beyond, hamfests are still good sources of equipment. However, additional sources may end up being a little different than the beginning surplus enthusiast might think. It used to be you could go to a hamfest like Indianapolis, or TRW (L.A.), or Orlando, and just walk up and down the aisles and grab all sorts of stuff for cheap. It wasn't really necessary to interact too much with the seller—the equipment and price made its own statement. Now the surplus supply is thinning out, and it's not so easy. You have to work for it a little harder. It turns out the best surplus may not be at the swap meets at all, but the people who know where you can find the surplus are. The best finds the author has come across recently are from contacts made at hamfests where he bought virtually nothing.

In the first instance, he'd volunteered to help at a friend's table at a Colorado swap meet. There wasn't much at the 'fest, but a passing attendee made a comment about the author's PRC-1099 which was there just for SWLing purposes during the day. The browser mentioned that he had an old military radio set and wondered if we could identify what it was. The equipment turned out to be a complete Wireless Set #19 that had been uncrated by the previous owner in 1946 but never powered up or used. We went to his place in Denver later that day to look at it, and the author ended up buying it, then reselling it and using the proceeds to buy more modern gear. There's no need to mention numbers, other than the buying and selling price differed by roughly an order of magnitude. In the second instance, the author found out about a very advanced receiver exciter, the Harris RF-5020 (see its own section), through a longtime friend and surplus expert. This was a fleeting, one-time chance to obtain a very excellent piece of gear at a very agreeable price.

These two transactions occurred several months apart, but the author likes to say he traded a radio built in 1942 even-steven for one built in 1991. Sometimes being in the right place at the right time is really a product of the contacts you make. Sooner or later, something good happens. In today's surplus market the best policy is to make as many contacts as you can at the hamfests. Spend a little extra time at the tables with any green stuff on them at all—even if none of it interests you. Talk with the sellers and get contact info, and follow up later to see if they've come up with anything else more interesting.

Don't give up on hamfests quite yet!

Surplus Dealers

Dealers used to be last on the author's list for places to look. That was before Epay...er, sorry, Edismay... oops, Ebay. But now the dealers have moved up a notch or two. For instance, the author was looking for a couple extra ST-138 carriers for his PRC-1099. He /PMs so often—see the HFpack Phenomenon section—he's probably the only surplus nut on the planet who's actually in danger of wearing one out. Fair had 'em for $35 in used good condition, but Mike Murphy at Murphy's Surplus had 'em in new condition in the wrap for $12. It pays to shop around. See the Appendix for a list of surplus dealers the author has done business with and has found to be straight-shooters.

Dealers are good sources for the gotta-have accessories like mast sections, handsets, cables and the occasional surprising deal.

The Internet

You can always try the newsgroups. In particular are

rec.radio.swap
rec.radio.amateur.equipment
rec.radio.amateur.boatanchors

The author has never had much luck with rec.radio.swap, which seems to have degenerated lately into a series of flame wars and auction announcements. The boatanchors site can be very amusing and informative. General Google searches can be productive, too. Still, the author has always preferred buying in person, or from someone he knows. Get plugged into the various Yahoo groups that cater to surplus. Try these groups for starters:

armyradios	Lots of surplus discussion, "for sale" or "buy" notices allowed
hfpack	Many of the members use mil pack sets, including yours truly
milpack	A group specifically for portable ops using military packsets

Hint: Get into the Yahoo Groups page and do a search on "military radios." You'll like the results.

Estate Sales

Don't count this one out. We're all gonna SK sooner or later, and surplus enthusiasts tend to leave a large amount (the author's group calls it tonnage) of strange, wonderful, technically improbable, and HEAVY stuff behind. That's part and parcel of the hobby. Often the wives are stuck with it. Remember, YLs live longer than us guys, on average. Unless the executors are hams, they won't have any idea what to do with it, either.

The author must leave exactly how to find the estate sales to the reader. One possibility would be to simply read the obituaries and run the names through QRZ.com. Use your best discretion here on contacting, of course. Another possibility is simple word of mouth.

Ebay

Ugh. A last resort for buyers in the opinion of the author. If you don't have a fast connection, you can't snipe (bid in the last few seconds of the auction). If you can't snipe, you won't win, at least on high demand items. Ebay has pushed the price of even the most basic surplus through the roof, because there is *always* someone somewhere willing to pay a ridiculous price. The author has had better luck on items a little off the beaten path like test equipment and accessories.

The Bizarre

While driving through a residential Michigan neighborhood on her way home from work, a friend's wife saw a big blue metal box sitting at the end of a driveway next to a full trash can. There were other full cans —either that day or the next was obviously trash pickup day. She stopped the car to take a look and ended up loading the thing into the trunk. She wasn't sure what it was, but it looked like some sort of radio and *that* was her husband Jim's favorite hobby. Well, when Jim pulled into his garage, he was somewhat (!) startled to see a mint condition Heathkit SB-220 linear amplifier sitting on a table. He was even more amazed to find the thing worked perfectly. He's been driving it with his Harris RT-1446 ever since.

And THAT is what the author means when he says, *"Surplus is where you find it."*

Maintenance and Troubleshooting or, When Things Go Blooie

U.S. military comm gear is among the best electronics made. The individual components have been especially manufactured to tighter specs, then assembled into equipment which, in turn, performs to tighter specs—including both performance and reliability.

These last two often conflict with each other. To get high performance it's often necessary to pursue a complex design with a high parts count. That lowers reliability. To accomplish both simultaneously gets very expensive! Many mil radios are made to the state-of-the-art for the particular time frame. When the PRC-74 was introduced in the early 1960s, it was a tremendous improvement over the GRC-9 it replaced. Not only did the PRC-74 herald a switch from AM to SSB for manpack gear, it was much lighter, more efficient, and had direct frequency readout.

But the PRC-74 also had 45-plus crystals, a lot of hand-picked components, and a pair of custom high voltage transistors in the power amplifier that pushed the state of the art of the time to the limit. It also had a very tedious and time-consuming alignment procedure to get the tuned radio frequency (TRF) stages exactly right and get all those crystals in the synthesizer on frequency.

The author did a rough count...the GRC-9 transceiver unit and PE-237 vibrator power supply together have 16 active devices (not counting a few crystals). The PRC-74B, which runs on 12VDC and has its own internal supply, has 93 active components (not counting all the crystals). Sooner or later one of these components IS going to fail. The author likes to sum the situation up thus:

KI0PF's Law of Critical Surplus Mass: Sooner or later, your surplus collection will reach a size at which something is always busted.

First Corollary to the Law of Critical Surplus Mass: If you don't troubleshoot what's busted WHEN it busts, pretty soon (sooner than you think), EVERYTHING will be busted; or at least it'll seem that way.

Second Corollary to the Law of Critical Surplus Mass: Refraining from operating surplus radio gear is NOT a valid way to avoid problems. Equipment busts just sitting there. (See PRC-47 section).

The above actually carries a fairly profound message. If you really want to get involved in surplus, and you don't have an utterly bottomless pit of cash, you're going to have to learn to troubleshoot your gear. The author began to realize this with his very first radio, an RT-70 he bought from Fair Radio in 1976. It was new-in-the-box, beautiful looking...and didn't work worth a darn. Upon inspection, he found a tube with the top "nubbin" broken off. He replaced the tube and the radio has worked for the ensuing quarter-century. That was an easy fix, and his first. Other fixes have been devilishly difficult (shorts in an R-390A's front panel wiring and an intermittently leaky high voltage cap in a linear amp's HV power supply come to mind).

Troubleshooting is about 30% science and 70% art. The only way to learn how to troubleshoot is to DO IT.

Tip: Don't wait for failures before learning the theory of your equipment. The technical manual (TM) should be thought of as a required accessory for the radio, just like handsets, antennas, battery boxes, etc. Part of the fun about surplus collecting is learning how it works, and this knowledge will really come in handy if (er...when) the radio fails. You'll also experience the satisfaction of repairing your own gear and getting it back on the air.

An illustrative story is in order. The normal idling plate current for the author's Technical Materiel Corporation PAL-350 linear is 160ma. That is, the bias pots are set to about 80ma per 4CX250B tube. The system developed a fault with a really strange symptom. When keyed with no drive, the plate current would abruptly zoom up to about 300ma, hang there for maybe a second, then drop back down to the correct 160ma. At that point the linear handled a suitable drive signal correctly. 300ma is the correct value for full output (350W) from the linear, so something was definitely haywire.

The author found he had to wait until the plate current dropped back before applying drive, or the current would bounce well above the 360ma allowable maximum and trip the over-current protection, kicking the system off the air. Even more interesting, he found that if the system were immediately keyed again after the plate current "bounce" happened, the idle current would go smoothly to160ma where it belonged. But if he waited a few minutes (such as in a normal QSO rotation) the symptom seemed to reoccur. Not always, but a lot more often.

Cleaning the HV relay contacts in the power supply proved ineffective. Suspecting a leaky HV filter capacitor, the author lifted each of the three in turn to no avail. After several months of mulling over the situation, a possible solution presented itself.

The "science" here is knowing from the schematic that a leaky cap isn't the only component in the screen supply circuits that can cause fluctuations in the screen voltage. If any one of the three series gaseous regulator tubes fails to fire, the screen voltage/plate current will soar and the transmitter will instantly kick off line.

Could there be some sort of intermediate failure in the regulator string? A partially firing tube? It seemed particularly suspicious that the meter seemed to prang more often when the transmitter was left unkeyed for more than a couple of minutes. This sort of theorizing is the "art" part of troubleshooting. Sure enough, swapping out the VR tubes in the screen voltage regulator string cured the problem! Clearly, one or more of the VR tubes was "soft firing" or partially failing to ionize for an instant. They all lit up okay visually. After the transmitter was unkeyed, there might have been residual ionization in the bad tube that helped it fire normally, if one didn't wait too long.

If it's the bottom of the ninth, tie score in the seventh game of the World Series, your team has a guy on third, and your pitcher hits a dumb busted-bat dying-quail bloop infield single, the result's the same as if your biggest McGwire type launched a homer to Alpha Centauri—you win. The author will almost certainly never know the exact physics describing the failure mode of that VR tube, but he fixed his power supply!

A roomful of advanced test equipment certainly makes it easier to troubleshoot, but it isn't strictly necessary. The author has a friend with a bench sagging under more fancy test gear than most corporate service facilities. In fact, collecting test gear is a whole 'nother hobby that could be the subject of a separate book. There are a few pieces of test gear you really need, though:

Digital Multimeter: The author has a Fluke 77 digital meter he's used for years. There are many others on the market now that probably have more features for the same cost.

Analog Multimeter: It's a good idea to also have a meter with an analog movement. Sometimes you'll want to measure voltages that are changing rapidly, and digital meters often don't respond fast enough.

Vacuum-tube voltmeter: For measuring RF voltages. The author uses an old TS-375.

Calibrated signal source: Some sort of signal generator. The author doesn't even own a formal sig-gen. He uses a Wavetek frequency synthesizer and a step attenuator to get the signals and levels he needs.

A frequency synthesizer may be used as a test signal source

Step attenuator: Any one of a number of varieties will work here. The author uses a 50dB (in 1dB steps) unit. You can generally find them at swap meets for a fraction of their original cost.

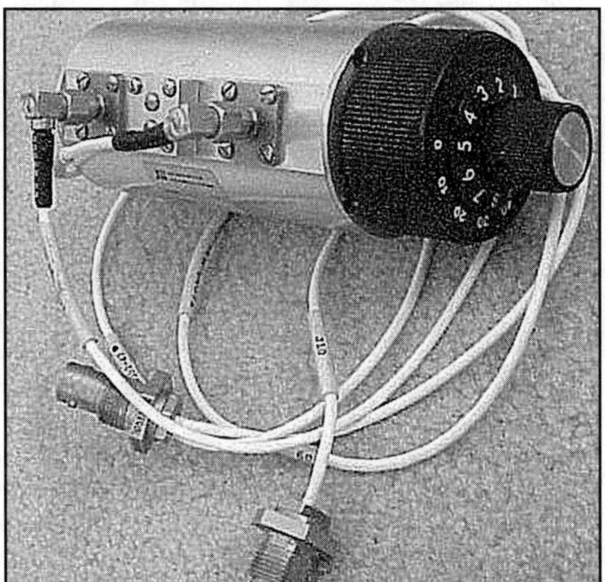

Use a step attenuator to set the test signal level.

Frequency counter: Again, nothing especially fancy is needed. The latest equipment will cost you a bundle. OR you can do like the author and pay forty bucks at a swap meet for an HP 5245L 50MHz counter with a 5253B 50-500MHz plug-in. Sure, it has a nixie tube display but that gives it charm. More important, the thing is very stable and it works just fine for most applications.

The venerable HP 5245L frequency counter.

Bench power supplies: Several units with different ranges. Constant-current and/or constant voltage capability is very helpful.

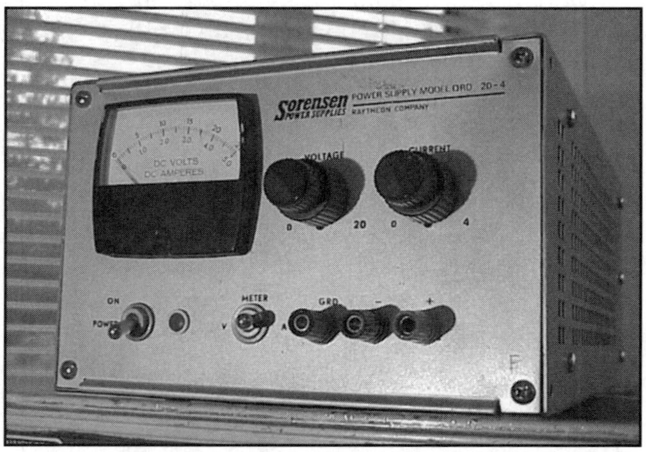

A typical power supply available at hamfests.

Oscilloscope: The author got along for many years without one. They sure help, though. 'Scopes are like PCs; you'll pay a bundle for the latest and fastest, but if you drop a few notches in performance they get much cheaper. He uses a USM-338 75MHz scope (basically a militarized HP 1707B).

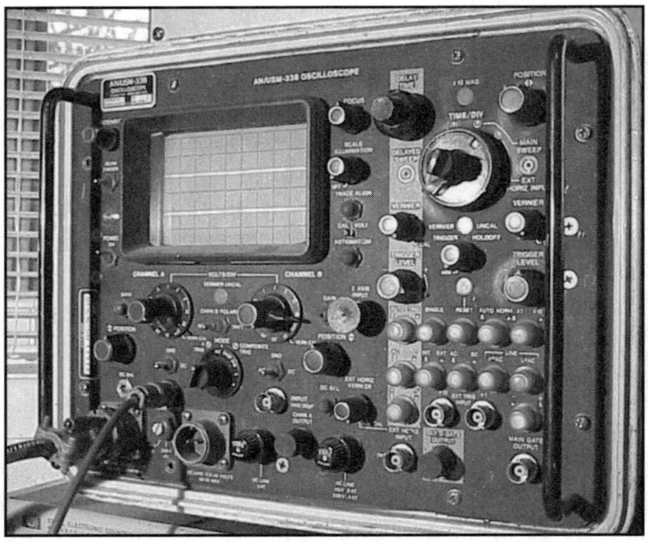

The USM-338 75mHz oscilloscope.

Miscellaneous: Coax adapters, clip leads (many!), dummy loads...you'll find yourself accumulating this sort of stuff as a matter of course.

Starting in the early 1960s, radios with specifications like the PRC-74 and PRC-25 were getting too complicated for field-level and lower echelon repairs and something had to be done. The answer was modular construction, where the fault could be easily and quickly traced to a removable module. This was done at first with specialized test sets and later with automated built-in-test (BITE) functions. The defective module is replaced from stock and sent back to a depot for repair. This system is great for the military but not so hot for the surplus enthusiast. Even if you manage to find the specialized test gear, simply throwing out a defective module from your PRC-74 is not an option!

If you acquire any amount of modern surplus you'll eventually troubleshoot a modularized radio. Extender cables probably won't be available—this is where your ingenuity and patience will really pay off. You'll need to *thoroughly* understand the theory of the equipment—especially in the suspected fault area—then correctly power up and supply signals to the individual module out of the equipment.

A word on tube testers: Don't always believe them! They're great for showing obviously dead, weak, or shorted tubes but sometimes fail miserably at finding subtle cases like interelectrode capacitance. Always check hollow-state devices by substitution if at all possible.

Field Expedient HF Antennas for Your Green Goodies

You can enjoy running portable from almost any location with a few simple, proven wire antennas. There have been entire books written on field expedient HF antennas, and it certainly isn't the author's intention to rehash them all. A few of his favorites will be presented here to aid the HF packset user in getting set up quickly (and in getting a relatively effective signal on the air).

A number of wire antennas, including dipoles, slant wires, verticals, and bobtail curtains can be constructed out of the following kit of basic supplies:

255 foot reel of lightweight multi-stranded bare copper wire
70 foot reel of lightweight multi-stranded bare copper wire
30 foot length of coax with suitable connectors each end
Dipole fixture (see text)
Connection adapters for your particular packset
100 feet or so of string
Throwing weight
Rubber bands or electrical tape

This is the deluxe kit that will build all of the antennas mentioned for frequencies down to 3.5 MHz. The list can be simplified considerably if fewer (or single) frequencies are expected. If you expect operation only on 40m and higher, you can cut the wire lengths in half. Insulators are shown throughout the examples but they can be replaced with simple lengths of string.

Tip: The author has the two wire lengths wound on bobbins—either on plastic units supplied with the radio sets, or simple reels cut out of thin plywood, plastic, or other material. The correct length of wire is unreeled and the rest secured with the rubber bands or tape (which shorts out the remaining length) and the whole reel is hauled up as need be. The thin stranded wire from the old Gibson Girl rescue transmitters works great. Standard 14-gauge stranded insulated wire also works okay, although you may find the length may vary from the theory somewhat due to the insulated wire left on the reel.

Tip: At HF frequencies and the usual packset power levels, RG-174 coax shows negligible loss and will lighten your kit load.

Tip: The dipole fixture is often home-built – just a bracket-mounted female BNC connector attached to some non-conductive material with wing nuts for attaching the wire ends. In fact, the author uses BNC-type connectors all around.

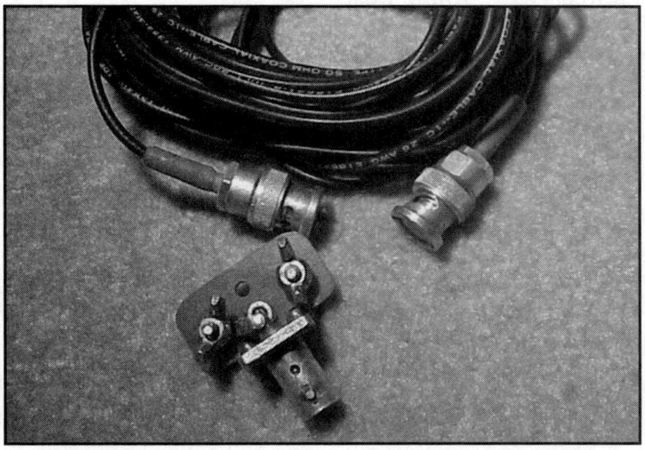

Author's simple homebrew dipole fixture and 30-ft coax feedline

For those wanting something fancier, Fair Radio Sales may still have the dipole fixture that originally came with the PRC-74. It includes 30 feet or so of Teflon-insulated twisted pair coax.

The coax is wound to one side in the illustration to show one of the fixture's binding posts.

PRC-74 dipole fixture

Basic Length Formula

The basic building block for the antennas, a quarter wavelength of wire, can be calculated by:

Length in feet of a resonant quarter wavelength = 234/frequency in MHz.

Example for 7.280: MHz: 234/7.280 = about 32 feet.

Vertical Antenna

The vertical antenna is a single quarter-wavelength of wire arranged straight up and down (hanging from a tree branch, etc). Example for 7.280 MHz: Throw a weighted string over a horizontal tree branch located higher than 32 feet up. Connect the 32-foot antenna wire (or reel) to the string and haul it up until the end can just be connected to the packset antenna posts or connector. Lay out a 32-foot counterpoise wire in the desired transmission direction and connect to the radio case or ground post.

This setup gives low-angle radiation, and is slightly more effective in the direction the counterpoise is pointing. The more counterpoise wires of the correct length you lay down with this type of antenna, the better it will work. But always include at least one!

Use the vertical antenna for maximum distance (generally) on the HF bands.

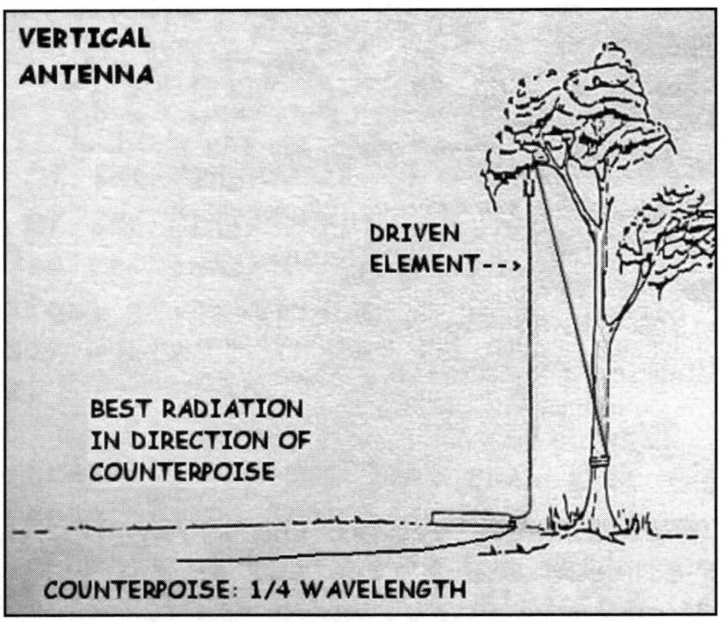

Vertical antenna built from the basic kit.

Slant Wire Antenna

The slant wire (or sloper) is a quarter wavelength of wire arranged with the feed end at ground level and the far end raised up high by a support.

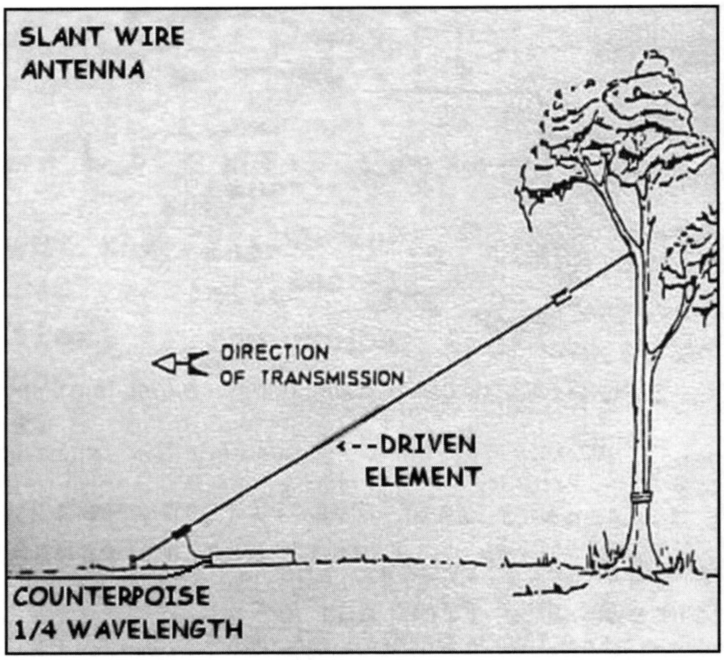

Slant wire antenna built from the basic kit

The slant wire is somewhat directional in the manner shown, so some thought should be given to where the string is thrown over the support. Example for 3.825 MHz: Throw a weighted string over a tree branch – the height will vary, depending on the length of the antenna; it isn't really critical. Connect the 61-foot antenna wire (or reel) to the string and haul it up until the end can just be connected to the packset antenna posts or connector. Lay out a 61-foot counterpoise wire in the desired transmission direction and connect to the radio case or ground post.

The slant wire is slightly directional and exhibits a somewhat higher angle of radiation than the vertical. Often it's hard to tell a difference between the two.

Dipole Antenna

The next two antennas require two supports, but are often more effective.

The first is the simple dipole, which consists of two one-quarter wavelength wires generally fed with coaxial line. Example for 7.280 MHz: Connect a 32-foot length of wire to each side of the dipole fixture and connect the coax. Throw two lengths of string over supports at least 66 feet apart (a few feet more than the total length of the dipole), and 20 to 30 feet up. Haul both ends of the antenna up into the supports. The inside conductor of the coax hooks to the antenna post or inside pin of the antenna connector, and the braid hooks to the ground post or radio chassis. Tune the antenna like any other 50-ohm load.

The dipole has an intermediate angle of radiation and is directional in and out of the illustration.

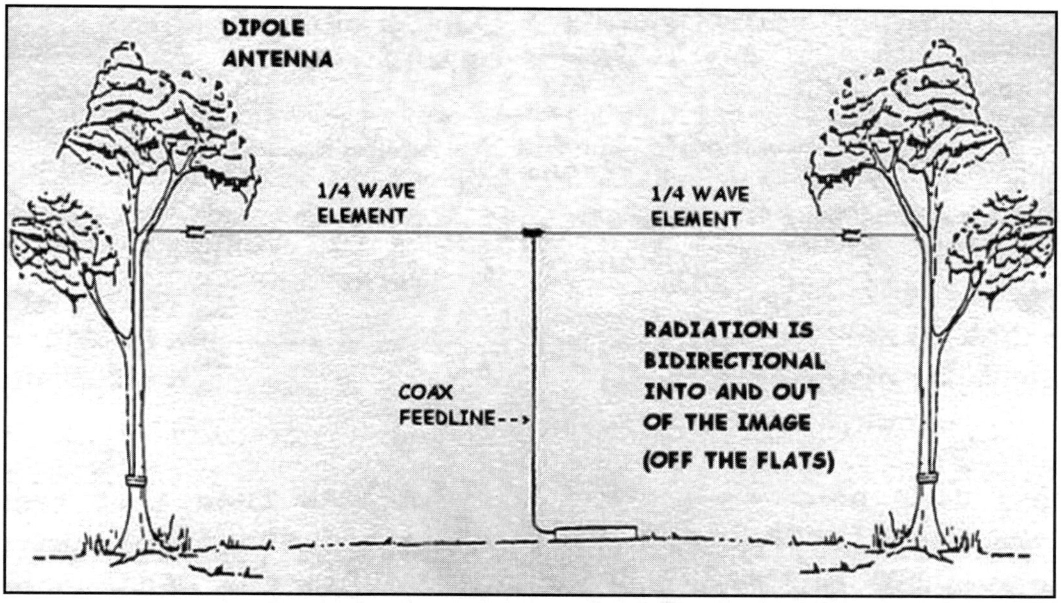

Dipole antenna built from the basic kit

Tip: The dipole's angle of radiation can be greatly elevated – to almost vertical – by dropping it to within a few feet of the ground. The efficiency is also greatly lowered, of course, which is not always a bad thing. This sort of NVIS (near vertical incidence skywave) operation can be of some use when transmission out of gullies, canyons, etc., is desired, or if transmission over short distances is required.

Curtain Antenna

This is an interesting expedient antenna consisting of a single end-fed wire that goes up vertically a quarter-wavelength, horizontally for a half-wavelength, then down vertically again for a quarter-wavelength. The half-wave counterpoise is located beneath the upside-down U of the main antenna.

The radiation pattern is relatively low angle, directional off the flats of the antenna, with some gain over the standard dipole. An example for 7.280 MHz: Measure off 128 feet of wire, making a kink 32 feet from each end. Throw two strings over supports at least 35 feet up and 66 feet or so apart. Choice of the supports is important, because this antenna is directional. Connect the two strings at the kinks and haul the wire up until the 32-ft end sections are dangling

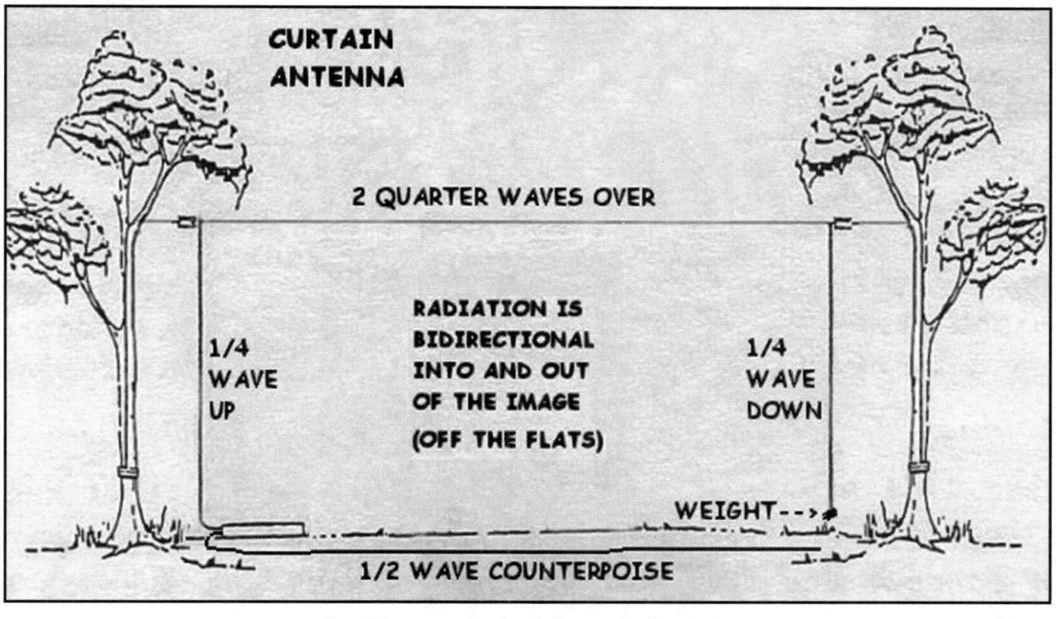

Curtain antenna built from the basic kit

within reach of the packset. Either end of the antenna can be fed – pick one and hook the wire to the antenna post (or center pin of the antenna connector). The far end of the antenna can be hooked to a string and suspended with a rock to keep it from flopping around. Hook a 64-foot length of wire to the radio chassis or ground post and arrange it on the ground the length of the antenna. Tune the curtain as with any other 50-ohm antenna.

Slanted Dipole Antenna – Variation on a Theme

If you only have one support, one end of a dipole can be dropped to nearly ground level as illustrated.

No counterpoise is needed with this antenna.

These antennas, or variations on the same, should be enough for most any field-portable situation. As with everything in this hobby, experimentation is the key!

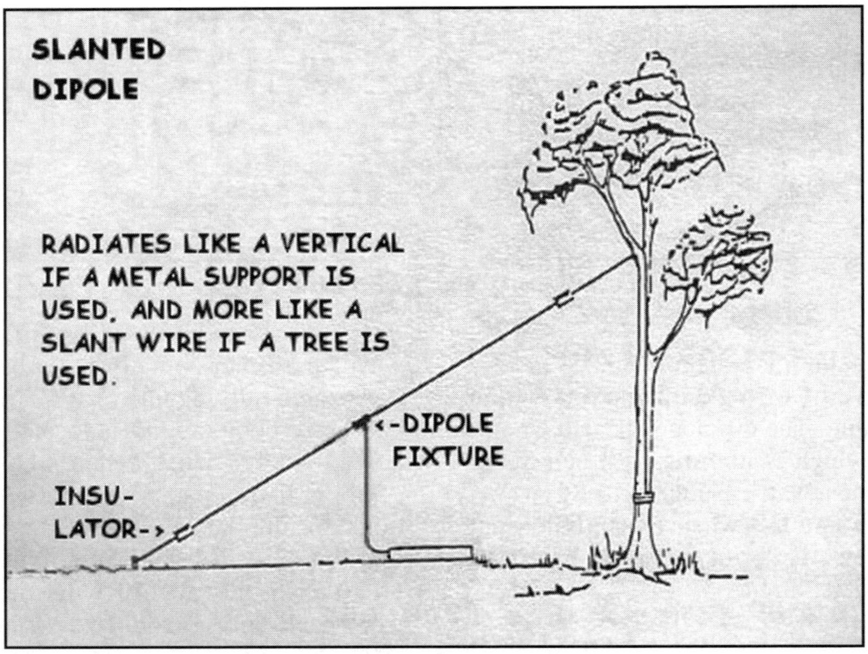

Slanted dipole built with the basic kit

The HFpack Phenomenon—Strap on that Radio!

In 2001 Bonnie Crystal, KQ6XA, thought there might…*might*…be twenty-five people worldwide interested in low power manpack and portable operation on the HF bands. She started a Yahoo group called "HFpack" to see if she could scare up some interest. By winter of 2005 there were over five thousand members in HFpack! What in tarnation happened? A whole lotta fun, that's what.

Here's how the author got hooked. In early 2002 one night on 40m a friend told him there was this thing called HFpack, a group of people messing around with low power HF manpack ops. The frequency he mentioned was 18157.5MHz USB. The author didn't do anything with the suggestion for a couple of weeks, then happened to flip to the frequency. He heard several stations, weakly. Then all of a sudden Budd Drummond, W3FF, came literally crashing in from northern California.

"*HEY!*" thought the author, *"I might actually have a chance…"*

So he scrambled around looking for the various pieces he needed to get the PRC-1099 going—the 9ft whip, the battery, the headset. There was no time for a ground. For all he knew, the opening was only gonna last another five minutes.

The illustration is an accurate recreation of the setup and location used for the author's very first HFpack QSO. He just ran out the back door of his place in Colorado and plopped the radio down on the ground, with no counterpoise. To his amazement, W3FF answered his call! Budd has been a great HFpack proponent, offering his Buddipole antenna design gratis to homebrewers on his website:

<www.qsl.net/w3ff/antenna_design.htm>

He also sells high quality pre-made versions of his antenna. Excuse me, but this is a shameless plug!

After that preliminary contact the author ran inside and dug out a dusty old ST-138 radio carrier, lashed on the radio, and rejoined the fray. Now, he was walking around talking to California (and soon after, a mess of other places) with a radio on his back! HF ops haven't been the same for him since.

The author's first HFpack QSO.

Running low power on the ham bands has always been tough. Up until recently, people trying to run five or ten watts have been pretty much stomped into mush. It was hard to get a word in edgewise. During 2002 and 2003 the sheer number of HFpackers on 18157.5 at any one time began to change that dramatically.

During the "formal" (as formal as this group gets, anyway) 2230Z weekend skeds, we often have one or more stronger base stations running a little interference to keep the freek open. QRO stations are always welcome because they can inform a low-power station he's being called, or help set up DX contacts. There are a number of JAs who enjoy looking for HFpackers, and west coast stations like W3FF can be a real help.

The Official HFpack Logo

Once a number of stations form an impromptu "net," the usual procedure is for the QRO station to ask each low power station to give a few "CQ HFpack" calls to see who they could reach. This often increases the net size. The QRO station pauses for new breakers, then moves to the next station on the list. With some luck we get enough low-power stations on frequency to reach critical mass and hold it.

While everyone is welcome, emphasis is on pedestrian mobile, bike mobile, rowboat mobile, in-line roller skate mobile (no kidding!), or any other low power stations. Just what the heck is this "pedestrian mobile" you ask? KQ6XA has proposed a new category for Field Day that comes as close to a formal definition as you'll find:

> **Proposed Field Day CATEGORY F (Pedestrian Mobile):** A singular radio operator shall freely move about, sit or stand while carrying and supporting the full weight of all radio equipment including power and antenna in the configuration used for operation. The total weight of all radio equipment, including antenna and power supply, shall be less than 15 kilograms (33 pounds) and the maximum antenna height shall be 33feet above surface level. Fixed or tethered metallic objects shall not be used as antennas, reflectors, directors, or counterpoises. Operator shall be located on a normal walking surface such as a natural earth surface or non-elevated road, sidewalk, or wooden walkway. This category includes an operator utilizing skates, skis, or riding an animal.

If this sounds like fun (and it is) check out the Yahoo group at

<http://groups.yahoo.com/group/hfpack>

and/or the site at

<www.hfpack.com>

The hot frequency up until winter of 2004 was 18157.5, but the sunspot cycle downturn has lowered the MUF so much that 17m is no longer reliable. Even with a transition to 14342.5, HFpack activity has fallen off considerably. We may have to wait through the lowest part of the cycle, several years, before we experience the same great conditions. One of the newest HFpack frequencies is 5371.5 USB, taking advantage of the new 5meg band channels. 17m has been surprising, though, and has periodically offered up some good conditions in spite of the cycle downturn. There are still enough openings to merit calling "CQ HFpack!" on 17 every once in a while. Despite the QRM on 20m, we still manage to have fun during the 2230Z weekend nets on 14342.5…so keep trying!

We always get questions about why most of the default channels are set up on half-kHz boundaries. The reasoning goes thus:

1) QRP operation on the HF bands is tough; we need all the help we can get. A QRP station on or very close to a QRO station will simply disappear. Non-HFpack stations generally seem to operate on even-kHz channels. The 500Hz difference often helps HFpackers to stand out. At times the conditions are such that answering the HFpacker will result in interference to the QRO station, so you have to hold off anyway. But at least you know there's activity and you can wait for the QRO station to finish. The author has found just about any station exactly on one of the default frequencies is worth checking out as a potential HFpacker.

2) If HFpackers are set up in QSO and start getting hammered by QRO stations on "standard" channels either .5kHz above or .5kHz below, we're still close enough for them to hear our "frequency in use" admonition.

HFpack Technical Considerations

By far the best thing for readers to do if they're interested in pursuing HFpack operations is to peruse the web sites, where KQ6XA has amassed a wealth of information. Here are some high points.

The advent of the FT-817 and its imitators (IC-703, etc.) has opened up a whole new aspect of the amateur radio hobby. They have made possible, and practical, the ability to carry a full-featured, synthesized HF radio set on one's person, creating the concept of pedestrian mobile or /PM.

Note: While the idea of low-power battery-powered man-portable HF sets is relatively new on the amateur scene, it has been a matter-of-fact part of the military scene for many years—at least as far back as the PRC-74 in Viet Nam where special forces routinely used it to get info back from recon patrols that were too far in-country for the use of VHF. Our commo people knew the value of expedient antennas, dragging counterpoises, and all that stuff. Yogi Berra was right: "It's déjà vu all over again." The author has no problem with that. Good ideas need to be rediscovered each generation!

Equations always have two sides, of course, and the other half of this one has been KQ6XA's HFpack concept—the method of organizing low-power HF enthusiasts into a coherent group. (Some say "incoherent" but that's a matter for debate. Ha!) HFpack has almost eliminated the situation in which a bunch of QRPers wander around aimlessly in the ether trying to find each other.

You'll already find a wide variety of radio sets available for /PM ops; the ham equipment manufacturers are certainly aware of what's going on! There's the FT-817 and FT-897 from Yaesu, the IC-703 from Icom, the Kachina MP-25, the VX-1210 from Vertex, and others, as well as a mess of military manpacks including the PRC-70, PRC-74B, PRC-2000, PRC-138, and the author's PRC-1099. The tradeoffs in size and weight, price, capability, and ruggedness are considerable. You need to do your research carefully before picking a radio because your investment is likely to be considerable.

The author has used his PRC-1099 HF packset with great success on the HFpack frequencies. The 1099 covers 1.6 to 30 MHz SSB or CW in 100Hz steps, runs 5 watts low or 20 watts high power, and has a built-in wide range tuner and integral battery box. The whole setup weighs about 15 pounds, including the whip antenna and a 4.5 Ah NimH pack.

Author with PRC-1099, AS-1887A whip, and H-161 Headset/Chest switch.

Weight is one of those big tradeoffs. It all depends on how much you're willing to carry, and how you ultimately plan to use your radio. The 1099 also has a stout antenna connector so a whip can be connected directly to the radio. No awkward lashups; everything's self-contained. The radio was bought as a sales demo from the factory in 1991 for $1500. That sounds like a lot until you consider the following:

• The radio is complete as-is. Add together the cost of a new FT-817, a larger capacity battery setup, outboard tuner, outboard RF amp, and building/buying a carrying frame, and you'll probably be pushing a thousand bucks. And the result isn't ruggedized or weatherproof.

• The 1099 uses accessories the author already had for other radios—H-189 or H-250 handsets, H-161 headset/mike, AT-271 or AS-1887 whip antennas, and carrier.

• The carrying method is pre-built. The radio has exactly the same form factor as a PRC-25 or PRC-77 and fits the same ST-138 carry frame. The whip antenna screws directly into the top panel of the radio.

• The author figures he got at least $500 worth of convenience not having to mess around with all the outboard devices and wiring.

This is a great example of the versatility of surplus. The author was able to piece together a great /PM setup from equipment he had already, but hadn't had a chance to use in that role. From the time he first heard W3FF on 17m that first day, to the time he had a fully functional /PM setup was about twenty minutes. He's using virtually the same arrangement several years later.

The antenna system, including the ground, is the single most important element in a /PM setup. It's enough of a challenge getting out on the HF bands with low power without fighting antenna systems that approximate a dummy load! It is absolutely critical that you strive for the most efficient antenna possible, given your personal list of tradeoffs. In fact, you'll find that a great deal of discussion on the HFpack Yahoo group centers squarely on antenna design. There are nearly as many tradeoffs while choosing a /PM antenna system as there are buying your radio! The good side of this is that once you've acquired your radio, you can experiment ad infinitum with antenna systems. There are a few general observations that, in the author's opinion, will hold in any case:

1) Get as much metal up in the air as possible. When an antenna is electrically very short, the issue rapidly ceases to be so much about the configuration, as about how much radiating surface there is. In HF /PM ops, virtually any antenna used below 18MHz will be electrically short. Below 10MHz, they will be extremely short. Obviously, there are limits to this, depending on how much you want to haul around, surrounding environment, etc. Stiff wind, surrounding forest, and proximity of people are all factors.

2) Always use the best ground possible. You're somewhat limited here because you'll run afoul of the definition of pedestrian mobile if you physically hook up to something. HFpackers have found one of the best grounds is a drag-along counterpoise. The main concept is that counterpoise or ground radial lengths generally given in previous literature apply when they are above ground by 1/8 wavelength or more. When they are in contact with the ground (ie: dragging), they must be cut shorter.

The type of surface the counterpoise is being dragged over affects the performance of the counterpoise. More conductive soil requires shorter counterpoises. It's tough to tune a counterpoise constantly, and Bonnie has worked up the following compromises (used here with her permission):

Pedestrian Dragging Counterpoise Length:

With broadly tunable whips (such as MP-1, mobile whips, etc) I've found empirically that a pedestrian dragging counterpoise wire can be about 10% to 25% shorter than the standard quarter wave formula predicts, and the whip will usually tune to a fairly good 50ohm match. Insulated wire with low ohmic resistance should be used. For best dragging quality, Teflon or slick PVC-jacketed or oil-resistant multistrand wire is best. For safety while walking with a dragging counterpoise, some sort of slip-connector or an alligator clip should be used so that the connection will break apart if the wire is caught by a rock or vegetation.

KQ6XA's Rule of Thumb For Quarter Wave Dragging Counterpoise:

~10% shorter 20 to 30MHz.
~15% shorter 11 to 20MHz.
~20% shorter 6 to 11MHz.
~25% shorter 3.5 to 6MHz

Formulas For Dragging Counterpoise Length

20~30MHz:
210/MHz = Length (feet)
64/MHz = Length (meters)

11~20MHz
199/MHz = Length (feet)
60/MHz = Length (meters)

6~11MHz
187/MHz = Length (feet)
57/MHz = Length (meters)

3.5~6MHz
175/MHz = Length (feet)
53/MHz = Length (meters)

Dragging Counterpoise Lengths By Band

Band	Feet	Meters
10m	7.4ft	2.3m
12m	8.0ft	2.6m
15m	9.9ft	3.0m
17m	11.0ft	3.3m
20m	14.0ft	4.2m
30m	18.5ft	5.6m
40m	26.3ft	8.0m
60m	34.3ft	10.4m
75m	45.4ft	13.7m
80m	49.3ft	14.9m

NOTE: The above formulas and charts are approximate.

The author uses an 11 foot length of flexible rubber-insulated test lead wire for /PM operations on 17m. It seems to slip right through the rock outcroppings and indigenous cactus and scrub he has in his area. The full length is used because the soil in the high arid area west of Pikes Peak is extremely dry most of the time. With 17m dying due to the sunspot cycle going down the tubes, he's resigned to lengthening the counterpoise somewhat for 20m.

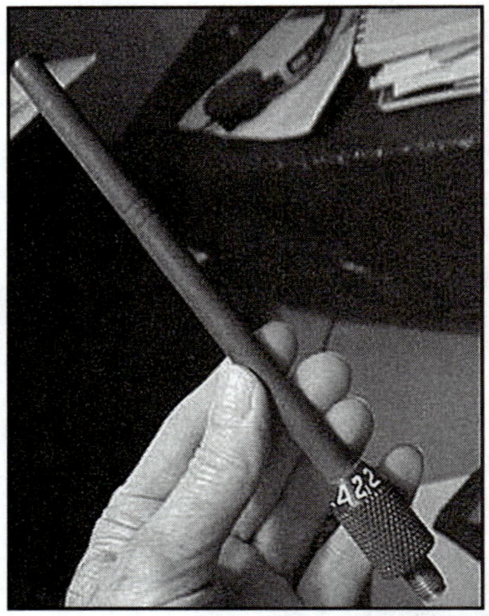

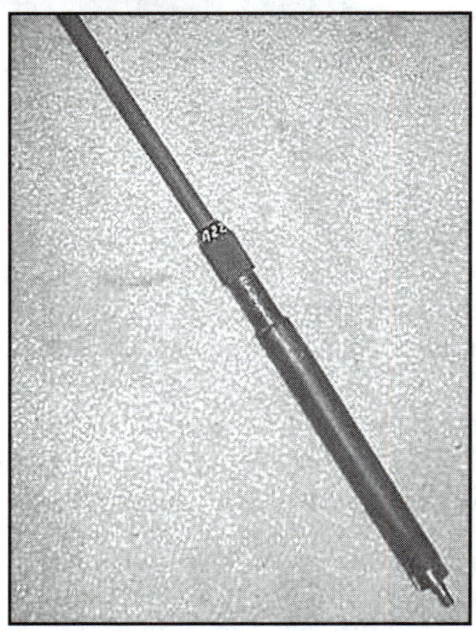

M-422 adapter.

The author has run a straight 10ft whip with great success on 17m. He uses an AT-271, which is the standard sectionalized "long" whip (as opposed to the "short" three-foot tape whip) used with the PRC-8/-9/-10 and PRC-25/-77 radio families (both of which have a section in this book). The AT-271 works well with the PRC-1099 above 7MHz. Below 7MHz, he switches to the AS-1887A PRC-74B whip, which has a center-mounted adjustable loading coil.

Alternatively, a really *long* long whip can be made out of stacked Korean War-vintage MS-116, MS-117, and MS-118 mast sections, an M-422 adapter, and AB-591 rubber base, and a reinforcing tube.

This gimmick screws into the top of a standard AB-591 spring base shown, and accepts an MS-116 mast section. Start with any number of stackable MS-116 sections, then finish the whip with an MS-117 and MS-118. Such a whip rapidly becomes too heavy for the AB-591 to support; some way must be found to reinforce the rubber base to keep the whole thing from flopping over. Of all things, the author found a length of thin wall tubing originally used in an old telescoping shower curtain rod. It was a near-perfect slip fit!

The author typically uses three MS-116s, one MS-117 and one MS-118 for a total length of 16 1/2 feet. If there's any wind, one MS-116 can be left out for a length of 13 1/2 feet. He used the longer version and a 45 foot counterpoise on the 1099 to set the 5 watt 80 meter SSB /PM world record with Ken Chong, WB6MLC. That was Colorado Springs to San Francisco, pedestrian mobile to pedestrian mobile, 200 miles per watt!

There's lots more info on the HFpack site, including an elaborate HF portable antenna shootout to help you decide how to configure your /PM setup.

Reinforcing tube on looooooooong whip.

Equipment Data Sheets

BC-611

Application:

The BC-611 is a short-range low-power amplitude-modulated walkie-talkie -- designed, manufactured, and used by the U.S. military before that name even existed. The BC-611 has a lot of history; it was designed in 1940 by the Galvin Manufacturing Corporation, which eventually became none other than Motorola. The BC-611 landed on the Normandy beaches in the hands of the Signal Corps on D-Day, and served with distinction across Europe. It wasn't replaced by the PRC-6 until 1950.

Tech Specs:

Frequency Range: one crystal controlled channel in the range 3.500 to 6.000 mHz.

Type of modulation and power out: Amplitude modulated (AM), 80-100mW or so.

Transmit bandwidth: About 6 kHz

Size: 3.5" x 5.5" x 15.75"

Weight: 3.85 pounds without batteries, 5.5 pounds with original batteries.

Active devices: Battery-type low filament voltage tubes—1R5, 1S5, 1T4, 3S4 (2)

Power requirements:
Receive: 1.5 VDC 250mA; 103.5VDC 11mA
Transmit: 1.5VDC, 300 mA; 103.5 VDC 35ma
Originally supplied by A-battery BA-37 and B Battery BA-38

Audio input/output devices: The mike and earpiece are built-in, however later model BC-611s have the provision for connecting an external handset of the type used with the BC-620/659 and BC-1335.

Whip antenna: Integral 39" collapsing whip antenna (see text).

Tech Manuals: TM 11-235 "Technical Manual, Radio Sets SCR-536-A, SCR-536-B, and SCR-536-C". The author's manual is dated May 14, 1943. There are versions of the BC-611 extending up to at least BC-611-F. Author's unit pictured above is a BC-611-C.

Construction Style: The components are wired point-to-point and on tube sockets on a single pull-out chassis. The internal construction bears some resemblance to that in the RBZ receiver. This is a repairable set, however care should be taken due to the very compact construction. (Compact, at least, for 1940!)

Description:

Amazing is the only word for this radio. The BC-611 was, as far as the author knows, the first truly hand-portable fully self-contained radio transceiver. It was, in fact, a bona fide engineering marvel. If you were to suggest the idea of building an HF transceiver out of just five active components in today's world of surface-mount and integrated circuits, you'd get laughed right into the looney bin. Yet not only did Galvin successfully build it, they saw it enter active military service in WWII and succeed in combat conditions. The design relied on the vacuum tubes performing two or more functions between transmit and receive:

V1: 3S4—Receiver RF amp; Transmitter RF power amp
V2: 1R5—Receiver converter and oscillator; Transmitter oscillator

V3: 1T4—Receiver I.F. amp; (not used in transmitter)
V4: 1S5—Receiver second detector, A.V.C., A-F amp; Transmitter mike amp
V5: 3S4—Receiver output audio amp; Transmitter modulator

This is the only military radio known to the author with the same active device acting as receiver front end and transmitter P.A.!

This is all managed by a very clever and extensive switching process using a long, multi-contact sliding switch actuated by the PTT button on the side of the radio. The slide switch essentially reconfigures the entire circuit from a five-tube single conversion superhet receiver into a four-tube single conversion plate-modulated AM transmitter.

You can see the switch in the illustration with its contacts mounted on a long bakelite strip. Note "GALVIN" on the audio output choke at lower right.

Operation:

You'll need to make a battery pack for the radio. Both of the dry cell batteries originally used with the set are long gone. There are a couple of different power alternatives. The first is making a B+ source out of eleven 9V alkaline batteries, giving around 100VDC, and an A+ source (filaments) using a D-cell alkaline. This is best if you don't plan on using the radio a lot. The 9V batteries are expensive, but will last quite a while with their long shelf life and the low 30mA maximum draw.

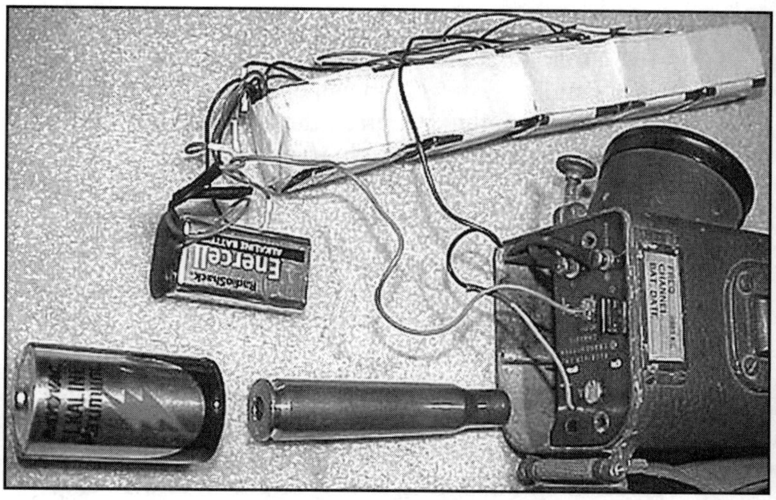

Improvised battery pack for BC-611

The B+ pack can be made from five pairs of batteries taped together and connected in series using standard 9V battery connectors. The eleventh battery fits in atop the rest of the pack once the pack is pushed into the B+ battery compartment.

The reader will doubtless wonder what the heck a shell casing is doing in the picture. (No, the author is not advocating shooting the radio, although there might have been a G.I. who would have liked to shoot his radio at some point!)

Tip: Long ago some enterprising radioman found himself short a BA-37 battery and discovered a spent .50 machinegun case was exactly the right size spacer to allow use of a single standard D-cell flashlight battery as an expedient. The old-time carbon-zinc battery probably lasted fifteen minutes, but it got him back on the air long enough to call in that mortar barrage. Today's alkaline technology makes this little trick more practical.

The A+ connection is effected by the contacts in the lid. The B+ positive connection can be soldered right to the round pad labeled "B+" on the radio chassis. The author ran the negative side of the B+ to a good ground on the chassis.

The other power alternative is to buy one of the pre-built DC-DC converters offered on the web. Try a Google search on "BC-611 power supply". This is not meant to be any sort of endorsement, since the author has never tried them.

Once you have power hooked up, operation is simple. Remove the protective cap from the whip and move it to the retaining

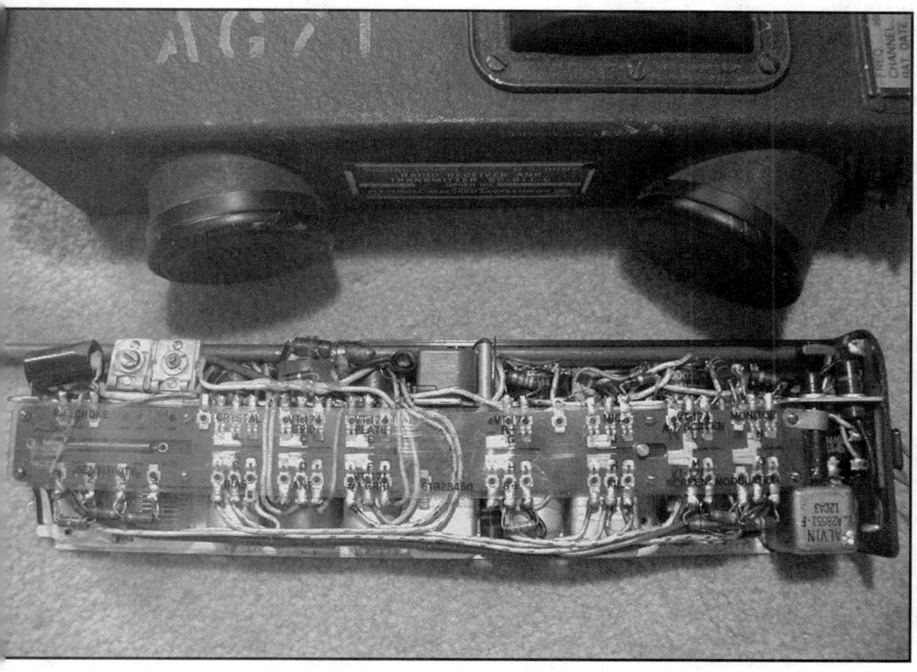

BC-611, chassis removed.

threads provided on top of the radio. Extend the whip fully -- this turns the radio ON via an internal switch. You should now hear atmospheric noise or stations, especially at night. Depress the PTT switch on the side and jabber in the mike to transmit. That's it!

Additional Comments:

Getting these radios to work again after 60+ years is an art as much as a science. Incredibly, many examples work right off, at least on major functions. Most units will be set up on the default frequency of 3885 kHz. While many were simply used as-is and never re-crystaled, this is not always the case. So you need to find out what crystals are actually in your set. For operation on 3885 kHz, you'll have a 4340 kHz crystal in the receive socket and a 3885 crystal in the transmit socket.

Receive Xtal: FT-243 type, 455 kHz above the incoming signal carrier frequency. Transmit Xtal: FT-243 type, equal to the desired transmit carrier frequency.

Aligning these sets to a new frequency or peaking them up is a hefty challenge. The whip, being about a meter in length is a one eightieth-wavelength antenna. The only ground available for this antenna system is the case of the radio itself and...you guessed it...you. The BC-611 has the endearing (or frustrating) characteristic of being the only radio the author knows of that must be aligned while the technician is holding the case. You may ask how in the world this can be done, because adjustment points are all on the chassis inside the case.

The answer is that there's a special case used during alignment with holes drilled in it over the required adjustment points! There were also some specialized jigs and test gear that are generally no longer available. However, assuming you have the TM and have read up on the alignment procedure, you can accomplish it without the special drilled case using a simple expedient:

Tip: Once you've removed the chassis from the case, carefully wrap it in a layer of newspaper, followed by a layer of tin foil. Use scotch tape to keep the wrapping on. Don't forget to screw the antenna-end of the foil down to the top part of the chassis to simulate the actual ground at that point. Then cut access holes through the foil to the adjustment points. The author's group has found this arrangement simulates the outer case well enough to get a decent alignment. Remember to hold onto the wrapped chassis with your hand to simulate operation!

Tip: The AVC in these radios is easily swamped. If you're close in to the other transmitter (like across a hamfest), the receiver can overload like crazy. You can de-sense it by holding onto the base of the whip. Let go before you transmit of course -- with a 1/80th wave antenna even a few hundred milliwatts can generate some pretty serious RF voltage on that whip!

What To Do If It's D.O.A.:

Make sure you have the correct voltages at the radio chassis. This may seem obvious, but it's easy to make mistakes with home-made battery packs. The internal power switch activated by the base of the antenna has a tendency to get "out of synch" and not actuated correctly. If this happens, the set won't turn on.

When any piece of equipment sits around for half a century, it's bound to get a little crusty. A little Caig De-oxit D5 on those switch wipers may do the trick.

Tip: Do not use standard tuner cleaner on switch contacts! It leaves no lubrication, and the blasted things will wear out even faster. De-oxit is the best stuff -- this advice comes from experience.

Serious symptoms like oscillation, erratic sensitivity, rotten sounding transmit audio, etc., demand more extensive troubleshooting and require the TM listed above.

Accessories:

Late model BC-611s have the provision for connecting an external handset of the type used with the BC-620/659 and BC-1335.

Also, the Army had problems in below-freezing temperatures with breath vapor freezing the mouthpiece holes shut. They came out with a water-repellant fabric doodad with a drawstring that fits over the mike. Of course, it had its own designation: "Microphone Cover M-367". This would have been a must during the winter of 1944-1945, the worst in Europe in thirty years.

GRC-9

Application:

The GRC-9 is low-power portable transceiver used in a wide variety of fixed and vehicular applications that operates in the lower half of the HF band.

Tech Specs:

Type of modulation and power out: CW, MCW (modulated CW), and AM. Power out 7 watts phone, 15watts CW (less with hand crank generator)

Frequency range: 2-12 mHz, continuous tuning, in three bands. Calibration via a built-in crystal oscillator with calibration points every 200 kHz. Up to six crystals (FT-243 at half the desired operating frequency), two per band.

Transmit bandwidth: AM about 6 kHz

Size: 12" x 16" x 8"

Weight: 30 pounds, basic radio and cover only

Active devices: Vacuum tubes; Receiver: 1L4(2), 1R5(3), 1S5, 3Q4; Transmitter: 0C3, 3A4(3), 2E22,

Power requirements:
Vibrator Power Supply PE-237 (vehicular battery 6, 12, or 24V)
Hand Crank Generator GN-58
Gasoline Engine Driven Power Supply PE-162
Dynamotor Power Supply DY-105 (vehicular battery 24V only)
Dynamotor Power Supply DY-88 (vehicular battery 6, 12, or 24V)
Battery BA-48 (receiver only)

Individual voltages and currents required by the transceiver are as follows:

Receiver
1.4V at 450ma (max), 105V at 20ma (max)
4.5V at negligible current (audio bias battery)

Transmitter
Voltage outputs from the PE-237 Vibrator Power Supply:

Function	Early Models	Late Models
Transmitter plates	475V at 90ma	580V at 100ma
Transmitter filaments	6.5V at 2amp	6.6V at 2amp
Receiver plates	105V at 42ma	120V at 45ma
Receiver filaments	1.35V at 450ma	1.5V at 500ma
Keying relay	6.0V at 500ma	6.9V at 575ma

Audio input/output devices:
Microphone: T-17 (or carbon type terminated with PL-68 plug, such as RS-38, etc.); also T-45 lip mike or T-30 throat mike with cord CD-318-A
Speaker: LS-7 (4K ohms impedance, PL-55 plug)
Headset: HS-30 in-the-ear unit

Antenna connection: Binding posts

Whip antenna: Mast sections MS-116A(3), MS-117A, and MS-118A and insulator IN-127; used with Counterpoise CP-12/CP-13 (see text)

Other antennas: Wire antennas AT-101 and AT-102 (see text)

Antenna tuner: Built in wide-range antenna tuner capable of handling whips, dipoles, or long wires

Tech Manuals: TM 11-263 "RADIO SET AN/GRC-9," description, operation, theory, troubleshooting, repair, alignment, and schematics for the basic radio set. Also covers the GN-58, PE-237. The author's copy is dated March 1951.

TM 11-263 "RADIO SETS AN/GRC-9, AN/GRC-9A, AN/GRC-9X, and AN/GRC-9Y," same as above, but a later version (author's is dated June 1956) that also covers the DY-88.

Construction Style: Receiver and transmitter are on separate panel/chassis units retained by latches in a common case. The front of both units is protected by a distinctive heavy wire cage arrangement. Internally, the wiring is point-to-point, typical of the era. Most components are easily reached and all are replaceable/repairable.

Description:

The GRC-9 is one of the more interesting U.S. military radio sets, and is a knob and button lover's dream. You really have to work to get your 15 watts CW or 7 watts AM out of this beast!

GRC-9 transmitter chassis

GRC-9 transmitter chassis, opened

In the transmitter chassis view, the 2E22 final tube is visible at center. One of the six sockets used for optional crystal control is visible between the oscillator and modulator tubes at left center. Note the large slugs partially inserted into the antenna tuning coils at top center. They operate off the ANTENNA TUNING control via a rack and pinion arrangement. The big relay behind the OC3 at center right is the T/R relay. They sure don't make 'em like this any more!

The bottom compartment of the transmitter chassis hinges open for access to the tuning capacitor and other components, which are fairly tightly packed into this engineering marvel.

Operation:

Of the many ways possible to power these sets, the author has two -- the DY-88 Dynamotor Power Supply and the GN-58 Hand Crank Generator. Setup and operation for these will be described here. The pin-outs at the POWER connector on the transmitter section are given here in case you have no power supply at all and wish to build your own:

Pin 54: +6.3 Regulated, filaments
Pin 57: +105 regulated TRANSMIT
Pin 56: +1.4 STANDBY (receiver and transmitter filaments)
Pin 46: +1.4 TRANSMIT (transmitter filaments)
Pin 53: COMMON GROUND (all voltages)
Pin 47: +500 (PA plate)
Pin 58: +105 STANDBY (receiver and transmitter)
Pin 45: +6.3 KEY (voltage here is used for keying when PTT or key closed)
Pin 55: SEND CONTROL (not a keyline—see note below)

Note: Pin 55 is grounded when the OFF-SEND-STANDBY switch on the transmitter section is set to SEND. This is what energizes the transmitter start relay in the PE-237, and dynamotor power supplies. No voltage is required here.

Pin-outs for the BATTERY connector on the transmitter section are as follows:

Pin A: +1.4 receiver filaments
Pin B: +90 receiver B+
Pin C, D: COMMON GROUND

Note: This connector powers the receiver only.

For operation with the DY-88 Dynamotor Supply, you'll need a source of 6VDC (22A), 12VDC (12A), or 24VDC (7A) to operate the set on high power transmit. The cable that interconnects the various supplies and hand generator is Cord CD-1086. Here are the pin-to-pin connections if you don't have the cable:

Cord CD-1086 Pin-outs:

Supply or Hand Generator connector pin	Transmitter POWER connector pin
20	54
21	57
22	56
23	46
24	53
32	47
33	58
34	45
35	55

Caution: The TM warns about possible faults in some CD-1086 cables and specifies that all cords be inspected before use and repaired if necessary:
a. Ground of cable shield may be connected to Pin 57 instead of Pin 53 in Plug PL-279 (female end).
b. Cable shield may not be dressed properly at ends, resulting in the sharp edges of the shield cutting into the cable connectors.
c. Cable connector inserts may be cracked or split.

The author's cable showed symptoms b. and c. and needed repair.

The cable that connects DC power to the supplies is Cable CX-2031. Here are the pin-outs for the POWER IN jack on the DY-88 if you don't have the cable:

Pins A,B: POSITIVE
Pins C,D: NEGATIVE

Warning: The window in the DY-88 panel at lower center must display the correct operating voltage ("6", "12", or "24") before applying power. If it doesn't, remove the chassis from its case and correctly set the internal switch. (See DY-88 section).

Here are the "condensed" operating instructions from the TM:

TO TURN ON RADIO SET

1. Turn switch E on transmitter to OFF.
2. Turn on the power supply.
Note. Be sure the key or mike switch is not closed.

TO RECEIVE

3. Turn switch E (on xmtr) to SEND when communicating with another set. For reception only, turn E to STANDBY (except when using the generator power supply).
4. Turn switch L to the type of reception desired.
5. Turn switch M to the band position desired.
6. Turn control N to the frequency desired and lock the TUNING dial. For CW reception, control N must be turned slightly above or below the transmitted signal frequency.
7. Turn controls O and P for a comfortable listening level in the headset. Rotating these controls counterclockwise reduces the volume in the headset.

TO TRANSMIT

8. Turn switch D to the type of transmission desired.
9. Turn switch A to the highest numbered position of the antenna used.
10. Turn switch F to the band position desired.
11. Turn control I to the dial setting corresponding to the frequency of transmission as given on the calibration chart. Lock the tuning dial in place.
12. See that the red dots on indicator B lenses are adjacent to each other.
13. Turn switch E to SEND
14. Turn control C while pressing the key down or closing the mike switch and TUNE FOR MAX INDICATOR GLOW on INDICATOR lamp (B). Lock the switch knob in place.
15. For PHONE operation (switch D), wait 2 seconds after closing the mike switch before speaking. When operating the mike, speak in a loud and clear voice.

TO TURN OFF RADIO SET

16. Turn switch E to OFF
17. Turn the power supply switches OFF.

Whew! You could get downright kilt waiting for your radioman to tune up so he can cancel off that off-target mortar barrage!

Quite a few important little details are glossed over in these condensed instructions, the most important of which is getting on frequency. This is no small feat with a separate receiver and transmitter and no direct frequency readout. The tuning chart will get you close, but you'll have to rock the receiver some to find your exact frequency.

Net Operation. The NET position of control L allows the transmitter to be tuned exactly to any frequency on which the receiver is operating. This is the situation that often arises when all transceivers in the net are to be tuned to the same frequency. For net operation, proceed as follows after the radio is set up for operation:

a. Turn switch E to SEND or STANDBY position depending on power supply.
b. Receive the desired signal with the receiver TUNING control L on either PHONE or CW and lock the tuning dial.
c. Observe the frequency of the net control station and, referring to the transmitter calibration chart, adjust the transmitter FREQ. CONTROL dial I to the approximate frequency.
d. Turn switch L on the receiver to the NET position.
e. Turn switch D on the transmitter to CW.
f. Tune the transmitter FREQ. CONTROL I for zero beat.

Note: Don't press the key or mike, because the transmitter will transmit full power and block out the received station.

g. After zero beat is found, lock the FREQ. CONTROL knob.
h. Reset controls D and L to the correct modes of transmission/reception.

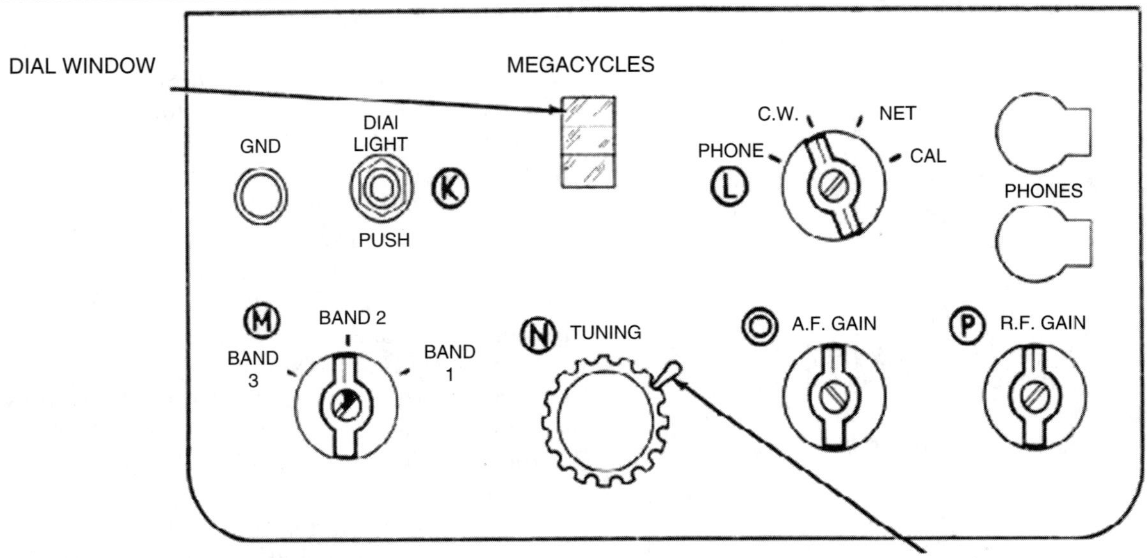

GRC-9 receiver front panel

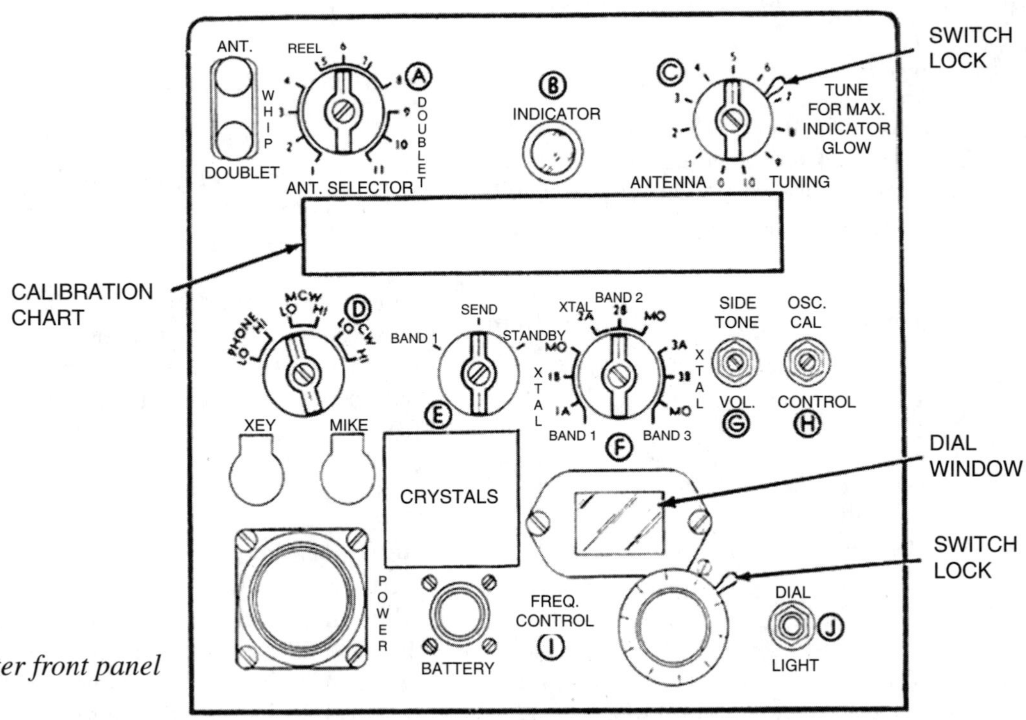

GRC-9 transmitter front panel

Transmitter Calibration. Since the GRC-9 receiver is zeroed to the transmitter, some way must be found to ensure the accuracy of the transmitter. To do this, the receiver must first be calibrated to the closest 200 kHz point to the desired operating frequency. Assuming operation on 5835 kHz is desired, proceed as follows:

a. Calibrate the receiver at 5800 kHz by setting control L at CAL and tuning the receiver for zero beat at or near 5800 kHz.
b. Turn control F to MO (master oscillator) for the band being used.
c. From the transmitter calibration chart, determine the dial setting corresponding to the calibration check frequency (in this case 5800 kHz) and turn FREQ. CONTROL knob I to that dial setting.
d. Turn control L to NET.
e. Set control D on the transmitter to CW.
f. Turn control E to SEND.
g. Turn AF GAIN and RF GAIN to about mid-position.
h. Adjust OSC. CAL. CONTROL H on the transmitter with a screwdriver for zero beat. Calibration for the check frequency of 5800 kHz is now correct, as well as for the rest of the 200 kHz range to next check frequency (5800-6000 kHz).
i. Move the FREQ. CONTROL knob I to the correct value for operation on or near 5835 kHz.

Note: According to the calibration chart on the author's radio, the closest frequency to 5835 given is "5800 +40kc" or 5840. You have to interpolate to get between indicated settings.

j. Return control L to CW or PHONE for normal operation.

Additional Info on ANTENNA TUNING control C. Sometimes it can be difficult to get a resonance indication, or the indication may be indistinct. From the detailed tuning instructions in the manual at the point the mike switch or key is depressed:

(9) Rotate control C until INDICATOR B glows at its maximum intensity.
(10) If the indicator does not glow through a complete sweep of knob C, turn ANT. SELECTOR knob A to the next lower number. Rotate knob C again until maximum glow is seen. If there is still no glow, move A down another setting.
(11) If the indicator glows with more than one setting of knob A, always use the highest numbered position of this switch at which the indicator will glow. When using a long-wire antenna it is sometimes difficult to see any indication of resonance on INDICATOR B when control C is tuned through resonance. In that case, temporarily remove the antenna lead from the antenna binding post; adjust control C to give maximum INDICATOR glow, then reconnect the antenna lead and proceed with normal operation.

[The GRC-9 is the only radio in the author's memory whose tune-up procedure in some circumstances involves transmitting without an antenna! A caution then appears:]

Caution: When tuning for an indication of resonance, never leave the transmitter on for more than 15 seconds with the antenna lead disconnected from the ANT. binding post.

No kidding!

For operation with the Hand Crank Generator GN-58, proceed as follows [have an ambulance ready]:

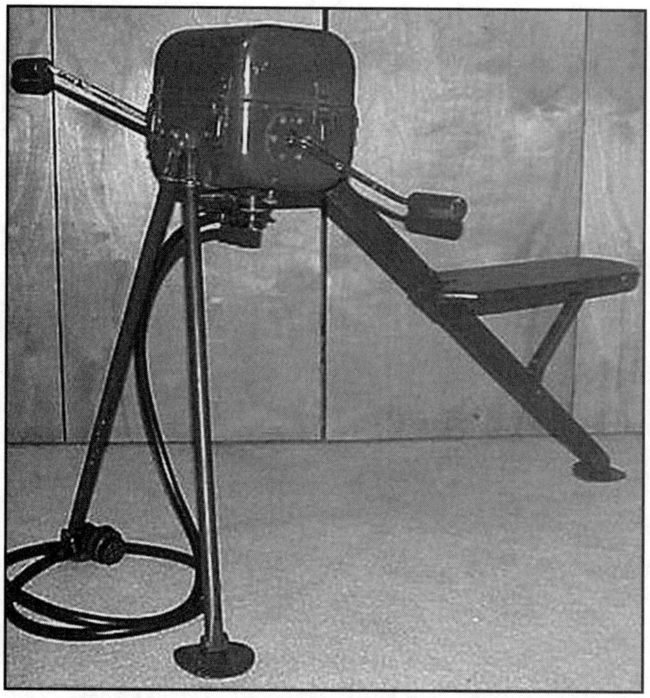

Hand Generator GN-58/GRC-9 a.k.a. thinly disguised torture machine

There's an old saying "What doesn't kill you makes you stronger." Running the GRC-9 off this thing will give you a chance to test that out. Operation is the same as with the power supplies above, except there's no power switch -- as soon as you start cranking, power can be turned on at OFF-SEND-STANDBY control E. Start by moving through SEND to STANDBY which will run the receiver only. Crank clockwise as viewed from the right hand side as you're sitting on the seat, at about 1 rev/sec.

When your operator switches to SEND you'll really feel the difference when the always-on portion of the transmitter filaments kicks in. On key-down, the load becomes noticeably greater when the set draws filament current and 500V B+ for the 2E22. You will never work so hard for seven watts of AM! Running the PRC-74 off the G-77 12VDC generator is a cakewalk in comparison, and the 74 delivers twice the power in SSB mode! The GRC-9 was a popular and widely-used radio, but it's a cinch the troops appreciated the advent of the PRC-74B because, if for no other reasons, of the reduction in packing weight and cranking effort!

Additional Comments:

The GRC-9 has all sorts of interesting quirks that can't be expressed in prose -- you just have to get one and use it for a while. Various tips from a friend with extensive GRC-9 experience:

Tip: The 2E22 final has a life of 200 or so hours. It will still light, but with little filament emission, as the filament is of the instant heating kind for a short delay using push to talk operation. They are getting a little pricey. Fair Radio Sales gets 20 bucks for used one, and 40 for new, old stock.

Tip: Make a two-foot length of RG-58 or RG-59 coax with a PL-259 connector on one end and tinned lead on the other for connection to the radio. With the addition of a barrel connector, dipoles and HF mobile antennas with PL-259s can be used.

Tip: A 40-watt clear light bulb and socket is just right as a dummy load for this radio.

Tip: There are two bias battery types in the receivers; one is soldered in and the other plugs in. With either one, you can cut them apart and install either three N-cells or three large mercury-style camera batteries (total 4.5V).

Caution: Don't rough-handle the IN-127 antenna insulator. They are fragile and have broken in the past.

Tip: Try using a carbon mike element from an old dial phone, like a Western Electric Model 500. This will raise the modulation level to almost 100%.

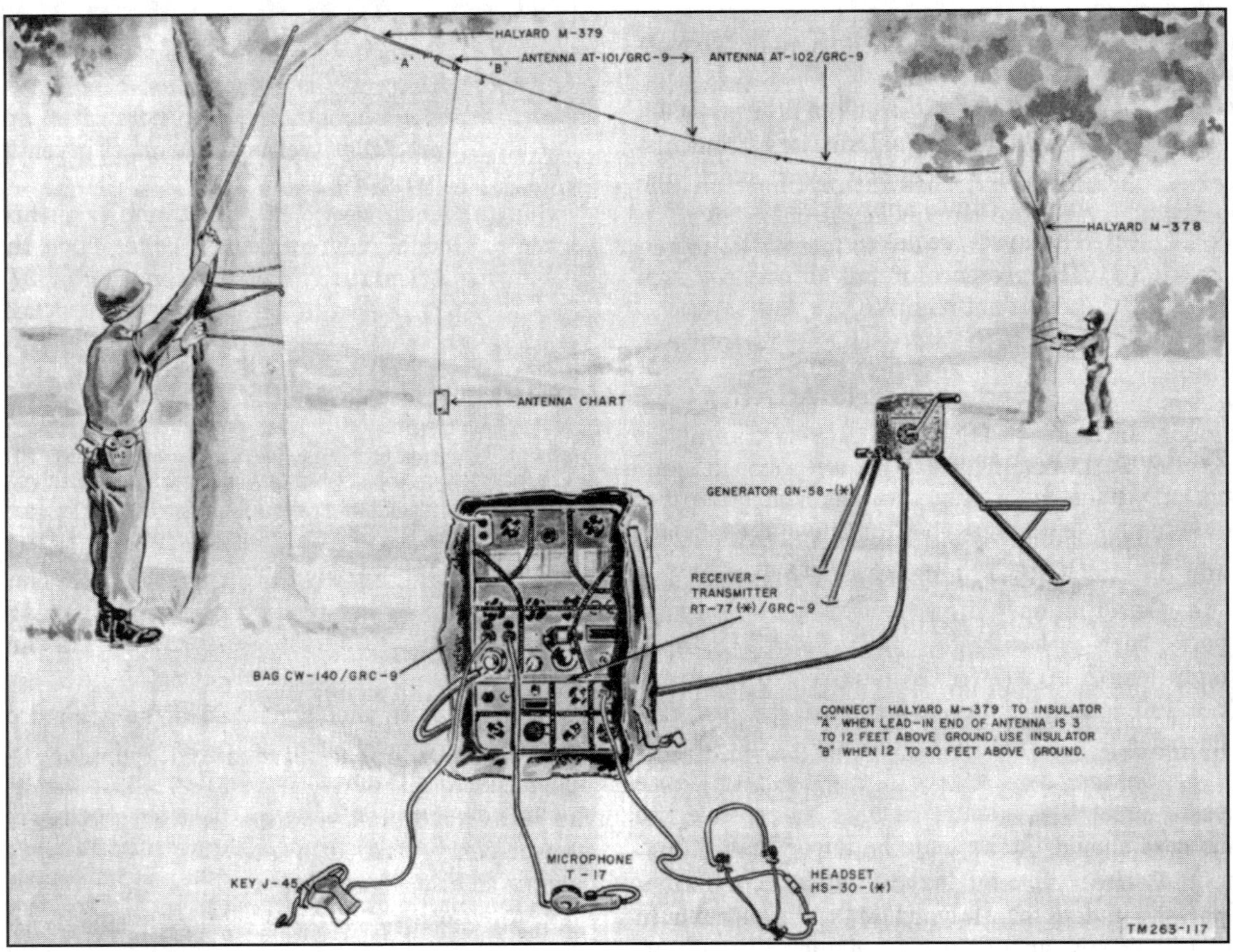

Complete GRC-109 field setup with AT-101 and AT-102 antennas, and GN-58 generator

Tip: The author's friend suggests always tuning up these radios with a field strength meter. After reading about the weirdness that can happen during tune-up, it sounds like a good idea! The military surplus answer is the Meter, Field Strength ME-61/GRC (see the ME-61 section).

What To Do If It's D.O.A.:

Even though there are an astounding number of buttons and knobs, this is really a fairly simple radio in concept. Most faults can be traced to a bad tube.

Tip: Tube testers are fine for general functionality tests, but nothing is as good as straight substitution if at all possible. This is mentioned here because of a problem one of the author's friends had with his GRC-9. It was exhibiting a really terrible CW chirp -- these radios aren't so hot in that category even when they're operating correctly, but his problem was exceptionally hideous. He spent many hours tracing, troubleshooting, re-timing T/R relay contacts and all sorts of other stuff. When he finally found the problem, it was by substituting another oscillator tube.

If you experience distorted audio out of the receiver, the audio bias battery located on the left rear of the receiver chassis may be bad. The author found his was missing altogether! Older receivers have soldered-in batteries; later production receivers have plug-in batteries. The bias is 4.5V negative, so make sure you hook the positive side of whatever solution you decide on to chassis ground. The bias current is tiny -- less than 1 mA.

Accessories:

The GRC-9 accumulated an amazing number of ancillary items, power supplies, and antennas over the years it was used by U.S. and NATO forces.

The antennas consist of wire sections interconnected with porcelain insulators that can be jumped in or out of the total length. A chart affixed to the antenna reel shows which insulators to jump in/out for operation on different frequencies.

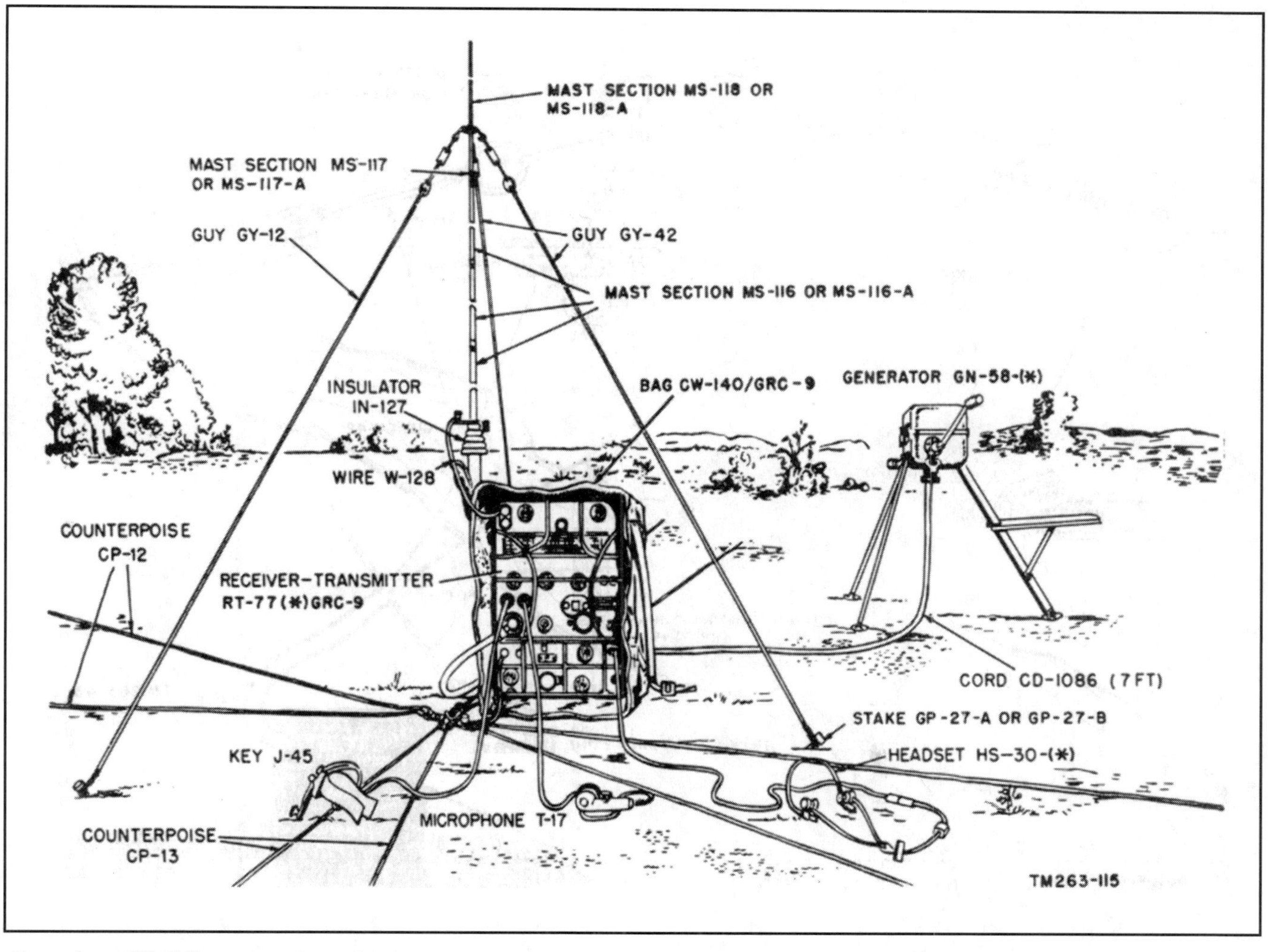

Complete GRC-9 setup using whip antenna

Note: The author's TM dated June 1956 has a "Note" covering the AT-101 antenna, reproduced verbatim here: "Some antennas were wired incorrectly and require the addition of jumper wires. Refer to MWO SIG 11-263-2, Modification of Radio Set AN/GRC-9 to Correct Error in Production of Antenna AT-101/GRC-9." No further data is given, so it's not possible to determine if they meant that only the jumpers were left off the insulators, or the entire insulator was left off.

Counterpoise CP-12, a group of 8 radials connected by a common wing nut, comprises the ground system. Guy Wires GY-12 and GY-42 meet at a metal clamp ring on the mast to steady the whip and radio. Pretty slick design.

A doublet (dipole) antenna is mentioned in the TMs and is a selection on the ANT. SELECTOR switch, but there is little mention on how to actually set it up physically. They do mention it should be fed with 50 to 72 ohm transmission line.

GRC-109

Application:

Compact, portable, weather-resistant CW radio station often used in clandestine or Special Forces roles. Shown above are Radio Transmitter T-784A (left), Radio Receiver R-1004A (right), and Power Supply PP-2685A (top). Not shown here but covered later is Power Supply PP-2684A. The set is shown assembled for operations using a common antenna system for the receiver and transmitter.

Tech Specs:

Transmitter T-784A

Frequency Range: 3 to 22mHz in four bands (3-6, 6-10, 10-17, 17 to 22mHz)

Type of modulation and power out: CW, 10 to 15W, depending on frequency

Type of frequency control: FT-243 crystals are generally used, at the carrier frequency, 1/2 the carrier frequency, or 1/3 the carrier frequency. Operation at 1/4 the carrier frequency is possible with degraded output.

Size: 9.75" x 8" x 5.5"

Weight: 11 pounds

Active devices: Vacuum tubes, 6AC7, 2E26

Power requirements: Usually supplied by Power Supply PP-2685A or Power Supply PP-2684A which produce 450VDC at 100mA and 6.3VAC or 6.3VDC at 1.2A

Antenna connection: push-type binding posts

Antenna: Single horizontal wire 25 to 75 feet long specified, depending on frequency

Other antennas: numerous other antennas possible

Antenna tuner: integral tuning and loading controls support a wide range of antenna impedances

Receiver R-1004A

Frequency Range: 3 to 24mHz in three bands (3-6, 6-12, 12-24mHz)

Modes received: AM, CW, MCW

Type of frequency control:
Continuous tuning—with VFO
Fixed-frequency—with FT-243 crystal used in local oscillator.

Size: 9.75" x 8" x 5.5"

Weight: 12 pounds

Active devices: Vacuum tubes, 1T4(4), 1L6, 1U5

Power requirements: Usually supplied by Power Supply PP-2685A or Power Supply PP-2684A which produce 108VDC regulated at 20mA (90V nominal will work fine)
1.5VDC regulated at 300mA

Antenna connection: push-type binding posts

Antenna: Same as transmitter, or can be set up independently

Antenna tuner: none in receiver

<u>**Power Supply PP-2685A**</u>

Power in: 75 to 260VAC, 40 to 400Hz

Voltages out: As listed for R-1004 and T-784 above

Active devices: 0B2 regulator

Size: 9.75" x 8" x 5.5"

Weight: 19 pounds

<u>**Power Supply PP-2684A**</u>

Power in:
75 to 260VAC, 40 to 400Hz
6VDC at 13A key down, 5A key up

Voltages out: As listed for R-1004 and T-784 above
Also 6VDC at 3.8A for battery charging

Active devices: 0B2 regulator

Size: 10.5" x 8.5" x 5.5"

Weight: 25 pounds

Tech Manuals: TM 11-5820-474-14 "OPERATOR'S, ORGANIZATIONAL, DIRECT SUPPORT, AND GENERAL SUPPORT MAINTENANCE MANUAL, RADIO SET AN/GRC-109"; description, operation, theory, troubleshooting, repair, schematics.

TM 11-5820-474-24P TECHNICAL MANUAL, ORGANIZATIONAL, DIRECT SUPPORT, AND GENERAL SUPPORT MAINTENANCE, REPAIR PARTS AND SPECIAL TOOLS LISTS FOR RADIO SET AN/GRC-109"; detailed parts list only, no schematics.

Construction Style: Heavy-duty die-cast front panel and chassis assembly slips into die-cast metal outer case in all units. Point-to-point wiring, 1940s and '50s-style. All units are repairable. Reaching components can get tricky due to very compact construction, especially in the receiver.

Description:

Much good detail has been written about the various GRC-109 configurations and the history of the set (do a web search on "GRC-109" or "RS-1"). The goal here is to get you on the air as quickly as possible, so the detail will be limited to mentioning that you'll find GRC-109s in three basic configurations. The first is the one pictured above, the lettered or "A" models. These units have heavier die-cast cases with protected hasp fasteners for the covers, and spring-loaded carry handles.

The second variety of GRC-109 is the unlettered model, consisting of the R-1004, T-784, PP-2684, and PP-2685. These units have covers retained by thumbscrews.

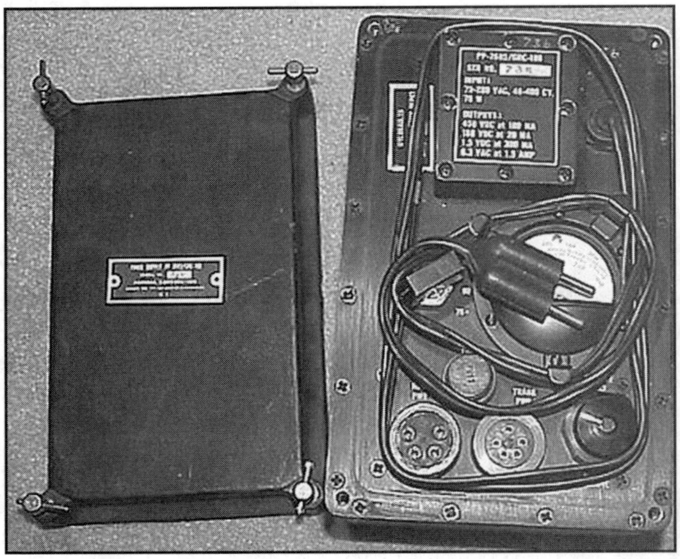

PP-2685/GRC-109, unlettered version, showing thumbscrew cover fasteners

The supply illustrated is the author's only unlettered GRC-109 component. Like the T-784A, the T-784 has the jack for the GRA-71 Burst Coder. (Covered in its own section)

The third version of the set was the RS-1, designed for the CIA in the late 40s, and chronologically the first set in the family. It's essentially the same as the unlettered GRC-109 except it's missing the jack for the GRA-71. The author also has a single piece from this set, the RR-2B receiver.

The RS-1 consists of the RR-2B receiver pictured, the RT-3 transmitter, and RP-1 and RP-2 power supplies. The individual units in the RS-1 are "sanitized" —no date codes on any parts, except (sometimes) the tubes. And tubes wear out so they may not be original. There is also no indication of manufacturer. R-1004 and R-1004A internals are virtually identical. These little units are beautifully made and probably represent the ne-plus-ultra in 6-tube HF receivers.

RR-2B Receiver

RR-2B internals, chassis covers removed, showing compact construction.

RR-2B internals, top view.

Operation:

As far as the author has determined, the various power supplies/receivers/transmitters all interchange. In case you have only the receiver or transmitter with no power supply at all, here are the power pin-outs, shown directly on the connectors:

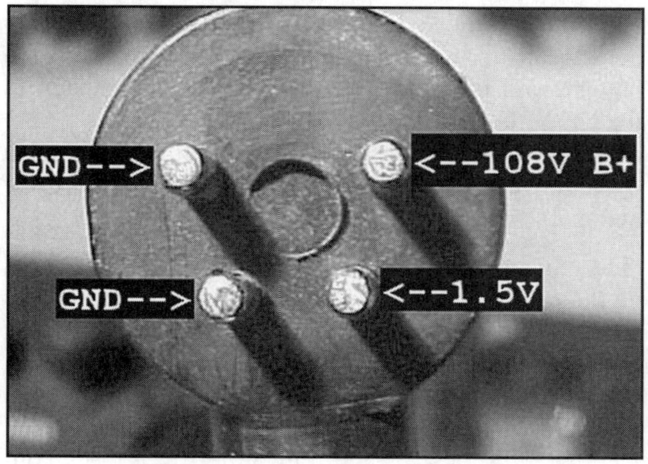

Receiver power connector pin-outs.

Transmitter power connector pin-outs.

Selected text from the operating instructions in the TM, author's comments in []:

16. Damage from Improper Settings. Take the following precautions when setting the power supply controls:

a. Before connecting either power supply to the ac source, be sure that the power selector switch is in the OFF position.

b. If the A.C. VOLTS meter indicates a voltage value that is between markings on the power selector switch, always use the higher numbered switch setting. Example: If the A.C. VOLTS meter reads 120 volts, turn the power selector switch to the 130 position. Never advance the power selector switch to a setting lower in numerical value than the A.C. VOLTS meter indication. If the switch is set to a value lower than the voltmeter indication, the 2-ampere fuse will blow. [Or you may damage the radio with overvoltage…]

c. Do not turn the CHARGE-OPERATE switch on the large power supply to the CHARGE position when operating from a 6-volt storage battery. To charge a battery, the power must be available from the ac source. The transmitter and receiver cannot be operated with the large power supply during charging. For charging, turn the power selector switch to the proper position as indicated on the A.C. VOLTS meter, connect the battery leads to the proper terminals on the battery, and turn the CHARGE-OPERATE switch to the CHARGE position.

d. Preliminary Connections, Common Antenna System. [Ie: same antenna connected to both receiver and transmitter].

(1) Connect the H-65/U to the PHONES binding posts on the receiver.
(2) Connect the antenna lead-in to the ANT binding post on the transmitter.
(3) Connect a length of rubber-covered wire between the ANT binding post on the receiver and the RCVR ANT binding post on the transmitter. Keep this connection as short as possible. [The setup in the image at the top of this section works fine].
(4) Connect a wire between the GRD binding post on the receiver and the RCVR GND. binding post on the transmitter. Keep this connection as short as possible.
(5) Connect the ground system to the GND binding post on the transmitter.

e. Preliminary Connections, Separate Transmitting and Receiving Antennas. Separate transmitting and receiving antennas may prove to be more effective under poor transmission conditions, particularly at lower frequencies. Separate transmitting and receiving antennas are recommended when using an external high-speed keyer.

(1) Connect the headset to the PHONES binding posts on the receiver.
(2) Connect one lead-in to the ANT binding post on the transmitter.
(3) Connect the other antenna lead-in to the ANT binding post on the receiver.
(4) Connect the RCVR GND binding post on the transmitter to the GND binding post on the receiver with a short wire.
(5) Connect the ground system to the GND binding post on the transmitter.
(6) Connect the REC ANT binding post and the REC GND binding post on the transmitter with a short wire.

15. Unit Interconnections, Primary Power Options

This paragraph contains instructions for connecting the transmitter and the receiver to the optional power sources.

Caution: Do not make connections to primary power sources until all units are interconnected.

Note: When not using the hand-cranked generator, be sure that the cover is in place on the HAND GEN INPUT receptacle on the large power supply.

a. Ac Line Operation.

(1) Plug the transmitter interconnecting cable into the TRANS. PWR. receptacle on the large or small power supply.
(2) Plug the receiver interconnecting cable into the RCVR PWR. receptacle on the large or small power supply.
(3) Turn the power selector switch on the power supply to OFF.
(4) Connect the power supply power cord to the ac source.

b. 6 Volt Storage Battery Operation.

(1) Plug the transmitter interconnect-ing cable into the TRANS. PWR. receptacle on the large power supply.
(2) Plug the receiver interconnecting cable into the RCVR PWR. receptacle on the large power supply.
(4) Turn the CHARGE-OPERATE switch to OPERATE.
(5) Connect the red battery lead to the positive terminal of the battery.
(6) Connect the black battery lead to the negative terminal of the battery.

OPERATION INSTRUCTIONS

18. Preliminary Starting Procedure Refer to the installation procedures before operating; operating procedures differ with various power sources.

a. Plug in the proper transmitting crystal for the desired operating frequency. The crystal frequency may be the operating frequency, or it may be one-half or one-third the operating frequency. Crystals equal to one-fourth the operating frequency may also be used, but the transmitter output power will be reduced.
b. Consult the tuning chart on the transmitter and turn controls (1), (2), and (3) to the settings indicated on the chart.
c. Set the RANGE switch on the receiver to the band that includes the frequency of the signal to be received.
d. Adjust the receiver TUNING control until the operating frequency appears under the hairline on the frequency dial.
e. If the receiver is to be operated with crystal control, plug in the crystal that will produce the desired operating frequency. The proper receiving crystal determined by adding or subtracting 455 kilocycles (kc)

from the operating frequency to obtain the control frequency. The crystal can then be selected for one-half or one-third the control frequency. For convenience in operation, mark the receiving crystals "receiver" and stamp them with the receiving frequency.

19. Starting Procedure, Power Supplies
With the controls set as given in the preliminary starting procedure, perform the steps described below.

a. For operation from an ac source, observe the indication on the A.C. VOLTS meter, and then advance the power selector switch on the power supply to the setting that corresponds to the voltmeter indication. If the voltmeter indicates a value between switch settings, set the switch at the higher numerical switch position.
b. For operation from a 6-volt storage battery, turn the CHARGE-OPERATE switch on the large power supply to the OPERATE position. Turn the power selector switch to the BAT. position.
c. For hand-cranked generator operation with the large power supply, leave the power selector switch in the OFF position.
d. Allow several minutes for the equipment to warm up.

20. Starting and Tuning Procedure, Radio Transmitter T-784/GRC- 109
[**Note:** The control number in the procedure corresponds to the control number on the front panel].

a. Depress the telegraph key.
b. Adjust control (2) for maximum brightness on the exciter lamp. If maximum brightness is obtained at more than one setting of the control, turn the control to the nearest figure indicated on the tuning chart.
c. Adjust control (3) for maximum brightness on the power amplifier lamp. If maximum brightness is obtained at more than one location of the control, adjust the control to the nearest figure indicated in the tuning chart.
d. Adjust control (4) slowly, first to the left then to the right; at the same time, readjust control (3) to obtain maximum brightness of the antenna lamp. Maximum brightness of the antenna lamp is subject to wide variation, depending on the antenna length and the operating frequency. If the illumination of the antenna lamp is perceptible, make the initial antenna adjustment by turning control (4) for maximum brightness of the power amplifier lamp. Adjust control (3) to give maximum brightness of the power amplifier lamp each time control (4) is adjusted. Continue this procedure, alternately adjusting controls (3) and (4) for maximum brightness of the power amplifier lamp, until a perceptible glow is observed in the antenna lamp; then adjust controls (3) and (4) for maximum brightness of the antenna lamp.

[In practice, the transmitter hardly ever tunes up exactly as described. Setting the controls for absolute brightest indication on all lamps is not necessarily the best thing to do. Often when this is done, the indicator associated with Tuning Control (4) may go out altogether, indicating no (or weak) power out. Moving the Oscillator Control (2) slightly off max will usually fix this. You'll get more of a feel for this with experience.]

e. When further adjustment of controls (3) and (4) does not yield an increase in antenna lamp brightness, adjust control (2) very slightly for maximum brightness of the antenna lamp. This final adjustment of control (2) is very important, especially if the crystal being used is one-fourth the operating frequency.

[The TM is definitely correct here when it says the Oscillator setting (2) greatly affects the transmitter operation. The tuned circuits here are very touchy.]

f. Release the telegraph key; the transmitter is now tuned.

21. Starting and Tuning Procedure, Radio Receiver R-1004/GRC-109

a. Advance the GAIN control clockwise until a rushing (hissing) sound is heard in the headphones.
b. If both the receiver and the transmitter are operated on the same frequency, use the transmitter signal to tune the receiver. Under these conditions, depress the telegraph key on the transmitter and slowly rock the receiver tuning dial above and below the approximate frequency setting until a squeal is heard. Release the telegraph key; the squeal should stop at the same instant. If the squeal does not stop when the telegraph key is released, the receiver is not tuned to the transmitter frequency. Depress the telegraph key and continue to slowly turn the tuning dial; check each time a squeal is heard by releasing the telegraph key until the receiver is tuned to the transmitter frequency.

22. [Receiver] Operating Procedures
For operating convenience, a logging scale and a vernier scale are provided on the receiver. These scales can be used to accurately record the frequency of a received signal for future reference. Note that ten units on the vernier scale correspond to one unit on the logging scale.

a. Continuous-Wave Reception. Start the equipment as instructed in paragraphs 18, 19, and 21.

(1) Rotate the TUNING dial to tune in the signal.
(2) Adjust the GAIN control for the desired sound level in the headset.

(3) Adjust the BEAT OSC control until the received signal has the desired pitch.

b. Amplitude-Modulated (AM) Reception.
Start the equipment as instructed in paragraphs 18, 19, and 21.

(1) Turn. the BEAT OSC control to OFF.
(2) Rotate the TUNING control to tune in the signal.
(3) Adjust the GAIN control for the desired sound level in the headset.

Note: SSB reception is also possible. Set up as for CW reception and adjust the BFO as necessary. Operation will be very touchy since this radio was not designed with the degree of stability necessary for true SSB reception.

c. Transmission.
Start the equipment as instructed in paragraphs 18, 19, and 20.

(1) Telegraph key. After the transmitter has been tuned (para 20), transmit by operating the telegraph key.
(2) External telegraph key. If operation is desired with an external telegraph key, connect the leads from the external key to the bind-ing posts marked EXT. KEY and GND.

Note: The front-panel key and the external key are now connected in parallel; the front-panel key is still operative.

(3) High-speed keyer. To operate with a high-speed keyer, remove the dummy plug from the connector marked KEYER, and plug the high-speed keyer into the connector. [This connector won't be present in the RT-3 version transmitter].

23. Operating Precaution

Overheating may occur if the radio set is operated in an enclosed space without adequate ventilation. Always try to provide ventilation when operating the equipment.

26. Stopping Procedure

a. When operating with the large or small power supply, shut down the entire radio set by turning the power supply power selector switch to the OFF position.
b. To remove the power from the receiver only, rotate the receiver GAIN control counterclockwise to the BAT. OFF position.

HAND CRANK GENERATOR SETUP/OPERATION

The hand crank generator specified in the TM for the GRC-109 is the G-43/G. The author's TM on the G-43/G states it is used to power the GRC-9 and "is connected in the same manner as Generator GN-58, the hand generator issued with the [GRC-9]. It also states "the operation of the [G-43] is identical to that of Generator GN-58." It appears the two generators are interchangeable.

To use the GN-58 or G-43, you need either the large power supply (PP-2684) or Voltage Regulator CN-690. The CN-690 is nothing more than a small breakout box with receptacles for the receiver and transmitter power plugs, and an 0B2 regulator tube for the receiver B+. Fair Radio Sales still had these available as of Summer 2003.

a. Hand-Cranked Generator Operation with Large Power Supply.

(1) Connect the transmitter interconnecting cable to the TRANS. PWR. receptacle on the large power supply.
(2) Connect the receiver interconnecting cable to the RCVR PWR. receptacle on the large power supply.
(3) Connect the hand-cranked generator to the HAND GEN. INPUT receptacle on the power supply with the 7-foot power cable.
(4) Turn the power selector switch to OFF.

b. Hand-Cranked Generator Operation with Voltage Regulator.
(1) Connect the transmitter interconnecting cable to the proper receptacle on the voltage regulator.
(2) Connect the receiver interconnecting cable to the proper receptacle on the voltage regulator.
(3) Connect the interconnecting cable on the voltage regulator to the receptacle on the hand-cranked generator.

GRC-109 connected to GN-58 via CN-690 regulator.

The illustration shows the hand-crank generator ready for torture...er...radio operation. This is definitely a two-man operation, and hopefully you will be the one on the radio!

Additional Comments:

The author finds the suggested radio installation in the TMs to be rather humorous:

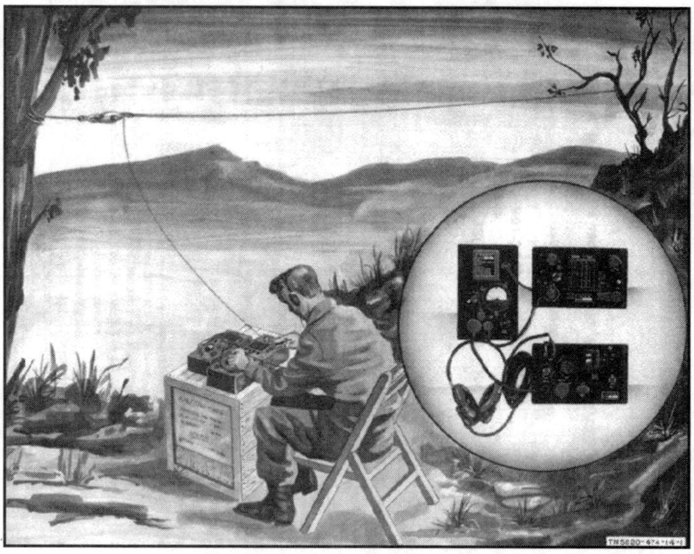

Suggested GRC-109 installation from the TM.

Somewhere in the picture, presumably behind the inset, there has to be another individual cranking away like crazy on a G-43/GN-58. Either that, or there has to be a big six volt storage battery inside that GRC-109 shipping crate the operator is using for a desk. In fact, how the heck did he get that crate out in the sticks? So, there's a couple of guys sitting around out in the open up on top of a hill, with a vehicle close by. Let's hope the good guys are maintaining complete air superiority!

What To Do If It's D.O.A.:

No indication on the power supply voltage meter when connected to the AC mains and in the OFF position indicates a probable blown fuse.

If you have both power supplies, you have some flexibility. Try both to see if one is defective. In any case, verify that the supply voltages match the tech specs.

Troubles can often be traced to bad/weak vacuum tubes. Testers will give you a good idea about the health of a tube, but the ultimate test is substitution. Hopefully, the 1L6 converter in the receiver isn't the culprit, because it's relatively rare and in demand by Zenith Trans-Oceanic fans. The author has been impressed in the past with how reliable these radios are, even after many years in storage.

Accessories:

The author has never seen the antenna reels or the 7-foot power cable.

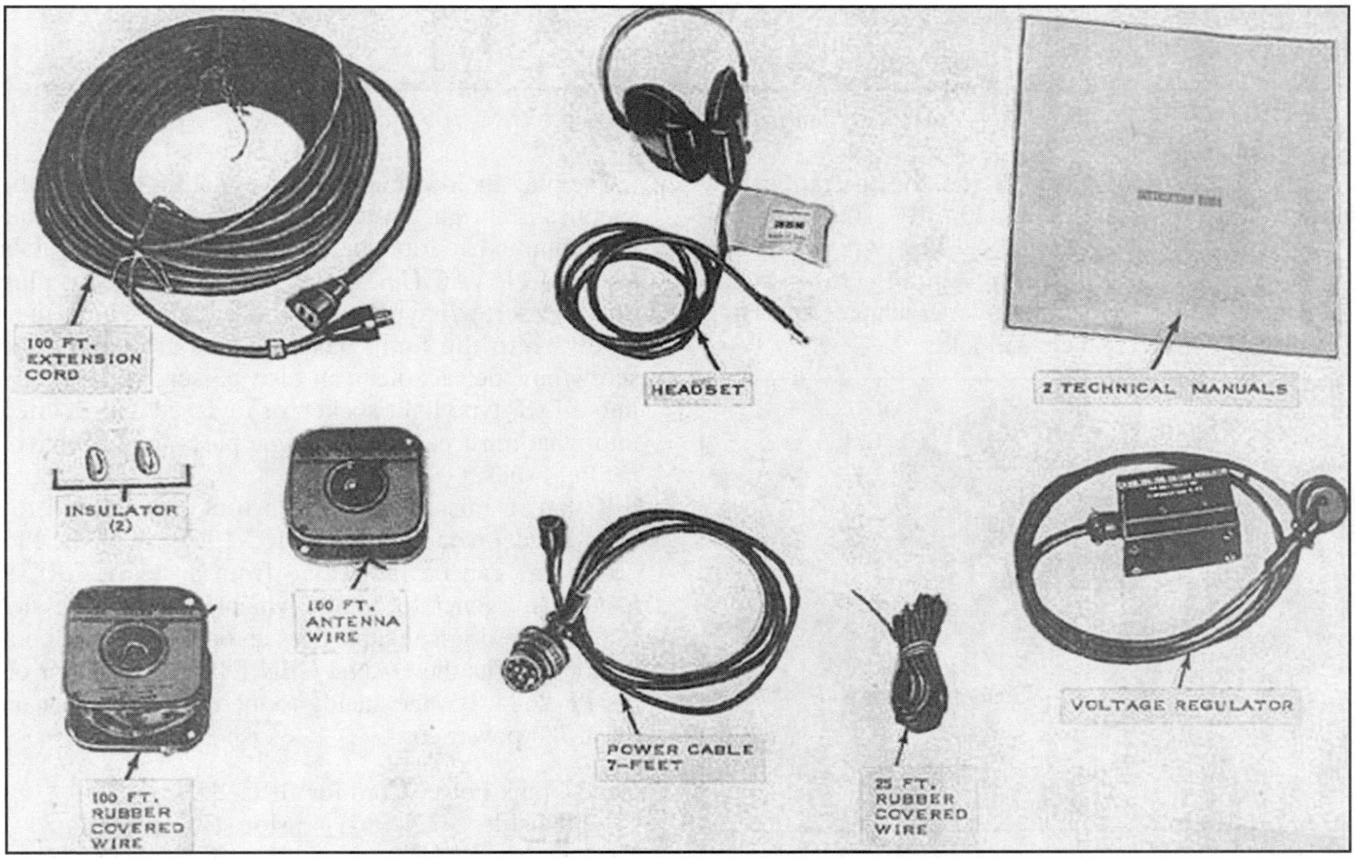

Accessories originally delivered with the GRC-109, part 1

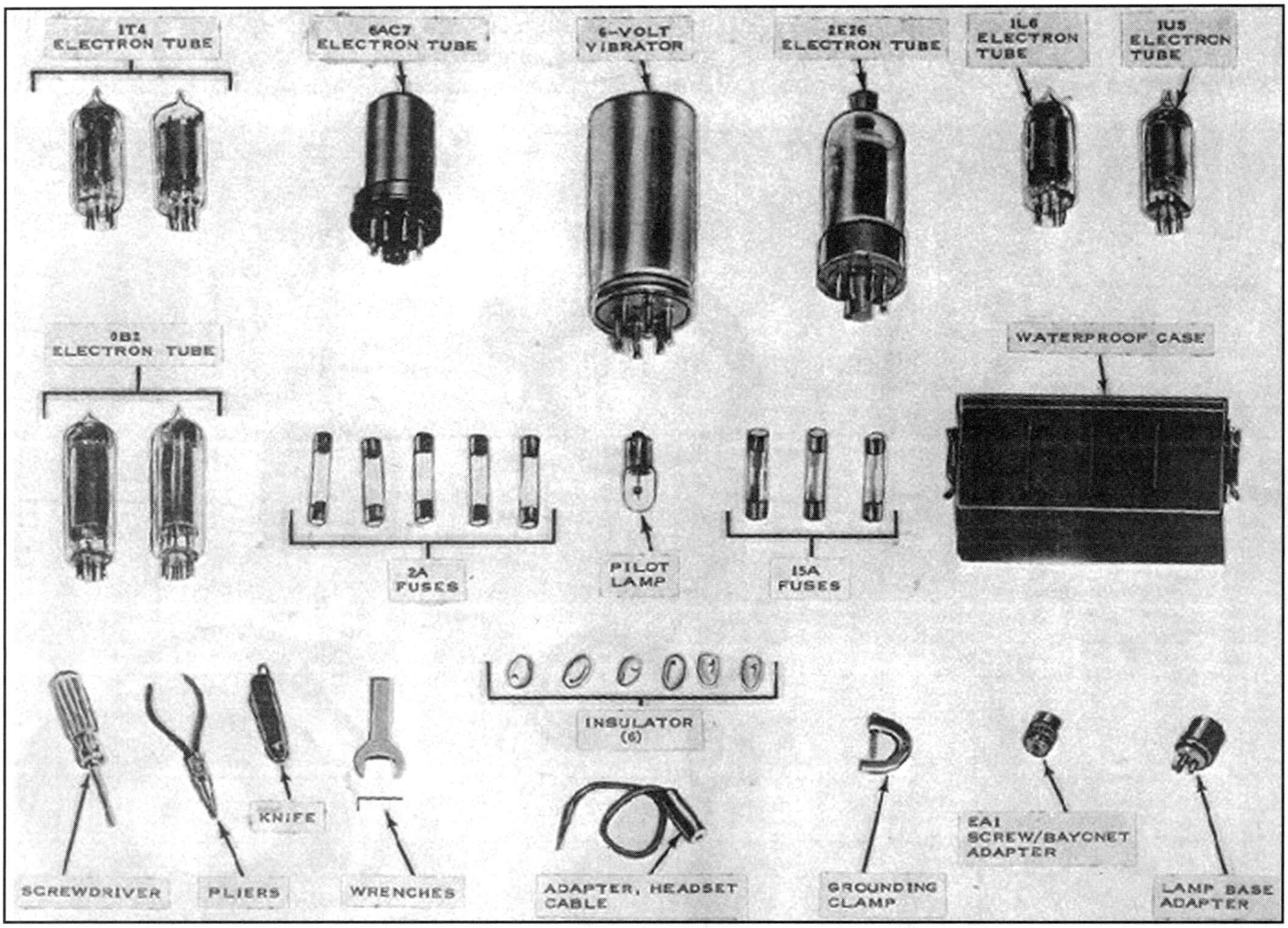

Accessories originally delivered with the GRC-109, part 2

The "waterproof case" is the Maintenance Kit, Electronic Equipment MK-833/GRC-109, and contains the above accessories. The wrench is for removal of the cap over the regulator tube on the power supplies. The screw/bayonet adapter and lamp base adapter deserve a closer look…

Power adapters, showing options

The plug molded into the power cord on each of the supplies is at right. Note how the prongs are reversible to accommodate foreign 220VAC outlets. For use in a standard 115VAC U.S. outlet, the smaller prongs plug into the screw/bayonet adapter at left, which then screws into the lamp base adapter at center. The screw/bayonet adapter can also be screwed directly into a U.S.-type light socket, or reversed and inserted into what must be a foreign-type push-and-turn bayonet light socket.

If you're missing the seven-foot power cord for hand-crank operation of the GRC-109 through the PP-2684, one can be fabricated from an extra GRC-9 hand-crank generator cable, which is probably easier to find. The original cable uses an oddball female connector that fits the HAND GEN INPUT connector on the PP-2684. It's the analog to the male connector on the T-784 power cord.

Hand Crank Power Cord for GRC-109:

PP-2684 side	GN-58/G-43 side	Function
Pin 1	24	GROUND
Pin 2		---

PP-2684 side	GN-58/G-43 side	Function
Pin 3	32	+450VDC
Pin 4		---
Pin 5	20, 34	+6VDC

Connector pin-outs for hand generator cable. Left: Power supply, Right: GN-58/GN-43

Tip: The author has had success in the past making connectors like the one for the power supply side in the following manner.

1) Fit three of the female socket inserts used in those nylon push connectors (available at Radio Shack) onto the pins so they hold, but aren't too snug. See the left image. Note these haven't been trimmed or soldered to the cable yet. You need to trim all the extra tabs off so there's no chance they'll short.
2) Cut a piece of index card, then roll and fit it into the receptacle to form a cylinder the size you want for the connector. See center image. A little trial and error may be necessary here. Index cards are cheap.
3) Tape a couple of layers of tin foil around the protruding part of the cylinder.
4) Remove the cylinder and female socket pins, solder the socket pins to the cable leads in the usual way, ensuring the cylinder top will reach above the cable insulation. Replace the cylinder and reattach the sockets.

Note: the cylinder in the author's demo is a little shorter than what you'd actually want, in order to better illustrate the method.

Making a connector for the PP-2684 to GN-58/G43 cable

5) Fill the cylinder with RTV, and let it cure thoroughly. Make sure the cable is centered during the curing process.
6) Pull out the result and remove the cylinder material.

Voila, connector!

TRC-77, TRC-77A

Application:

Six channel crystal controlled HF CW transceiver. The TRC-77A, pictured, is identifiable from the unlettered model by the new family push-and-turn audio connector instead of the original phone jack.

Tech Specs:

Type of modulation and power out: CW transmit only, 10 to 14 watts. Radio will receive SSB and AM signals.

Frequency range: Six crystal controlled channels in the range 3.0 to 8.0 mHz. Radios are typically supplied set up with the following CR-18A/U crystals:

Default TRC-77 Channel Frequencies:

Channel	Transmit (mHz)	Receive (mHz)
1	3.000	3.455
2	4.060	4.515
3	5.100	5.555
4	6.000	6.455
5	7.000	7.455
6	8.000	8.455

Receive and transmit frequencies are independently selectable.

Size: Basic radio—4.5" x 10.5" x 10"; Radio with battery box—4.5" x 10" x 17"

Weight: Basic radio—TRC-77: 12.9 pounds; TRC-77A: 11 pounds; Radio with battery box and original wet cell battery -- 25 pounds

Active devices: solid state, except for tubes 3B4 (transmitter oscillator), and 2E24 (transmitter P.A.). Older mid-60s vintage semis, like 1N3015, 1N457, 2N1039, etc.

Power requirements: 12VDC nominal, 14V gives operation to spec. (16V absolute max).

Current drain: Receive—17 mA typical at 14VDC, 11 mA at 9VDC (see text); Transmit—1.5A at 14VDC transmit mode, idle, 3.0A at 14VDC transmit mode, key down for 15 watts out

Audio input/output devices:
TRC-77—headphones with standard 1/4" phone plug (or PL-55)
TRC-77A—any of the headphones with U-229 connectors, H-140, H-251, etc.

Key supplied: J-45 leg key, with PL-55 plug

Antenna connection: Push-type binding posts

Whip antenna: not used

Other antennas: Originally supplied with AT-1098 wire antenna, consisting of 25, 40, and 57 foot wire reels terminated in pin plugs for insertion into the ANT binding post, and a counterpoise consisting of two 25 foot wires connected together at a common pin plug for the GRD post.

Antenna tuner: Internal tuner, adjusted by ANT. TUNE multi-position control

Tech Manuals: TM 11-5820-473-12 "RADIO SET AN/TRC-77 and AN/TRC-77A ORGANIZATIONAL

MAINTENANCE MANUAL", description and operation only, no theory or schematics.

TM 11-5820-473-35, "RADIO SETS AN/TRC-77 and AN/TRC-77A, DS, GS. AND DEPOT MAINTENANCE MANUAL", theory, troubleshooting, repair, and schematics.

Construction Style: Panel and chassis assembly fits into a slip-on metal case. Hybrid construction with most components on PC boards. All are replaceable/repairable. DC-DC converter hinges out allowing access to the component side of the receiver PCB.

Special equipment supported: Burst Coder Group AN/GRA-71, using the Special Cable CX-11389, or equivalent (see text)

Description:

The TRC-77 is an interesting artifact of the mid-60s, when conversion from vacuum tubes to solid state occurred. Like the PRC-25, PRC-41, PRC-47, etc., it incorporated semiconductors for much of its circuitry. But when it came to the RF-producing circuits, semis just didn't exist that could handle the job. So use of a DC-DC converter and vacuum tubes resulted.

This radio is very solidly constructed and clearly designed to survive rigorous field use. This fact, along with the possibility of split-frequency ops and compatibility with the AN/GRA-71 Burst Coder definitely points to clandestine ops.

Most radios the author has seen are missing battery boxes. This is likely due to the fact that the original box (BB-447A) consisted of 10 BB-418/U wet cell nicads containing potassium hydroxide electrolyte, a caustic base. Virtually all of the boxes seen by the author have been ruined by corrosion. The author has heard wailing in the past about the non-availability of the battery box. The solution: just make one!

This one was made out of scrap 1/2" plywood, latches from wherever, and some elbow grease. Note the decorative brass corners (painted O.D. of course) at the corners to protect the wood. The trick is cutting the wood assuming the radio opening is a sharp rectangle. Then you file the corners to fit the rounded radio opening corners. The new family bulkhead connector (U-183) is for charging. The box was made to

Author's homemade TRC-77 battery box.

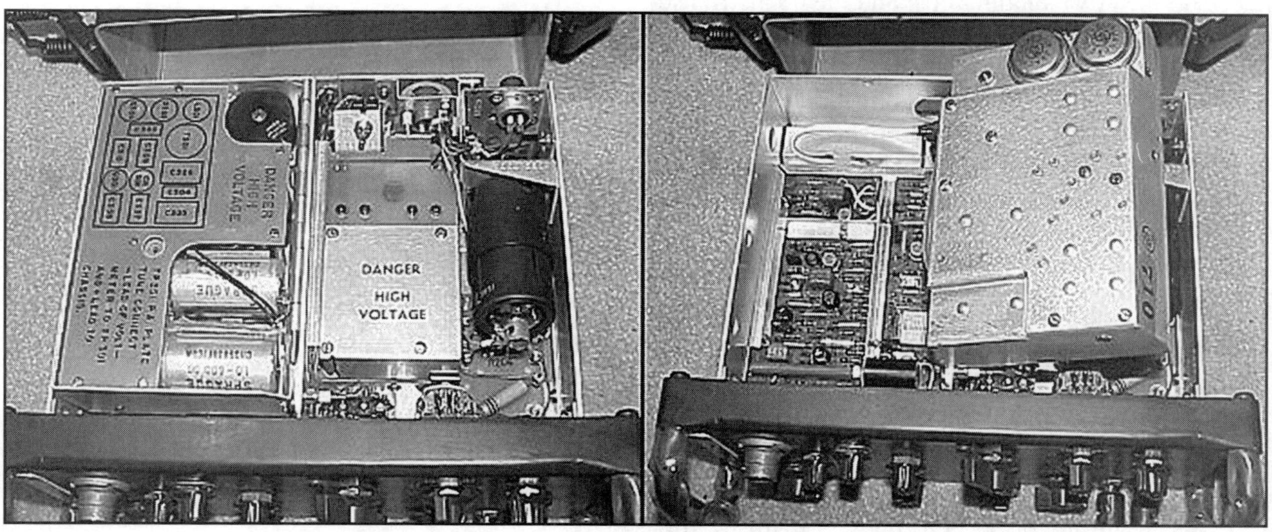

TRC-77A internals, with DC-DC converter opened for access to receiver components

duplicate the dimensions of the original. With the modern battery technology available, it could probably be made half the size.

Operation:

(See Additional Comments section for details on recrystalling the radio)

Assuming crystals are installed and the receiver and transmitter are tuned up, the pin-outs at J301, the power jack, are as follows:

TRC-77: Pins 2,3,4 +12VDC; Pins 5,6,7 NEGATIVE

TRC-77A: Pin A: +12VDC: Pin B: NEGATIVE

If you have any question about which pin is which, the red(+) and black(-) wires to the jacks are visible inside both radios.

Antenna setup with the supplied wires is as follows (from the chart given in the TM):

Suggested Wire Antenna Lengths For TRC-77:

Antenna Length	Channel	Frequency (mHz)
57 feet	1	3.000
40 feet	2	4.060
	3	5.100
25 feet	4	6.000
	5	7.000
	6	8.000

Note: This shows the wire length the tuner will match on the default frequencies generally supplied with the radios. Use this to determine which reel to use with your own frequencies. Of course, it's unlikely you can go wrong with a quarter wavelength from the formula 1/4 wave = 234/frequency in mHz. The antenna wire connects to the ANT binding post and the counterpoise is connected to the GRD post and spread out as shown on this page.

There's been some debate in the author's group about which direction of radiation is best for this type of slant wire. A friend did an analysis of the antenna with MININEC and the results show it's nearly omnidirectional, with possibly a slight advantage to the soldier's back, in the direction of the counterpoise.

Basic operating instructions from the TM (Author's comments in the []):

Types of Operation

The radio set is operated locally. It can receive both amplitude-modulated and continuous-wave signals. The transmitter can be operated by a manual telegraph key or by an automatic keying device at up to 300 words per minute.

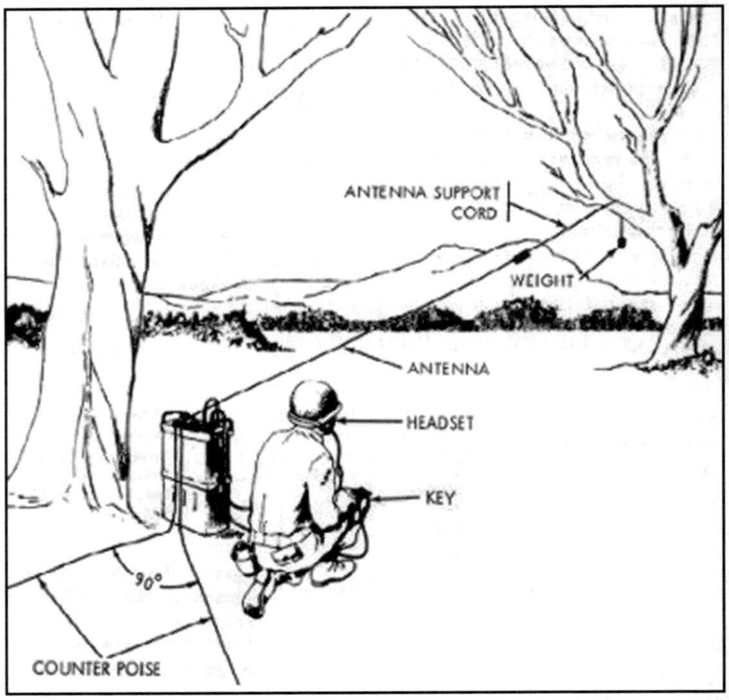

Antenna setup shown in the TM

Preliminary Starting Procedure

Perform the preliminary operations listed below before operating the equipment.

a. Set the OFF-RCVR-XMIT switch on the receiver-transmitter front panel to OFF.
b. Set the INDICATOR RANGE switch to OFF.
c. Set the XMTR CHANNEL switch to the desired transmitter channel.
d. Set the RCVR CHANNEL switch to the desired receiver channel.
e. Set the ANT. TUNE switch to position 1.
f. Set the AM-CW switch to the desired receiver mode.
g. Set the BFO PITCH control to 0.
h. Set the RF GAIN control and the VOL control to the approximate center of their ranges.
i. If the automatic keying device is installed, disconnect the automatic keying device from the KEY jack and connect the telegraph key to the KEY jack.

Operating Procedures

Warning: The following procedures require the breaking of radio silence imposed by any command.

a. Transmitter operation.
(1) Set the OFF-RCVR-XMIT switch to XMIT and allow 2 minutes for the radio set to warm up before proceeding.
(2) Set the INDICATOR RANGE switch to 1.
(3) Close the key (for not more than 30 seconds) and

rotate the ANT. TUNE switch until the brightest indication is obtained on the ANT. TUNING INDICATOR.
(4) If the ANT. TUNING INDICATOR does not glow, set the INDICATOR RANGE switch to 2 and repeat the procedure given in (3) above.
(5) If the ANT. TUNING INDICATOR still does not glow, set the INDICATOR RANGE switch to 3 and repeat the procedure given in (3) above.
(6) Set the INDICATOR RANGE switch to OFF.

[This is important, because the lamp robs a not-insignificant part of your power!]

(7) If the automatic keying device is to be used, disconnect the telegraph key from the KEY jack and connect the automatic keying device to the KEY jack.
(8) Operate the transmitter by the use of the key or the automatic keying device.

b. Receiver operation.
(1) Set the OFF-RCVR-XMIT switch to RCVR.
(2) Adjust the VOL control for the desired audio level at the headset.
(3) Adjust the RF GAIN control for the desired signal level at the headset.
(4) If the receiver is being operated in the CW mode, adjust the BFO PITCH control for the most desirable audio tone.

Additional Comments:

The frequencies supplied with the radio are useless to the ham. You'll want to crystal it up on some better ones. The best candidates might be the ones decided upon by the Yahoo ArmyRadios group at

<http://groups.yahoo.com/group/armyradios>

They've standardized on the crystals listed in the table below.

Specs of the modern crystals used are: CR-58A/U (HC-17 holder), .093" diameter pins at .486 pin spacing, parallel resonant, 32pf load, .002% accuracy.

Tip: The author remembers having some success with HC-6 crystals from the junk box. Also, FT-243s will plug in directly if you omit the ceramic socket adapters. Some FT-243s may have trouble starting, but the author has used them in the past.

Here are the transmitter and receiver alignment procedures to be followed after installing new crystals -- the author's comments appear in []:

[***Tip:*** The shafts on the slug-tuned oscillator coils sometimes are difficult to turn or get stuck altogether. To avoid breaking the slotted ends in the particularly difficult instances, try the following method.

Author's method of getting more leverage on the TRC-77 tuning shafts

Screw on small nuts a few threads, then solder them. The smallest nuts you can get away with are best. The coil may need to be adjusted farther down than the nut will allow and you may need to open up the hole with a Dremel® tool. The author used 1/4" nuts on three of the six coils and this gimmick saved his bacon on his TRC-77A].

Transmitter Tuning [alignment] Procedures
Warning: Do not use a vacuum-tube voltmeter for tuning and alignment. The common lead of the vacuum-tube voltmeter connected to the metal case of the AN/TRC-77(*) places the case at + 120 volts dc.

a. Oscillator Tuning.
(1) Remove the chassis from the case.
(2) Remove the transmitter crystal retainer holddown plate located below V201.
(3) Insert six transmitter crystals in ascending order in their crystal holders.

AN/TRC-77 Channel Plan (CR58A/U crystals):

Channel List	TX xtal	RX xtal (CH+455 KHz)
Channel 1 (80m QRP calling frequency)	3560 KHz	4015 KHz
Channel 2 (Alternate for above)	3565 KHz	4020 KHz
Channel 3 (40m QRP calling frequency)	7040 KHz	7495 KHz
Channel 4 (First alternate for above)	7041 KHz	7496 KHz
Channel 5 (Second alternate for above)	7042 KHz	7497 KHz
Channel 6 (40m Novice frequency)	7141 KHz	7596 KHz

(4) Replace the transmitter crystal retainer holddown plate.
(5) In the AN/TRC-77, position the power amplifier tuning range links for each transmitter channel frequency. A pictorial diagram of the link position settings is located below the power amplifier tuning range link board. The AN/TRC-77A does not have tuning range links.
(6) Connect the test battery or bench power supply to TEST POWER J302. Connect the positive (+) lead of the TS--352(*)/U to TP202 and the negative lead to TP201.

[The TS-352 is a common multimeter, like a Simpson, etc. DVMs will work fine].

(7) Set the TS-352(*)/U to the 10-volt range. Set TUNE OPERATE switch S204 to TUNE and the XMTR CHANNEL switch to the channel to be tuned.
(8) Turn the power switch to XMIT.

Note. If the chassis is laid flat on a bench, the TUNE OPERATE switch will snap back into the OPERATE position. To prevent this, support the rear of the chassis on a small wooden block.

[The TUNE OPERATE switch is a miniature toggle located on the chassis bottom at the rear].

(9) For each channel, rotate the oscillator coil tuning slug and capacitor fully clockwise.
(10) For each channel, in turn, rotate the oscillator coil tuning slug and capacitor in small increments counterclockwise until the TS-353(*)/U meter pointer dips to a minimum. It may be necessary to set the TS-352(*)/U to the 3-volt scale to observe minimum dip.
(11) Turn the power switch to OFF.

b. Power Amplifier Tuning. Before tuning the power amplifier, be sure that the antenna is disconnected.

[The above is not a misprint. Moving the TUNE OPERATE switch to TUNE reduces the output power of the transmitter while tuning the PA plate and antenna circuits. They truly don't want an antenna connected].

(1) Insert the negative lead of the TS-352(*)/U into TP301 and connect the positive lead to the case. Set the TS352(*)/U to the 10-volt range.
(2) Connect Key J-45 to KEY.
(3) Rotate the power amplifier tuning capacitors 1/2 turn clockwise (two capacitors for each channel).
(4) Turn power switch to XMIT.
(5) Close Key J-45.
(6) For each channel, rotate the power amplifier tuning capacitors, in small increments, counterclockwise until the TS-352(*)/U meter pointer dips to a minimum. It may be necessary to set the TS-352(*)/U to the 3-volt scale to observe minimum dip.
(7) Turn the power switch to OFF.
(8) Disconnect the TS-352(*)/U from TP301 and chassis.
(9) Disconnect Key J-45.

[Following a simpler method, one can simply leave the TUNE OPERATE switch in the OPERATE position, hook up a dummy load to the transceiver, and tune for maximum output. (First the oscillator stage, then the PA stage for each channel). The author has aligned the transmitters on both the TRC-77 and TRC-77A in this way several times in the past with excellent results].

Receiver Tuning [alignment] Procedures
Note. The RF and MIXER controls necessary for receiver tuning are located on the right-hand side of the chassis.

a. Preliminary.
(1) Remove the receiver crystal retainer hold-down plate.
(2) Insert 6 receiver crystals, in ascending order, into the receiver crystal holders.
(3) Replace the receiver crystal retainer hold-down plate.
(4) In the AN/TRC-77, position the receiver mixer and rf screw links for each receiver channel frequency. For receiver operation between 3 and 3.8 mc, position the channel links between the center screw terminal and the bottom screw terminal. To operate between 3.8 and 8 me, position the channel links between the center screw terminal and the top screw terminal. The AN/TRC-77A does not contain channel links.
(5) Connect the headset to the PHONES connector.
(6) Turn the RCVR CHANNEL switch to channel 1.
(7) Set the AM-CW switch to AM.
(8) Turn the RF GAIN and VOL controls fully clockwise.
(9) Connect the AN/URM-25 to ANT and GRD terminals and tune the AN/ URM-25 to the channel frequency.

[The URM-25 is a signal generator, and a rather simple one at that].

Note. The channel frequency is 455 kc below that marked on the channel frequency crystal.

(10) Turn the power switch to RCVR.
(11) Set the AN/URM-25 MOD SELECTOR dial to 1,000 cps and the % MODULATION control for 30 percent modulation.
(12) Adjust the RF and MIXER tuning capacitors for channel 1 to give maximum output in the headset.

Reduce the AN/URM-25 rf signal level as the RF and MIXER tuning capacitors are adjusted. With the circuit peaked, 1.0-microvolt output from the AN/URM-352/U, should produce a strong tone in the headset, and AN/URM-25(*).

[Another way to do this, if you are completely minus test equipment, is simply to hook up an antenna and tune for maximum night-time atmospheric noise. On HF especially down low between 3 and 8 mHz, once you are hearing atmospherics you can be assured the radio is working. Any more sensitivity is basically wasted].

(13) Repeat steps (10) through (12) above for channels 2 through 6.
(14) Turn the power switch to OFF.
(15) Disconnect the power supply used, TS-352/U, headset, and AN/URM-25.

Note. The TUNE OPERATE switch will automatically turn to OPERATE when the chassis is replaced in the case.

(16) Replace the chassis in its ease and lock the chassis to the case.

***

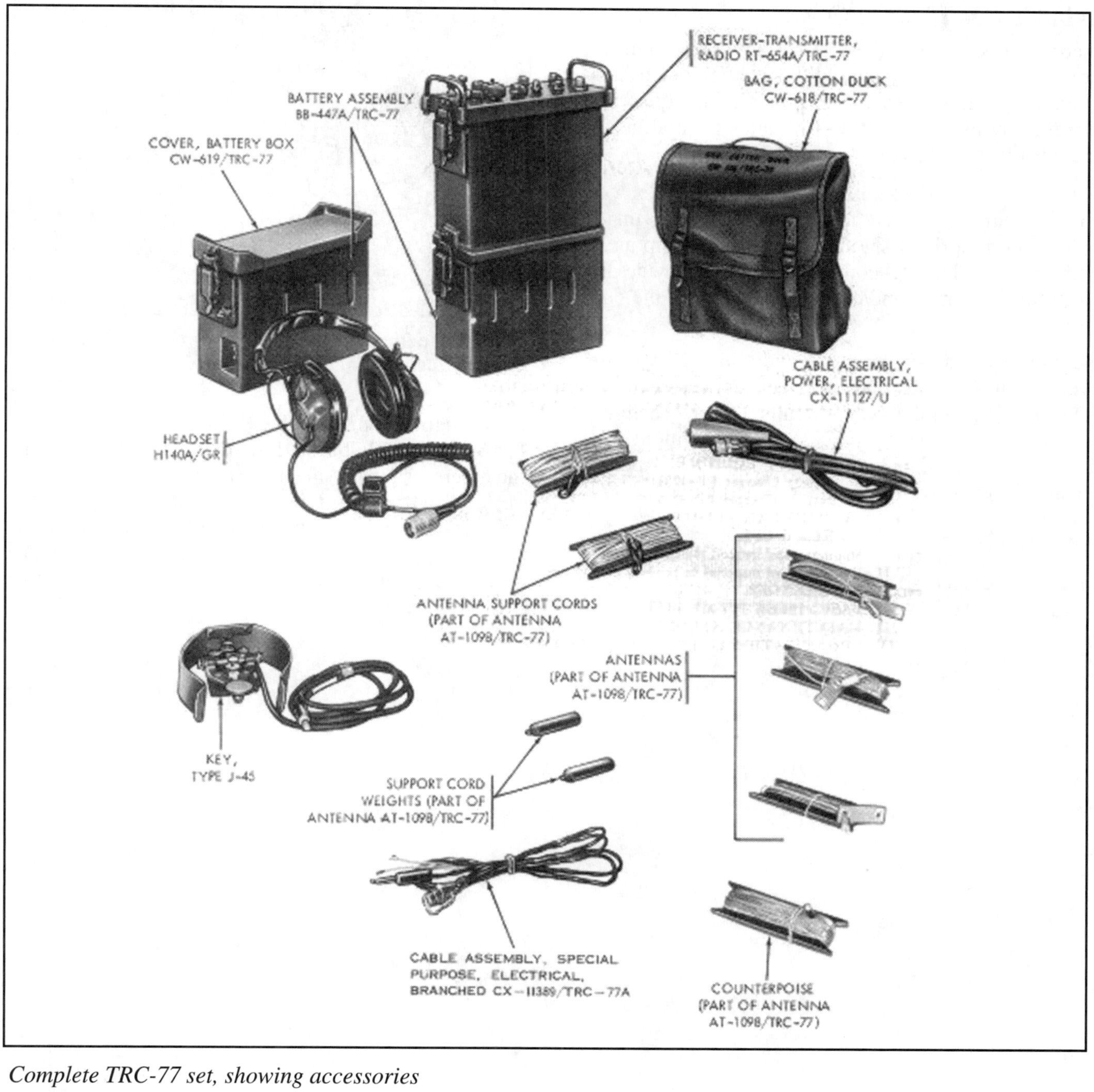

Complete TRC-77 set, showing accessories

Tip: Once you finish the receiver alignment, center the BFO using the BFO ADJUST access hole on the right side of the chassis. Set the BFO to "0", set the AM--CW switch to CW, input a signal on the channel frequency and adjust the BFO ADJUST for zero beat.

Tip: In a pinch, the receiver will run fine off a 9V battery at about 11 mA!

The TRC-77 was designed as the JAN system of semiconductor nomenclature was being adopted. The TRC-77A had quite a few semis replaced with their JAN equivalents.

What To Do If It's D.O.A.:

Incorrect operation in these sets -- low or no transmit power, rotten receive sensitivity -- is often due to misalignment. Units often come indiscriminately screwdrivered by previous owners. An attempt to align the radio is often the best way to troubleshoot.

When the radio is kicked into XMIT mode, you should hear the DC-DC converter squealing inside and the current drain should be around an amp and a half or so. No DC-DC converter action or a wide disparity in the current indicates problems with the converter.

No hum/hiss whatsoever in the audio in RCVR mode, even when the VOL control is fully clockwise indicates problems in the audio circuits. Often a seemingly dead receiver is just way out of alignment. Try injecting a hefty signal at the antenna on the expected receive frequency while trying to peak up the channel.

Accessories:

Most of the accessories are fairly conventional, but the CX-11389 cable deserves a closer look.

TRC-77A with CX-11389 cable and GRA-71 components, showing hookup

This cable connects the KY-468/GRA-71 to the TRC-77 for CW burst transmission. The terminal ends were originally meant to connect to the BB-447 wet cell battery box via a wing nut (ground) and a recessed push-type binding post (+12V) on the side of the box. This connection supplies power to the KY-468. For more info on the Burst Coder Group, see the GRA-71 section.

PRC-64

Application:

The PRC-64 (originally the Delco 5300) is a compact four channel, battery operated, crystal controlled HF transceiver.

Tech Specs:

Frequency range: 2.2 to 6.0 mHz (PRC-64, 64A), crystal controlled. The Delco 5300 covered 3 - 8.0mHz. The four receive and four transmit channels can be independent of each other. The unit is capable of receiving AM, SSB or CW signals, and transmitting CW or AM voice.

Type of modulation and power out: CW— 5W, AM —1.5W minimum

Crystal types:
Receiver - CR-78A/U, HC-25 holder (Actual channel frequency plus 455kHz)
Transmitter - CR-89/U, CR-32 holder (Actual channel frequency)
Both crystals are plug-in types

Receive bandwidth: selectable, .5kHz or 6kHz

CW keying speed:
Manual with built-in key: as fast as you can send; the TM says 12wpm
With external keyer: 300 groups per minute

Size: 5" x 4.5 x 10" (basic radio)

Weight: 5 pounds (basic radio); 7.5 pounds with SLA battery (see text)

Active devices: All solid state, mid-60s semis

Power requirements:
Power originally supplied by Battery BA-1509/PRC-64, (not available)

+28VDC	700 mA (CW), 300 mA (voice), transmit mode
+4VDC	30 mA (Receiver, no signal)
+12VDC	? mA (not given, but very low current)

Audio input/output devices: The PRC-64 was supplied with two headsets (earphones) and a mike that looks a little like a clip-on lapel unit. The author has never seen one of these mikes, but has found the Headset Electrical H-264/PRR-9 supplied with the PRR-9 squad radio works quite well as both a mike and an earphone. All audio jacks are common sub-mini types.

Audio output impedance: 600 ohms

Antenna connection: Push-type binding posts for ANTENNA and GROUND

Antennas: The PRC-64 originally used the following antenna setup:
96 foot reel of antenna wire, with frequency markers
96 foot reel of counterpoise wire, with frequency markers
35 foot reel of 75 ohm twin line with molded Y-connector

Antenna tuner: integral multiposition antenna LOADING control

Tech Manuals: The author has only the Repair Parts and Tools list, and Change 1 from the Depot Maintenance manuals...

TM 11-5820-552-24P, "ORGANIZATIONAL, DIRECT SUPPORT, AND GENERAL SUPPORT MAINTENANCE REPAIR PARTS AND SPECIAL TOOLS LISTS...FOR RADIO SET AN/PRC-64A", line drawings and listings of parts by National Stock Number.

Change No. 1 to TM 11-5820-552-15, "OPERATOR'S, ORGANIZATIONAL, DIRECT SUPPORT, GENERAL SUPPORT and DEPOT MAINTENANCE MANUAL, RADIO SET AN/PRC-64A", basic data and specs only, little else. HOWEVER, the Depot Maintenance TMs generally have theory and schematics so this TM is what you want.

Construction Style: Black anodized front panel/chassis in a hinged clamshell O.D. metal case. Components are on several circuit cards and construction is very compact. This radio looks to be a challenge to troubleshoot and repair.

Description:

The PRC-64 is a "spy" type set with a number of interesting features. The WHISPER MODE switch boosts the gain of the mike amp to allow the operator to lower his voice to almost nothing and still get good modulation. The capability for split-frequency operation is another clandestine-op giveaway, as well as the multi-pin KEY jack for connecting a GRA-71 or other similar code burst device.

Another pointer to the unit's clandestine CW function is the incorporation of the .5kHz bandwidth position with excellent filtering characteristics.

PRC-64A internals

The author believes the board at upper right handles power regulation and audio functions, the two at lower left are the receiver/exciter, and the one at upper left is the transmitter board.

PRC-64 power connections, looking down into battery box

Operation:

The receiver runs on 4VDC, which is sort of a screwball value. The author obtained this by simply putting three 1N4001 rectifiers in series with the 6VDC lead off any one SLA. Start with four or five rectifiers and remove them as necessary to get 4VDC while the receiver is on. There are a number of methods of getting batteries into this radio -- nicads, alkalines in holders, etc. The author decided on four small 6V sealed lead acid (SLA) batteries. Yuasa makes a 1Ah battery, four of which will fit perfectly into the battery compartment:

↑ *SLAs used in the battery pack, out of and in the battery compartment* ↓

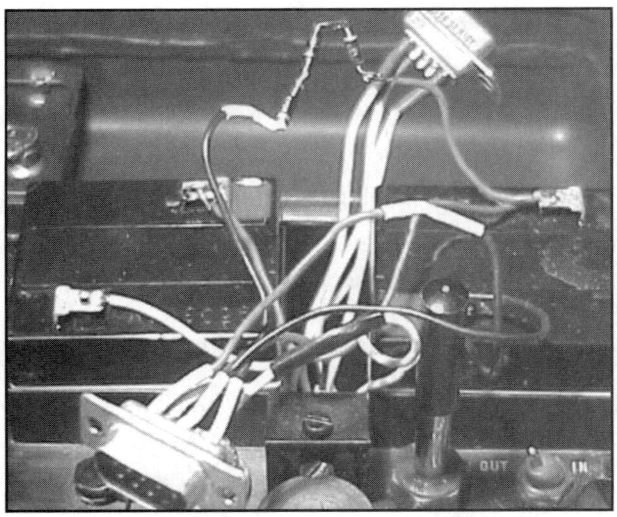

Normally when installed, the batteries are taped together on each side and the voltage dropping rectifiers (visible at top center) are taped to one of the batteries. The author used standard 9-pin DIN connectors to connect the pack to the radio, and to allow recharging. He had a 12V wall-wart charger, so he ran extra wires to allow charging the batteries in two separate sticks of 12V each, (one or the other but not both at once), when the DIN plug from the charger is hooked up. The lid will fit nicely over these. Don't forget to insulate the top SLA terminals!

Assuming you have the channels crystalled up as desired and aligned (see Additional Comments section), proceed as follows:

Antenna considerations. The radio is supplied with the wire reels outlined in the Tech Specs section. Antenna lengths for given frequencies are given inside the radio lid:

Antenna Wire Length for PRC-64

Frequency	Feet
2.2 - 2.5	93
2.5 - 3.0	75
3.0 - 3.5	64
3.5 - 4.0	55
4.0 - 4.5	49
4.5 - 5.0	45
5.0 - 6.0	37
6.0 - 7.0	31
7.0 - 8.0	27

These values are a little bit short for a quarter wave all the way down the line, which is consistent with the probable desire to tune the antenna with a some inductance at the radio end, ie: the LOADING control. The author has had good luck tuning these lengths in sloper and vertical configurations. (See the Field Expedient HF Antennas section).

Once your antenna is set up and connected, plug an earphone into either PHONES jack and turn on the receiver by rotating the GAIN control clockwise past OFF. Select the correct channel and move the BFO away from OFF to turn it on if desired. Adjust the GAIN CONTROL and select .5 or 6kHz bandwidth (BW) as desired.

To operate CW, check that the mode switch at far lower right is set to CW, and the TRANSMIT CHANNEL switch is set to the correct transmit channel. Set the IND SENS (indicator sensitivity) to "1" and set the REC--SEND switch to SEND. Depress the key and rapidly adjust the LOADING control for max brightness of the ANT INDICATOR lamp. If it's hard to see the lamp, move the IND SENS up a notch. After you get max brightness, turn the IND SENS switch to OFF. Depressing the key will transmit on the channel indicated by the TRANSMITTER CHANNEL switch. Releasing the key and returning the REC--SEND switch to REC will receive on the channel indicated by the RECEIVER CHANNEL switch.

To operate AM, set up and tune up in the same manner as with CW. After you get max brightness, release the key and move the mode switch at lower right to AM. An earphone plugged into the MIKE jack will work well. Moving the WHISPER switch to IN will greatly increase the mike gain.

Check the batteries by pressing the BAT. TEST button. If the lamp illuminates, the batteries have enough poop to run the radio.

The AN/GRA-71 Coder Burst Group may be used with the PRC-64 by connecting the KY-468 Keyer to the KEY jack with a suitable cable. The author made his own. Both sides are Winchester Electric M7P connectors (same as the Amphenol connector on the CX-10239/PRC-74).

Cable CX-10239 Pin-outs (cable is reversible):

PRC-64	KY-468	Function
A	A,F,C	GROUND
B	—	
C	A,F,C	GROUND
D	D	Keyline
E	—	
F	A,F,C	GROUND
H	H	+12VDC from PRC-64 to KY-468

Once you are set up for CW, the KY-468 will key the PRC-64 in SEND mode. See the GRA-71 section for operational details.

Additional Comments:

To align the channels, proceed as follows. The author's comments appear in []. The illustrations and text were scanned off the original PRC-64 alignment card:

RECEIVER RF ALIGNMENT

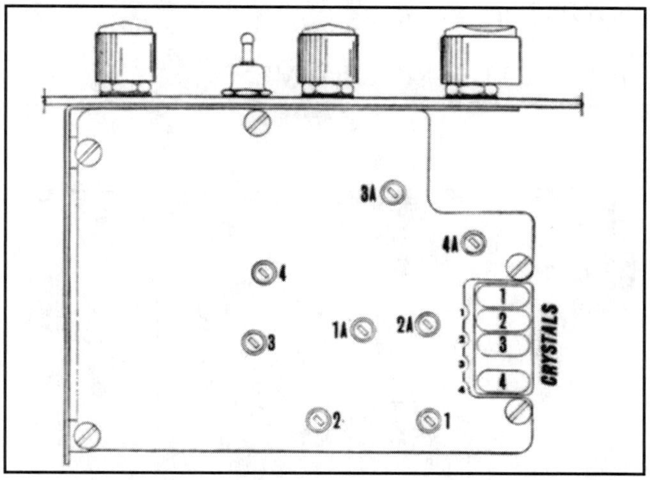

PRC-64 receiver alignment points

[*Note:* You must pull the panel/chassis from the case by removing the ten screws around the panel periphery to gain access to the crystals].

1. Insert the proper crystal for each channel into the receiver. (455 KC above channel frequency.)

[*Note:* The crystals for receive and transmit have different physical sizes and designations. See the specs above for details. Suitable crystals were originally obtained from Jan Crystals, but other vendors like International Crystal should also carry them.]

2. Connect a 600 ohm audio power meter to the receiver output by inserting the power meter input plug into one of the jacks designated "PHONES."
3. Connect an RF signal generator to the ANT and GND terminals.
4. Turn the SEND-REC switch to the "REC" position.
5. Turn the CHANNEL selector to position 1.
6. Turn the BFO control to the "OFF" position.
7. Put B.W. Switch in the 6.0 KC position.
8. Turn the GAIN CONTROL to the fully clockwise position.
9. Adjust the signal generator for approximately 1000 microvolts output using 30%, 1000 cps modulation.
10. Adjust the frequency of the signal generator to the frequency to be received using the power meter as an indicator and reduce the input to obtain about 1 milliwatt of audio power output.
11. Adjust the RF collector coil 1 and RF antenna coil 1A for maximum output indicated on the power meter as the signal generator output is decreased to keep the output indicates on the power meter under 3 milliwatts.
12. Turn the CHANNEL selector to position 2 and repeat steps 8, 9 and 10. Repeat for Channels 3 and 4.

[The author has gotten perfectly good results doing all the above by ear, without any test equipment at all. Just do the alignment at night and peak up on atmospheric noise!]

TRANSMITTER RF ALIGNMENT

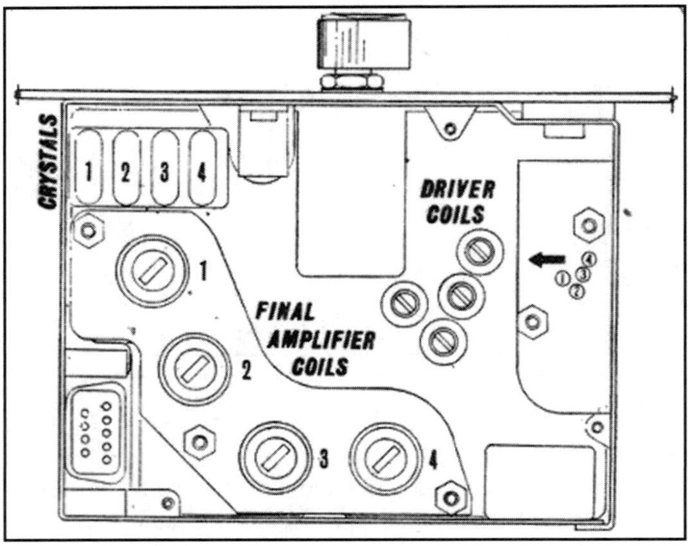

PRC-64 transmitter alignment points

1. Insert proper crystals for each channel.
2. Make sure the LOADING control is in position 1 (counter-clockwise) and the IND. SENS. is OFF.
3. Connect to the ANT and GND terminals a meter capable of measuring 15 watts of RF power.
[The assumption here is that it's also terminated in a 50-ohm dummy load!]
4. Turn the channel switch to Channel 1.
5. Turn the AM-CW switch to 'CW' and the SEND-REC switch to "SEND."
6. Depress the key and tune the two Channel 1 adjustments for maximum power as indicated by the meter.
7. Turn the AM-CW switch to "AM" and again tune the Channel 1 driver coil for maximum power output.
8. Switch to Channel 2 and repeat steps 5, 6 and 7. Repeat for Channels 3 and 4.
9. The power output on AM should be approximately one fourth that on CW. This may be adjusted by turning the potentiometer located on the speech amplifier circuit board.
[Don't overdo this or you'll just end up producing all sorts of distortion].
However, the correct setting is determined by the modulation envelope viewed on an oscilloscope.

What To Do If It's D.O.A.:

Check your power connections. See if you can get any of the four receiver channels to peak up. See if the channel crystals are actually oscillating by tuning a local receiver to the actual crystal frequency and loose-coupling a wire connected to the local receiver antenna. Except for radios that have been deliberately destroyed by demilling, the only radios the author has seen have been working units. They tend to be very reliable.

Accessories:

Here is the complete AN/PRC-64 set:

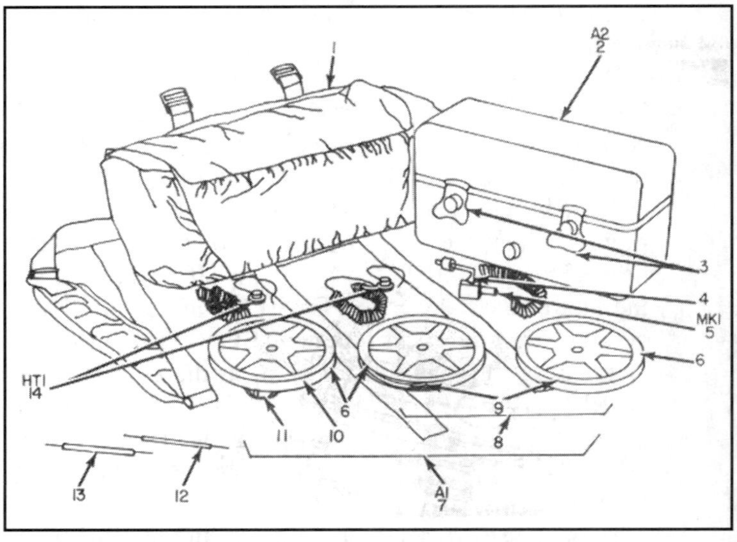

Radio Set AN/PRC-64, showing antenna reels (6), bag (1), tuning tools (12, 13), headsets, and microphones.

PRC-74

Application:

HF manpack Vietnam-vintage transceiver originally used by Special Forces. Designed by Hughes Ground Systems in the early '60s, it was the world's first synthesized HF packset.

Tech Specs:

Frequency Range: 2.000 to 11.999 mHz (PRC-74, PRC-74A); 2.000 to 17.999 mHz (PRC-74B, PRC-74C). All versions are synthesized in 1 kHz steps

Type of modulation and power out: USB voice, about 15 watts PEP

Transmit bandwidth: 2.7kHz

Size: 4" x 12" x 19" with standard metal battery box

Weight: 24 pounds (transceiver, battery box, and 4 Ah gel-cell)

Active devices: 60s-vintage discrete transistors. No chips.

Power requirements: 10.0-17.0VDC (100ma in receive, ~4A in xmit)

Audio input/output devices: H-250, H-189, H-138 handsets, H-161 headset/mike, LS-454 speaker (see LS-454 section), numerous others that fit the new family connectors.

Antenna connection: ANT and GND binding posts on front panel

Whip antenna: AS-1887A center-loaded whip with AB-955/PRC (or AB-591) spring-loaded base and MT-3613/PRC-74 antenna bracket

Other antennas: quarter wave slant wire with counterpoise, half wave dipole, any antenna of roughly 50-ohm impedance.

Antenna tuner: built in manual type, relatively narrow range

Tech Manuals: TM11-5820-590-35-1 (theory, troubleshooting, schematics); TM-5820-590-12-1 (operating info, no schematics or theory)

Construction Style: Removable, repairable modules.

Description:

The PRC-74 is a useful packset for amateur use, once it's been gone through and aligned properly. The

author has noted that hardly any examples he's seen are up to snuff as bought, although they often work on major functions well enough to get by.

The PRC-74 was a transition radio from older tube equipment like the GRC-9 and GRC-109 to the more modern chip-filled PRC-104. And that's not all bad! The 74 has all discrete components and is relatively easy to work on. The active devices are 35+ years old, though, so be prepared to scrounge some to find them.

Caution: The P.A. final transistors are 2N2887s in the very earliest 74s and Hughes house numbers in the later 74, 74A, B, and C variants. These transistors are generally unavailable. If you pop a final, you're out of luck (unless you can find another P.A.). There is some protection built in for the finals, but nothing like modern automatic power reduction on high SWR, etc. So tune up rapidly into an antenna that's already close to resonance. The tuner in the 74 is not broadband -- it's meant to tune wires and whips that are close to 50 ohms to begin with! The loading coil on the AS-1887A whip has taps that get you close, then the tuner matches the last bit.

Warning: Remember that if you do manage to find P.A. transistors, they must be installed as a matched pair. Otherwise distortion and noise will result on the output signal. The author has heard a rumor that the NTE72 device may work in this application, but he's never used them, or seen the device used in the PRC-74 P.A. You can go to the NTE site here...

<http://www.nteinc.com/>

...and click on "View Datasheets", enter "NTE72" and decide for yourself. A friend has mentioned he is looking into whether some type of modern transistor will work as a final.

Numerous mods to this radio exist, the more useful being addition of LSB, and the author's mod to replace the 17.000-17.999 mHz band with 18.000-18.999 mHz, and the 2.000-2.999 mHz band with 1.000-1.999 mHz.

Tip: Many PRC-74 mods are total hack jobs. Beware the radios where the owner has pulled the stock battery connector off the power supply module, indiscriminately drilled the front panel to add switches, etc. These units have most often been reduced to basket cases and are parts radios. Not always, of course (see author's mods below); just look over the radio carefully and pay accordingly.

Tip: The correct mating connector for the power supply connector on the bottom of the radio is a Deutsch 44007-7P. (See the Appendices for possible sources). Please try to find the right connector -- or better yet, the correct battery box -- before yanking the original connector.

The 74 has a crystal-mixed synthesizer with 46 (!) high-quality crystals in it to generate the 3.750 to 19.749 mHz (PRC-74B/C) output to mix with the 1750 kHz frequency standard. All of these crystals are adjustable in one way or the other and have to be very close to on-frequency for the radio to work to specs. Seeing a radio receiving WWV on 5 mHz at a swap meet tells you it works on major receive functions BUT it doesn't tell you much about the state of alignment on other frequencies. You'd need to check enough frequencies to have all four knobs touch all their positions at least once. This isn't generally practical, so the next best thing is to check several strategically picked frequencies.

Start by checking for WWV on 2.500, 5.000, 10.000, and 15.000 mHz (74B/C)

Ideally, check for good transmit power at 2.000, 4.000, 7.000, and 11.000 mHz (PRC-74/74A), or 12.000 mHz (PRC-74B/C). These are breakpoints between bands in the RF module.

Check for receive noise and (if possible) good power out on as many of the following frequencies as possible:

2111, 3888, 4222, 6777, 7333, 11666, 12444 (74B/C), 17555 (74B/C)

On all frequencies, check to see the CALIBRATE function works (see "Operation" below). Obviously, it's a good idea to take a small power meter, various cables and connectors, and some basic tools with you! Failure on any of these frequencies does not necessarily indicate a hard fault in the radio. One of the crystals may be off frequency. However, if any test frequency is completely dead, the corresponding crystal(s) may not be oscillating and that's more serious. For instance, if the crystal for the "4" position on the 10 kHz dial is dead, you will lose all frequencies with a "4" in the 10 kHz digit. 7040, 3842, 11649, etc...

If you don't have an original battery box, the power connections at the connector on the bottom of the radio are:

Pins 5,6	+10.0 to +17.0 VDC
Pins 2,3	Ground

Caution: Do not hook up power backwards. The radio is not reverse polarity protected and you'll just blow an internal fuse -- or worse. (Fuses are located in the power supply module).

Some of the accessories for this radio have become difficult to find, especially the whip antenna (AS-1887A) and clip-on antenna bracket (MT-3613/PRC-74). They aren't absolutely necessary as long as you use some sort of resonant antenna (eg: a slant wire 33ft long with a counterpoise for 40m, etc.) The most useful battery box for the PRC-74 series is the CY-6121/PRC-74 wet cell metal box.

The original wet cell nicads that went in the box are seldom encountered, and should be avoided in any case due to the caustic potassium hydroxide used as

PRC-74 with CY-6121 battery box attached

the electrolyte. There are far better battery solutions available now. Any battery setup capable of delivering 4 - 6Ah will work fine. The author uses gel-cells, and has shortened one of his battery boxes to accept a couple of 6-volt 4 Ah SLAs in series. Fair Radio may still have new CY-6121 boxes in stock.

Tip: There is a plastic battery box available but, in the author's experience the cover-to-box fit is very poor, and the latches tend to break off. Opinions vary, but the author believes the plastic box is a good accessory to avoid.

Operation:

These instructions are adapted from the operator's manual.

CONDENSED OPERATING INSTRUCTIONS FOR RADIO SET AN/PRC-74B

Receive Mode

a. Attach whip antenna mounting bracket to receiver-transmitter.
b. Install whip antenna and adjust to desired frequency range.
c. Attach lead wire from whip antenna to front panel ANT terminal.
d. Connect headset to either front panel AUDIO jack.
(1) Set front panel MC, 100 KC, 10 KC, and 1 KC selector knobs to the desired frequency.
(2) Turn OFF-ON-TUNE switch to ON position. [The "ANT IND" meter should move up to roughly half scale or so to indicate battery power.]
(3) Set RF GAIN control to maximum and adjust PEAK NOISE control for a definite, noticeable increase in noise heard in the headset.
(4) Adjust ANT LOAD and ANT TUNE controls for maximum noise heard in headset. [This is IMPORTANT. It roughly matches the antenna to the P.A. circuits so you minimize tune time.]
(5) Press in CLARIFY/PUSH TO CALIBRATE knob and adjust so that a zero beat condition is observed in the headset.
(6) Release the CLARIFY/PUSH TO CALIBRATE knob and reset the pointer to midscale.
(7) Set RF GAIN control for desired volume of received signal.
(8) Adjust CLARIFY knob so that quality of received voice signals is natural,

Transmit Mode

f. Perform steps outlined above, do not perform steps (7) and (8) if no received signal is present.
g. Connect either microphone or key (depending on mode of operation desired) to the remaining audio jack.
(9) Turn OFF-ON-TUNE switch to the TUNE position and listen for tone in headset.
(10) Adjust ANT TUNE knob to mid position. Adjust ANT LOAD knob until a maximum reading is obtained on ANT IND meter. Adjust ANT TUNE knob until a maximum reading is obtained on ANT IND meter.
(11) Release OFF-ON-TUNE knob and allow to return to ON position.
Note. Repeat-step h each time transmitting frequency is changed.

AUTHOR'S NOTE: Do steps 9-11 rapidly. You will find that adjustment of the TUNE and LOAD knobs is touchy. This is especially true with the LOAD knob which peaks, then drops off immediately if you go one increment too far. Do NOT try to tune the set without a reasonably resonant antenna connected!

i. Press microphone button and speak directly into the microphone. [The ANT IND meter should bounce to about half to three-quarters on voice peaks.]

Additional Comments:

Steps d.(5) and d.(6) calibrate the radio for operation on the selected 1 kHz step. It's also possible to move between 1 kHz steps by first calibrating to the selected step, then pushing in the CALIBRATE control and moving it off center. Moving CCW brings the receive/transmit frequency down roughly 500 Hz per division; moving CW brings it up roughly 500 Hz per division. When the CALIBRATE knob is pushed in, the receive and transmit frequencies are both changed. When the CALIBRATE knob is pulled out, it becomes a CLARIFY control and moves only the receive frequency (at a somewhat slower rate). Radios vary quite a bit here; you'll have to experiment with yours. You can be assured if you can hear a distant station correctly with the (pulled out) CALIBRATE knob centered, that station will hear you on frequency. Of course this is with a correctly aligned set.

If you have the 74B or C version and the radio is working to your satisfaction, (most critically at the high end at 17.999 mHz), you might want to replace the top mHz band with 18.000 to 18.999 mHz to allow ops on the 17m ham band. This is done by replacing the 23530 kHz HC-49 crystal in the 17 mHz position in the synthesizer module with a 22530 kHz unit. A suitable crystal can be gotten from International Crystal (#EX 49AH98, HC-49/U, third overtone). The result of this mod will replace the 17.000 to 17.999 mHz band with 18.000 to 18.999 mHz. No further alignment was necessary on the author's radio or either of the two others he knows about. Before doing ANY mods to these radios, get the dash-35 tech manual from Fair Radio Sales and read up on what you want to do.

If you've done the LSB mod, you may also want to similarly replace the bottom mHz band with 1.000-1.999 mHz. This is done by replacing the 38530 kHz HC-49 crystal in the 2 mHz position in the synthesizer module with a 39530 kHz unit. The author's 74B functions from about 1880 – 1999 kHz with good sensitivity and 15 watts out. Below 1880 the receive sensitivity and power output falls off very rapidly. For a challenge, try low power on 160!

Caution: Do not attempt use the AS-1887 whip antenna on 160 meters. It won't match that low and you run the risk of frying your finals. For 160, use only correctly cut resonant antennas.

Tip: The PRC-74 series will receive fairly well with input power all the way down to just a hair over 9VDC, and drawing about 70mA! As far as he knows, this is a record for any sort of serious receiver. This allows running the receiver directly off a solar panel, which might come in handy during an emergency. Note that solar panels often run well above 17VDC no-load, so it's a good idea to make a simple regulator circuit out of an LM340K (see Appendix B).

What To Do If It's D.O.A.:

If you turn the radio on and the ANT IND meter doesn't advance, check the fuses in the power supply module. This is a fairly complicated set and the author recommends that beginners start with one that's been demonstrated to be good at least on major functions.

There are control settings that can make the radio seem like it isn't working at all. In particular, the PEAK NOISE control is very critical, and sometimes the actual peak is not very apparent -- especially if the frequency is very quiet. Try messing with the TUNE, LOAD, and PEAK NOISE controls during receive.

Beyond these simple checks, you need the dash-35 manual and troubleshooting skills. There is a module test set called a PRM-31, merely a "go/no-go" tester that will tell you which module in the unit is probably bad.

Warning! Do not indiscriminately tweak adjustments to "peak up the radio". For instance, the adjustments on the RF module are extremely touchy and critical. It is not possible to align the RF module by "ear" -- it's a TRF (tuned radio frequency) arrangement in which each succeeding stage is highly dependent on the previous stage being exactly aligned. If the receive gain seems low or the power output is insufficient over part(s) of the operational range, chances are the RF module needs aligning. You need the dash-35 TM to do this (and a lot of patience). The slugs in the RF module transformers are extremely delicate and easily broken. They must be doused with MEK (methyl-ethyl-ketone) solvent per the TM, before attempting to move them. The author has worked on many PRC-74s and knows it is virtually impossible to align an RF module without breaking a slug, or encountering a stuck slug. *If your set is working with marginal, but sufficient, transmit and receive gain and you don't have any spare RF modules,* ***leave the radio alone!***

Tip: If you locate RF module spares, and have a choice, pick the ones with slugs retained by wax instead of glyptol. They're easier to work loose.

The author's personal favorite 74 has an LSB conversion, 17m and 160m capability, and the calibrate micro-switch wires run out to a pushbutton to allow sliding the RX/TX operating frequency between 1 kHz points while receiving.

A lot of thought went into placement of the new controls. The toggle between the "100 KC" and "10 KC" markings switches between USB and LSB. The button above the CLARIFY knob activates the PUSH TO CALIBRATE function. Both retain the water tightness of the set with rubber boots.

By separating the internal calibrate microswitch operation from the PUSH TO CALIBRATE knob, it's possible to push in the knob without energizing the CALIBRATE function and slide the operating frequency around while listening. This helps to get on

Author's favorite 74B, showing shortened battery box

Location of added controls on author's (carefully) modified PRC-74B

the distant station's frequency if it's between 1 kHz points. For regular CALIBRATE operation per the instructions above, one simply pushes both the knob and the button.

Tip: Information about the LSB and frequency slider mods should still be available from the author's friend Jim Karlow, KA8TUR. (See Appendix C).

The standard metal battery box has a lot more space than is really necessary given modern battery technology, so the author shortened one of his examples by about 3.25". This involves removing the corner pieces by drilling the rivets, shortening the box, moving the latches, etc. This involves a considerable amount of work, but the result is pretty snappy looking and brings the bulk of the set down considerably. For pack use, the legs can be removed, cutting the size down even more.

This radio ended up being the author's best-performing and worst-looking set. He started with the crummiest appearing set to test mods on, and ended up putting the most time into it.

Accessories:

The MT-3613 antenna brackets are tough to find anymore, but it would be relatively easy to make something to clip on the side of the radio. AS-1887A whips are becoming harder to find also -- they appear on Ebay from time to time. The AB-955/PRC-74 is nothing but the more common AB-129 (for PRC-10) with a

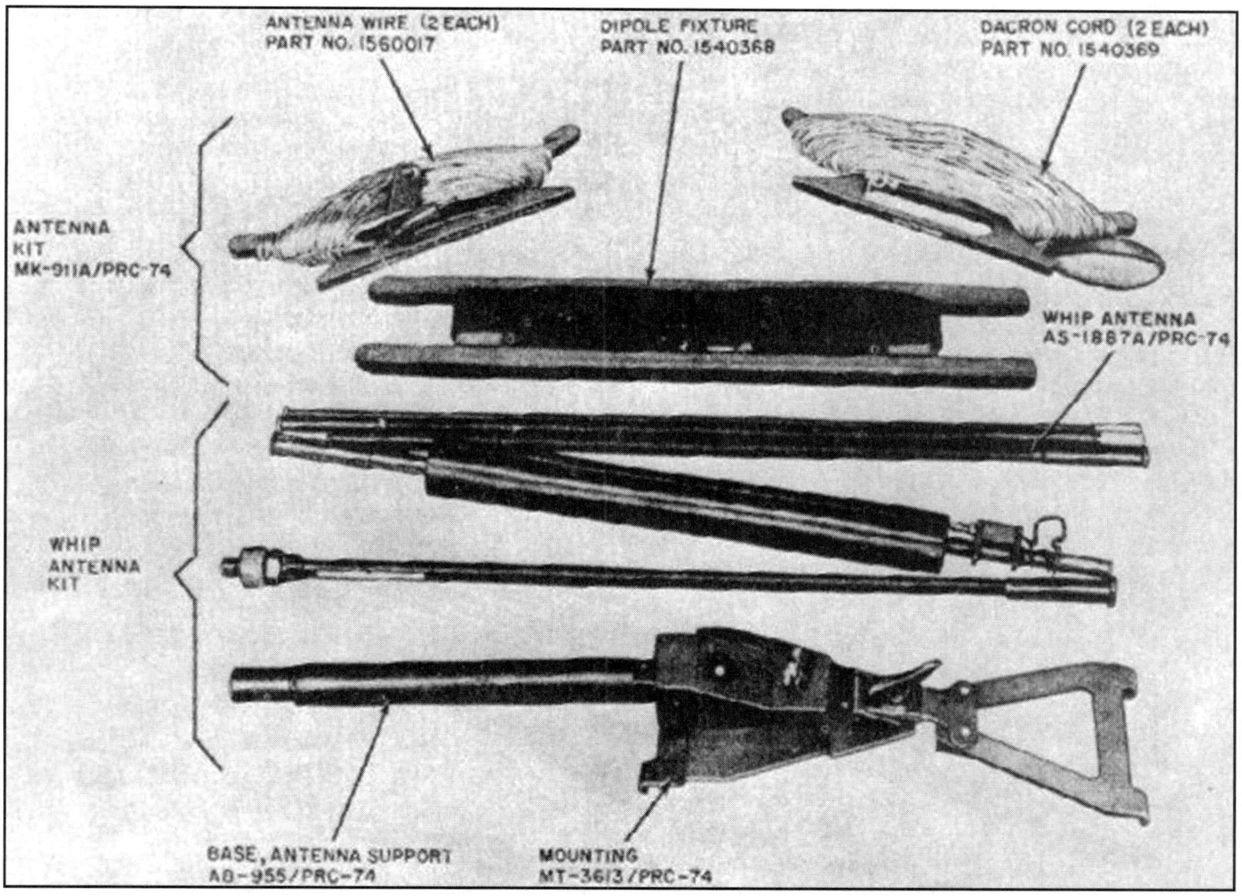

Basic PRC-74 antenna accessories

wire pigtail attached. Again, effective wire antennas can be easily made by simply setting up resonant lengths. The dipole fixture in the illustration is just a reel of coax with two binding posts. Remember that the 74 will NOT tune a straight 10ft whip like the PRC-2000, PRC-1099, PRC-70, etc. The whip ***must*** be correctly loaded. The taps on the AS-1887A coil accomplish this. Audio input/output devices mentioned in the Tech Specs are covered in their own sections.

The PP-4514/PRC-74 is a power supply/battery charger, which supplies basic power for the radio from 117VAC. For more, refer to its own section.

The G-77/G is the hand crank generator used to power the PRC-74. (Details in G-77 section). This is a truly useful item because the low cranking effort allows for long periods on receive. Tuning up for transmit takes one person on the crank and another operating the radio, but after that a single person can crank and operate the set if a headset/boom mike arrangement like the H-161 is used:

Once the radio is tuned up, the ICS constant-on position of the chest set toggle can be used to key the radio while cranking. Effort during transmit is far less than the GN-58 used with the GRC-9! And your talk power is more than doubled -- 15 watts SSB versus 7 watts AM.

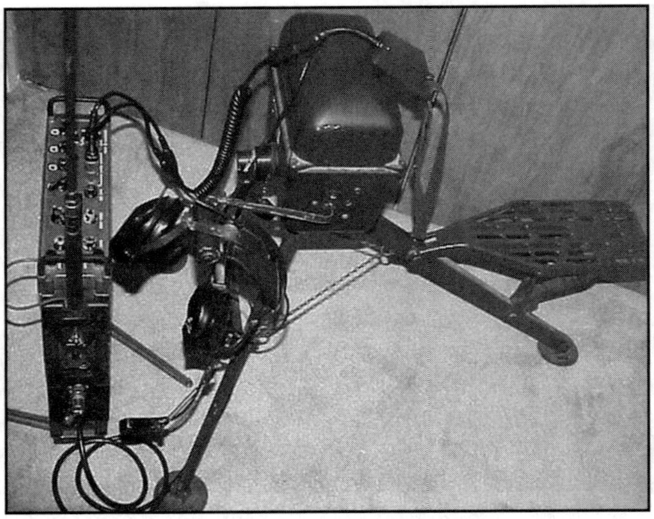

Generator G-77 set up with PRC-74 and H-161

PRC-70

Application:

Solid state multi-mode HF/VHF microprocessor-controlled transceiver used in manpack, fixed, and mobile service. Made by Cincinnati Electronics.

Tech Specs:

Frequency Range: 2.0000 to 75.9999 mHz in 100 Hz steps

Receiver sensitivity:
FM, 0.60 uV; SSB/FSK/CW, 0.375 to 0.50 uV (varies with frequency); AM 2.5 uV

Receiver signal to noise ratio: 10 dB at referenced sensitivity

Receiver selectivity:
FM, 32 kHz at 6 dB, 70 kHz at 60 dB; USB, CW, FSK, 2.8 kHz at 6 dB, 4 kHz at 26 dB, 6.0 kHz at 60 dB; AM, 6.0 kHz at 6 dB, 14.0 kHz at 60 dB

Type of modulation and power out:
High power setting:
USB—21-42 watts PEP
CW, FM, FSK—21-42 watts average (FM enabled from 30 - 80mHz only)
AM—7.5 watts carrier, 30 watts PEP at 85% modulation minimum

Low power setting: 3 watts, +2dB, -2.5dB

Size: 11.9" x 13.5" x 20.5"

Weight: 20.5 pounds (transceiver only); 27.5 pounds (transceiver and BB-542/U battery)

Active devices: 70s-vintage discrete transistors and ICs.

Power requirements: +20 to +32VDC. Usually supplied by 24V nicad battery BB-542/U.

Power consumption:
Receive mode, 7 watts max (250ma current draw after initial startup)
Low-power transmit mode, 50 watts maximum
High-power transmit mode (FSK, CW, FM, AM), 160 watts max; SSB, 115 watts max

Audio input/output devices: H-250 handset (see text), H-251 phones, H-161 headset/mike, LS-454 speaker, numerous others that fit the new family connectors.

Antenna connection (three types):
ANT connector (BNC)
GND and WIRE binding posts for random wire antennas
WHIP antenna connector

Whip antenna: AS-2974 whip; 6-foot section, 4 to 76mHz; 9-foot section, 3 to 76mHz

Other antennas: AS-2975 doublet antenna, 50-ohms, 2 to 30 mHz

Antenna tuner: built-in wide-range automatic type, connected to both the whip and BNC antenna connectors

Tech Manuals: TM 11-5820-553-10 "OPERATOR'S MANUAL RADIO SET AN/PRC-70": operating info for radio, PP-6148, and G-76, no schematics or theory)TM 11-5820-553-23 "PARTS, SPECIAL TOOLS ORGANIZATIONAL & DIRECT SUPPORT RADIO SET AN/PRC-70": exploded module diagrams, parts list, main chassis schematic, module/board block diagrams, NO detailed schematics)

DMWR 11-5820-553: Detailed module/board schematics, NO parts list (See text for more on the above situation)

Construction Style: Removable, repairable plug-in boards and modules

Description:

The AN/PRC-70 is a mixed-bag. It's a very capable radio that will get you on every HF ham band (USB) plus 6m USB and FM in 100 Hz steps at a very healthy power level for a portable/manpack set. The radio uses a fully coherent multiple phase-locked loop synthesizer referenced to a precision TCXO standard to give 100 Hz steps through the entire range. This is very much unlike the RT-524 and PRC-25/77 radios, which used incoherent indirect synthesis. That's the good news.

The bad news is the 70 is heavy for manpack use, nearly 30 pounds counting whip antenna and handset. It requires a relatively high capacity battery to get any sort of reasonable operating time. The stock battery pack has a fairly small volume, probably to keep the overall radio size from getting completely out of hand. This means replacement cells for pack rebuilds have to be high-capacity, 1/2-D-size and thus are expensive. Peak current draw on high power can be as high as 4A at 28V, which really hammers the batteries. Members of the author's group have had varying degrees of difficulty keeping 20+ cells going at the same time. More on the battery and power situation later.

The antenna tuners can also be problematic. Virtually everyone in the author's group has experienced some sort of tuner malfunction.

The construction is individual plug-in circuit boards and removable modules. Unless you have a lot of spare boards and modules for troubleshooting via substitution, you will need extender cables for any serious work. Even if you get a radio in working condition, you will eventually be troubleshooting it. The PRC-70 really pushed the envelope when it was designed in the mid to late 1970s. It was listed as "developmental" in the author's 1977 FM 24-24, "Radio and Radar Reference Data." As an exercise, try and think of any microprocessor-controlled military comm gear from that period.

Cincinnati Electronics was compelled to do a major redesign in the late 1970s after the first batch of PRC-70s released to the military came back with a variety of problems. This may have contributed to the 70's ongoing bad rep. Another factor is that most 70s encountered in the surplus world have problems to begin with that need to be ironed out. You can buy an RT-68 and expect it to work right off. Not so with a complex set like the PRC-70. Be advised.

Now that you're completely and totally terrified, it must be said that the PRC-70 is a great radio when it's working, and as long as you pay attention to the tips outlined here. For those who experience failures and do not wish to do their own troubleshooting, there are individuals who can help out. See Appendix C.

Operation:

Caution: Do not attach antenna to this unit if high power transmitters in the range 2-76 mHz are being operated within 200 feet. At least one member of the author's group has lost the mixer in his 70. It's a really good idea to yank the antenna in a high-RF environment.

Caution: Handcrank Generator G-76/G must be producing power BEFORE the AN/PRC-70 is turned on. Failure to do this may damage the radio.

Tip: The original manpack whip used with the PRC-70, the AS-2974, is a fairly rare bird. The PRC-25/77 long whip works well. It's the AT-271 attached to the AB-591 rubber base. The AS-2975 is the factory doublet (dipole) antenna and includes a balun at the feed point. It's often available through surplus dealers.

The PRC-70 has what the author considers a serious design flaw—the radio can actually be damaged by low input voltage. At supply voltages below 19V or so, certain DC-DC converters can hang up, draw large currents, and take out the associated voltage regulators. The above caution addresses the issue of having the generator up to speed before turning on the radio. This suggests that if the cranker has occasion to slow down or stop for any reason without warning the operator (a situation that may certainly happen), the radio could be damaged.

In fact, there is a circuit in every battery pack that senses supply voltage. When the voltage drops below 19 volts or rises (for whatever reason) above 28V, the circuit simply cuts power to the radio. This will generally occur during a transmission, usually in mid-syllable, with no warning whatsoever. Since this board was manufactured by the Frezzolini Electronics company, the author's group naturally refers to it as that "#$@#% Frezzolini board."

If you decide to power the radio with an external supply or a battery of your own construction, Do NOT…it will be repeated, DO NOT allow the input voltage to drop below 19 volts with the radio turned on.

Jim Karlow KA8TUR has recently worked out a mod to the radio that takes advantage of an unused feature in a regulator chip. It seems the regulator had a low voltage sense feature all along that was never implemented by the PRC-70 designers. Jim's mod allows low voltage cutoff to be accomplished inside the radio, thus eliminating the requirement for low voltage detection in each battery. Eliminating the Frezzolini Board will free up space in the battery box. See Appendix C for contact information.

Power connections on the Power Connector are Pin 5: 24VDC, and Pin 1: negative. 24V is the nominal supply voltage but the radio will operate correctly from about 20 to 30VDC. Maximum current has been measured at 7.5 amps in high power transmit, and should never exceed 8.5 amps for a correctly functioning radio.

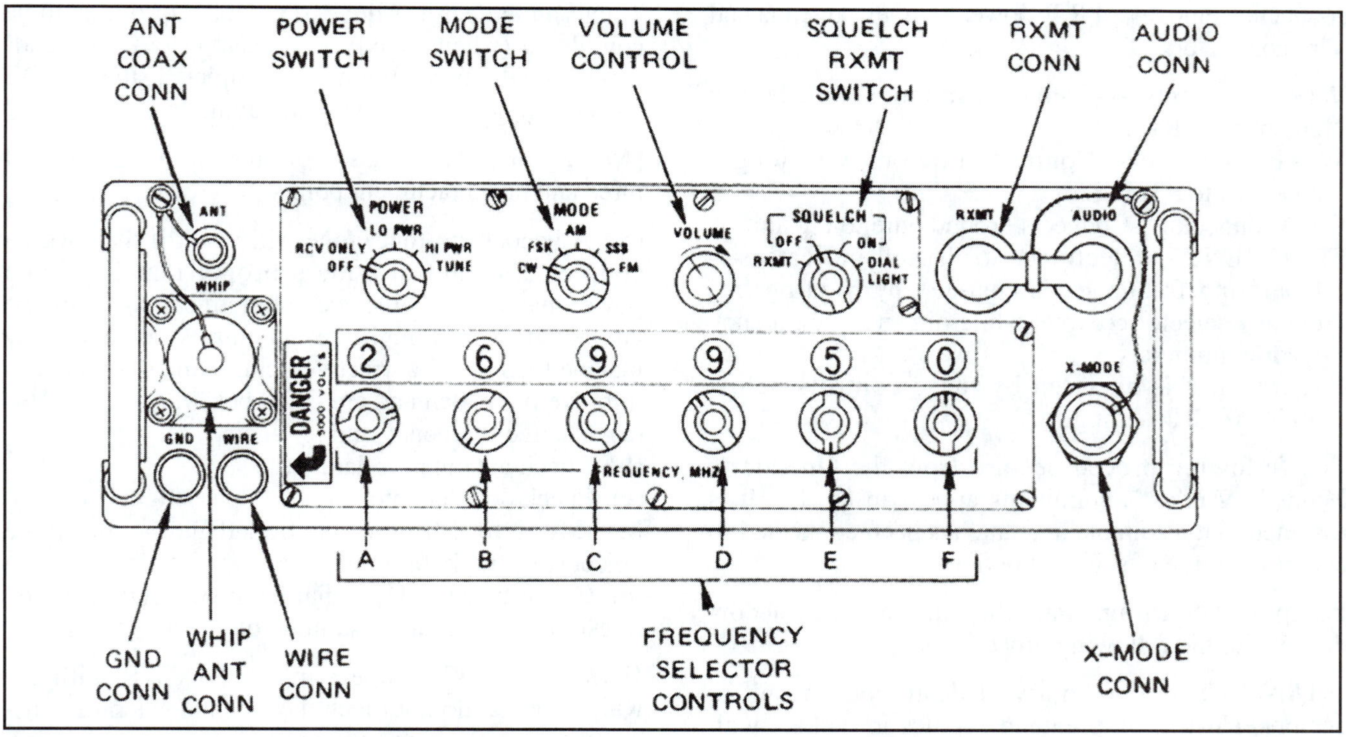

PRC-70 front panel controls and connectors

POWER switch—Sets major functions of RT-1133/PRC-70 as follows:

Positions:

OFF—Removes all power from RT unit.

RCV ONLY—Applies power to receive circuits only.

LO PWR—Allows transmission in low-power mode when transmitter is keyed. Output is approximately 3 watts.

HI PWR—Allows transmission in high-power mode when transmitter is keyed. Output is approximately 30 watts.

TUNE—Starts automatic tuning sequence in unit to tune antenna coupler. Switch is spring loaded in this position and returns to HI PWR position when released.

MODE switch—Selects operating mode of unit. Modes are CW, FSK, AM, SSB, and FM (only in 30.0000 to 75.9999 MHz range).

VOLUME control—Sets volume of received audio.

SQUELCH switch—Controls operation of receiver squelch and unit dial lights as follows:

RXMT—Actuates switching circuit which controls second RT-1133/PRC-70 in retransmit mode.

OFF—Disables squelch circuit.

ON—Enables receiver squelch in voice modes. Squelch is disabled if MODE switch is set to CW or FSK.

DIAL LIGHT—Momentary switch position; applies power to dial lights. Switch returns to ON position when released.

RXMT connector—Connects unit to second RT unit RT-1133/PRC-70 in retransmit mode. Connects CW Key KY- I 16/U for CW or FSK operation.

AUDIO connector—Connects Handset H-250/U to RT unit for normal voice operation and Headset H-251 /U during CW or FSK operation.

X-MODE—Connects voice security applique to unit connector for secure voice operation.

Caution: Dummy X-Mode shorting caps for AN/PRC-70 and AN/PRC-77 are not interchangeable!

Note: Shorting cap must be connected to connector at all times when voice security applique is not being used or power to RT unit will be interrupted.

ANT connector—Connects feedline for Doublet Antenna AS-2975/PRC-70 to RT unit.

WHIP connector—Connects Whip Antenna AS-2974/PRC-70 to RT unit.

Note: Installing the coax connector of the doublet antenna to the antenna connector removes all R.F. power from the wire connector post.

WHIP connector—Connects Whip Antenna AS-2974/PRC-70 to RT unit.

Note: Installing the antenna swivel base to the whip

connector removes all R.F. power from the antenna and wire connectors.

WIRE connector—Connects random wire length antennas to RT unit.
WIRE connector—Connects random wire length antennas to RT unit.
GND connector—Connects ground wire to RT unit.
FREQUENCY selector controls and indicators—Operating frequency is changed by rotating frequency selector controls while observing frequency readout numerals.
FM transmit is disabled by these controls below 30.0000 MHz setting.

The following procedures are from the Operater's Manual. Author's comments appear in the []. It is assumed that a suitable antenna has been connected to one of the three attachment options.

Equipment Starting. Start the equipment in accordance with the following procedures:

WARNING: Serious injury or death could result to personnel if the whip antenna comes in contact with power lines.

(1) Set RT unit to desired frequency.
(2) Set RT unit POWER switch to RCV ONLY.
(3) Adjust RT unit VOLUME control for suitable audio output level.

[***Tip:*** Use the military H-250 handset. Other handsets such as the H-138 and H-189 have the correct connector but have been known to malfunction due to differences in wiring, mike elements, and earpieces. One of the author's friends who has repaired many PRC-70s has had several units come in that were perfectly good —the owner had just been using a handset other than an H-250.]

Note: The POWER switch is spring loaded in the TUNE position and will return to HI PWR when released. When RT unit is in the coupler tuning mode, radio silence is broken.

When the POWER switch is placed in the TUNE (momentary spring loaded) position, the coupler unit will automatically tune to match the power amplifier to the antenna being used. If a good match is made, the tune cycle will stop and the RT will be ready for use. If a good match cannot be made within 17 seconds, the coupler will stop the tune cycle. A no-tune condition will exist and is indicated by a series of 2000 Hz beeps when the RT is keyed. In SSB mode, the operator must talk into the handset to detect a no-tune condition.

(4) Momentarily set RT unit POWER switch to TUNE and release. Allow 10 seconds for tuning to be completed (when TUNE tone disappears from handset). Once coupler tuning is completed, the MODE switch can be changed without having to retune.

[***Note:*** The PRC-70 autocoupler is engaged to all three antenna attachment points.]

(5) A beeper (no-tune) tone will result if the antenna coupler does not provide a proper match. Some power may be available and communication should be attempted even though the antenna is not optimally matched. Several corrective actions may be taken to improve the match and are listed below:
(a) Change frequency and try again.
(b) Check antenna for damage.
(c) Check doublet antenna for proper length.
(d) Move to different location if near some large object (tree, rock, building, etc.).
(e) Check battery. Hold dial lamp switch on and key transmitter. Dial lamps should not go out.

WARNING: When operating in HI PWR with the whip antenna, do not touch the antenna when in transmit mode-an RF burn can result.

A potential RF radiation hazard exists when Radio Set AN/PRC-70 is operated with the whip antenna, either the 6ft or 9ft configurations. This hazard is increased when the radio is operated in the portable (backpack) mode. For protection against these hazards, observe the following safeguards:
a. Use a dummy load when operating the radio, if possible, such as when performing Preventive Maintenance Checks and Services (PMCS) to determine equipment readiness/availability.
b. Operate the radio in the off-the-back (field-emplaced) mode, if possible, using either the doublet or the long-wire antenna.
c. If a whip antenna must be used for transmissions in the field-emplaced mode, use a microphone cable long enough to maintain a distance of at least 50 centimeters from the antenna.
d. If the radio must be operated in the portable (on-the-back) mode, do not transmit for longer than 30 seconds out of any 6-minute interval.

[This last is very interesting. It seems unlikely this rule was ever followed.]

(6) Set RT unit POWER switch to LO PWR.
(7) Set RT unit SQUELCH switch to ON (if desired).

Note: If RT unit FREQUENCY selectors are reset, momentarily set POWER switch to TUNE and release. After retuning is completed (within 10 seconds), LO PWR position of POWER switch may be selected.

b. Voice Mode Operation. Transmit in any of the voice modes by pressing the H-250/U Handset push-to-talk button and speaking into the microphone.
c. Standard CW Operation. Use Headset H-251/U for CW operation. To transmit in the standard CW mode, key the RT unit with CW Key KY-116/U.

d. Miscellaneous Operating Notes.
(1) Changing settings of RT unit controls should not be done while transmitter is keyed.
(2) To read FREQUENCY selector dials in poor light, set SQUELCH switch to DIAL LIGHT. The switch is spring-loaded in this position and will return to ON when released.
(3) The radio set will not operate well with FM sets having 25-kHz channelization unless the RT unit has been modified for this mode. Such modification can be performed by depot support maintenance personnel.

[This statement is probably referring to the difference in deviation between the RT-524, PRC-25/77 etc., and the PRC-70. The "modification" is likely adjusting the PRC-70 deviation upward for compatibility with these radios.]

e. Equipment Shutdown. To shut down the equipment set the RT unit POWER switch to OFF.
f. Special Tuning Procedures. Certain frequencies and loading conditions may affect tuning. If several attempts at tuning fail, the operator should try the following steps:
(1) Change the location of the antenna. Objects near the antenna (including the operator, other people, and the handset cable) affect the impedance of the antenna. The operator can try to tune the radio set after doing one or more of the following:
(a) Change his position relative to the antenna.
(b) Move the handset cable away from the antenna.
(c) Move the radio position if it is sitting close to a tree, post, or some other conductor.
(2) Change frequency.
(a) Change the frequency of the radio set in 100 kHz steps to a frequency at which it will tune.
(b) Change frequency as in (2)(a) then, change back to the desired frequency. This will often solve the problem.

[The antenna tuner in these radios can often be finicky, and may malfunction at certain frequencies. That's what they're talking about here -- if you can get the radio to tune a few hundred-Hz steps off your desired frequency, the tuning will be close enough that the radio won't issue the tune error beeps and allow you to transmit. This is not "solving the problem," strictly speaking, because the radio is refusing to tune on a certain frequency. You're simply force-tuning the set on some close by frequency to get the tuner to operate. Other tuner problems have been encountered and will be covered shortly.]

g. Field Checkout. There may be times when the operator wants to know the operational condition of the unit without breaking radio silence. The following steps will provide a partial check and may lead to the corrective action required:
(1) With POWER switch OFF, attach battery.
(2) Set FREQUENCY to 10.997 MHz, MODE switch to SSB, and SQUELCH switch to OFF. Do not connect any antenna to the radio.
(3) Turn POWER switch to RCV ONLY and adjust VOLUME control to obtain a suitable level of tone which should have a frequency of approximately 3000 Hz.
(4) Change the FREQUENCY to 10.998 MHz and observe that tone changes to a frequency of approximately 2000 Hz. This test indicates that power supplies, audio, and IF circuits, volume control, and handset or headset are operable. If no audio is heard, turn VOLUME control fully clockwise. If audio is still not heard, try a new handset and/or new battery.
(5) To check squelch operation, set FREQUENCY to 10.898 MHz and turn SQUELCH switch to ON. The radio should mute (no audio or noise unless strong traffic exists on that frequency). Set FREQUENCY to 10.998 MHz and observe that squelch breaks and 2000 Hz tone is heard.

[***Note:*** This has got to be the first instance the author can recall of an operational test based on the presence of a birdie in a synthesized radio set! The PRC-70 is pretty clean throughout the tuning range, but a friend checked and found a birdie on 11.000 mHz, exactly as suggested by the above check. It appears all 70s have the 11.000 mHz birdie.]

(6) A battery check can be accomplished by setting controls as given below:
(a) FREQUENCY at 10.998 MHz.
(b) Turn SQUELCH switch to DIAL LIGHT and observe light level. While holding switch in DIAL LIGHT position, change FREQUENCY to 14.998 MHz. The bandswitch motor will run and lights should not dim (motor drains about 200 mA from battery).
(7) Transmitter check can be accomplished either hot, into an antenna, or cold, into itself.
(a) Cold or silent tune check.

Note: Do not use antenna for this check.

1. Set FREQUENCY to 7.100 MHz and MODE switch to a voice mode (SSB, FM or AM).
2. Turn POWER switch to TUNE and note presence of a tone in handset while coupler is tuning.

3. Note that tune sequence finishes in less than 10 seconds and the tone shuts off.
4. Key transmitter and talk into handset. Observe that voice sidetone is heard. If a series of beeps (a 2000 Hz tone coming on at approximately a 1 second interval) is heard in background, a no-tune condition exists. Repeat test at 8.1, 9.1 or 10.1 MHz. If beeps continue, RT unit is defective. It should be noted that some radiated power may be available even though beep is activated during a transmission.

[*CAUTION:* Tuning the radio without a suitable load seems like a really stupid thing to do. The author cannot imagine why one would want to perform the above "silent tune check" and recommends that the procedure not be attempted. The pre-driver and driver transistors are rather sensitive in this radio. In low power, they are what are actually connected to the antenna, bypassing the P.A. Even thought the radio tunes at the 2-watt level and there is an SWR-sensing circuit, it just isn't worth the chance of popping them. ALWAYS transmit into a reasonable load. The predriver and two drivers will set you back about a hundred bucks—why risk it?]

(b) Hot tune check can be made as given below.

Note: Use antenna for this check. Radio silence will be broken.

1. Set FREQUENCY to assigned frequency.
2. Turn POWER switch to TUNE and note presence of a tone in handset while coupler is tuning. (Radio silence is broken.)
3. Note that tune sequence finishes in less than 10 seconds and that tone shuts off.
4. Key transmitter and observe that voice sidetone is heard.
5. If a series of beeps (a 2000 Hz tone coming on at an interval of approximately 1 second) is heard, try the following procedure: change frequency, check antenna for damage or doublet antenna for correct length, move to a different location.if near some large object, tree, building, etc., or change battery if available.

(8) Changing batteries.
(a) Battery removal.
1. Position RT-1133/PRC-70 (RT unit) with battery assembly attached on a flat surface. Insure POWER switch on RT unit is in OFF position.
2. Unfasten battery side latches.
3. Lift RT unit away from battery.

(b) Battery installation.
1. Place a fully charged battery on a flat surface with latches pulled out and down.
2. Ensure POWER switch on RT unit is in OFF position. Place RT unit over battery and carefully mate battery connector with RT unit connector.
3. Press two units together and lock side latches.

Note: A temperature-sensing switch is located in the power amplifier module. If continuous transmission at HI PWR is required, during high temperatures the RT unit my switch to lower power (approximately 3 watts) after approximately 10 minutes of such operation. When the temperature is sufficiently reduced the RT unit will automatically return to the higher power level.

Caution: Do not change front panel controls while RT unit is transmitting.

Caution: The antenna base must be screwed down all the way until the antenna base meets with the connector. If a gap is left between the antenna base and the connector, the antenna base may break.

Additional Comments:

The antenna coupler is a fairly wide-range unit and is connected to all three of the antenna attachment options: BNC jack, whip, and wire posts.

If you use the battery long enough for the undervoltage sensor in the battery to trip, the relay in the battery will be reset when you apply normal charging voltage of about 32V from the Charger/Power Supply PP-6148/U or other source.

The squelch is a syllabic signal-to-noise type that actually quite effective on SSB, since it senses voice signals that are above the noise level. Good sideband squelches are seldom encountered, but the PRC-70 example is one of the better ones. The same circuit is used for FM, however, and this results in a compromise. In FM mode the squelch is less sensitive than you would expect, and the squelch tail tends to be long due to the use of the same AGC (and the same release time constant) as in SSB.

The antenna coupler is a novel design that uses no rotating variable inductors or capacitors and hence does not use motors in the coupler circuits themselves. It's essentially a T-network with two series inductors and a shunt capacitor. The correct capacitor is selected via a bank of high-voltage sealed transmitting relays. The relays are latching types so no power is consumed after the tuning process completes. Each inductor, which consists of two windings on a high permeability powdered iron core, is varied using a very interesting scheme. During the tuning process, one of the windings is current pulsed to change the core permeability until the other winding reaches the correct inductance as determined by the sensing circuits. The core retains its permeability value after the current pulses cease, so no power is consumed during regular operation.

The coupler utilizes sensors for forward power, reflected power, resistance, and phase. This, along with the flexibility of settings available for the T-network components, allows matching a wide range of loads.

Batteries are a real issue with the PRC-70. The radio hogs a lot of current, especially on high power transmit, and the physical volume in the BB-542 battery is limited. It has just enough space for 20 1/2-D cells. The original cells had a capacity of 2200 mAh.

Caution: The cells in the original BB-542s (taxpayer cost: $800) were fancy 2.2 Ah fast charge batteries. The PowerSonic cell part number is 1/2DFT, and they cost $5 apiece. PRC-70 owners have been known to install the regular charge rate batteries, then charge them at the "gummint" rate of 3 amps with the G-76 or PP-6148. *This will destroy the standard charge rate batteries after a few charges!*

Recommended charging procedure for BOTH fast and standard charge batteries, determined by experience (thanks to Jim Karlow, KA8TUR for this info):

1. Release the BB-542/U battery from the bottom of the radio and set it on the table near the charger.
2. Plug in the charging power supply and set the current limit to 220 milliamperes. Set the charge timer for 15 hours.
3. Connect the negative end of the charging power supply to pin 1 of the 6 pin radio power output connector on the battery. Connect the positive lead to pin 5. Immediately a click should be heard as the electronic module latching relay closes (this relay opens when the battery output falls below 19.5 volts). Confirm the power supply ammeter still reads 220 milliamperes. It should remain at this level for the rest of the charge cycle. As charging continues, the voltmeter reading will rise to near 32 volts at the end of the charging cycle.
4. Charge the battery 15 hours with the power supply at a maximum charging current of 220 milliamperes. This is usually enough to bring the battery to full charge.
5. Disconnect the charger and reattach the battery to the bottom of the radio. The radio is now ready for battery operation.

Packs will generally give about 1.5 to 2 hours of service, assuming 9:1 RX to TX ratio and high power transmit. That life will decrease as the battery ages. Nicads are recommended for SSB ops because of their excellent peak current characteristics. 1/2D nicads are available now with nearly 4Ah capacity. Be advised -- you will pay upwards of $200 to re-cell your BB-542. Since most serious 70 users will want two packs, you can see it's an expensive radio to feed!

The TM (Technical Manual) situation for the PRC-70 is sort of unique, and a bit frustrating. There was never any dash-35 DS/GS maintenance manual issued for the radio because board level maintenance was handled by civilian contractors. There were four service stations in North America, at which were located a complete suite of specialized test benches for repair/alignment. The dash-23P manual has block diagrams and a complete parts list, but no detailed schematics.

The "DMWR" manual was not published by the military in the usual sense, and was not available through the usual channels. It was the manual meant for use by the civilian contractors and was probably a fairly low-volume publication. The DMWR contains detailed board and module schematics necessary for serious troubleshooting, as well as detailed descriptions of the specialized test fixtures. The DMWR has no exploded diagrams or parts list, so you really need both it and the dash-23P to work on the radio.

What To Do If It's D.O.A.:

While the author does not own a PRC-70 himself, he's seen several members of his group have problems of one sort or another with the radio.

Coupler problems. The coupler is an interesting design that uses relays and magnetically biased transformers to tune the antenna without any moving components like variable capacitors or inductors. It does tend to get a little weird from time to time, and has been know to fail both intermittently and totally.

One failure mode is inability to tune, causing very high current (6 to 8A) at the battery and a tripping of the undervoltage sensor due to low voltage. When this happens, the relay on the Frezzolini board must be reset be inputting a normal charging voltage of about 30-32V at the battery terminals. One group member has a 70 that periodically exhibits these symptoms, but hasn't hard-failed long enough to accomplish real troubleshooting. He generally takes along a hand crank generator or a small 12VDC-28VDC converter supply to help reset his Frezzolini board. Obviously, if you experience continuing failure of this type, there's something seriously wrong with your tuner or P.A.

Another failure has been intermittent sticking of one (or possibly more) of the high-vacuum high-voltage transmitting relays in the coupler. When this happens the coupler won't be able to find a match and will search for an extended period of time (tens of seconds) before giving up. In some cases the coupler may find alternate values of capacitance. A friend has this problem occasionally, and has noted that it happens much less often when the battery is fully charged (full voltage to the relay coil). It has been found that some of these relays can become "lazy" and fail to energize at battery supply voltages otherwise sufficient to run the radio.

Mixer problems. Mixers have been known to fail in these radios, whether from bleed-over in a coax switch, nearby lightning hits, or nearby high-power transmitters. If you lose your mixer, a valid replacement is the MiniCircuits SBL-1, at about $5.00.

Accessories:

There are quite a few accessories for the PRC-70. The more important ones will be covered here.

1) Six foot whip antenna (4 sections)
2) Nine foot whip antenna (6 sections)
3) Swivel antenna base
4) Handset H-250/U

Note: 4) above is actually incorrect. The handset pictured is the H-138. The H-138 is judged by the author to be the best sounding "new family" handsets. Unfortunately it's been out of print for many years now. If you find one at a swap meet, grab it.

5) Ground rod
6) Wire rope assy (2 ea)
7) Halyard assy (2ea)
8) Headset H-251/U
9) Five foot measuring tape
10) Balun assy with RG-58 coaxial cable and BNC adapter
11) RG-58 cable assy
12) CW key KY-116/U with cable CX-13101/PRC-70
13) Adapter UG-914/U

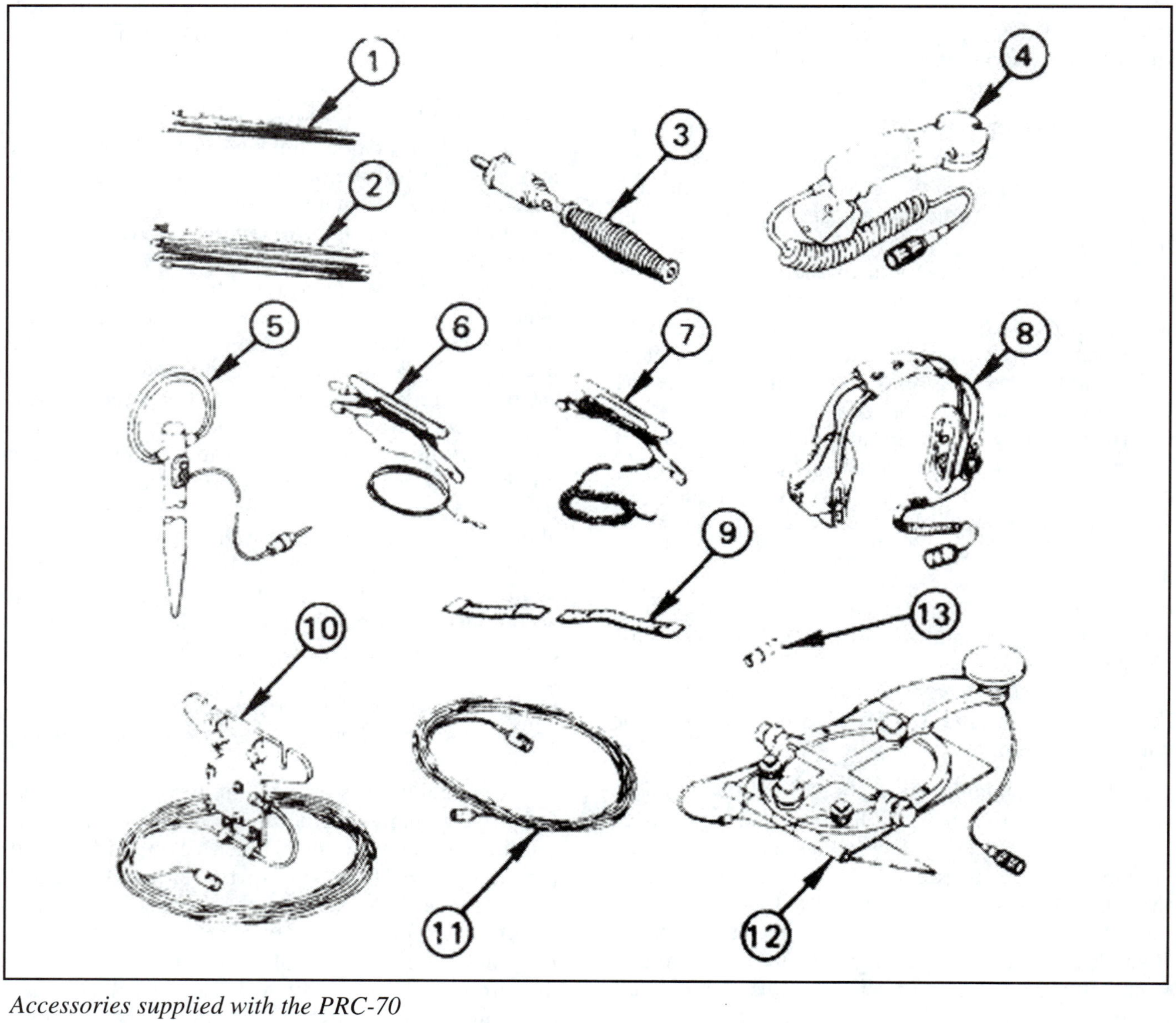

Accessories supplied with the PRC-70

These accessories come in an accessory carrying bag similar to the one for the PRC-74.

The hand crank generator for the PRC-70 is the G-76.

2) 6-foot Generator Cable CX-13175/G
3) 1-foot AN/PRC-70 Radio Cable CX-13176/G

The PP-6148 is the AC supply/battery charger used with the PRC-70. It looks like a heavily finned battery box about 2/3 the height of the R/T and, in fact, attaches to the R/T unit in the same manner as the BB-542 battery. This supply is quite useful and will charge 28V or 12V batteries. It can be set up as a constant current/constant voltage charger and has a CHARGE OVER light that will come on when the battery is fully charged.

Caution: Use the doublet (dipole) or longwire antenna when using the power supply on high power. This is because when using the whip on high power, RF may worm its way into the power supply and affect the regulation. The TM says this can be severe enough to cause an overvoltage condition and damage the supply or the radio. The designers probably assumed the radio would not generally be used on AC power with the whip antenna.

Tip/Caution: When running the PRC-70 on the PP-6148, do not latch the supply physically to the radio; use the interconnect cable instead. This is so the state of the power supply can be monitored. Make sure you set the voltage and current correctly on the PP-6148 to prevent a low voltage condition that could damage the radio.

The PRC-70 is compatible with the AN/GRC-71 Burst-Coder Group, which allows pre-recording Morse messages on magnetic tape and sending them at 300 words per minute. For more, see the GRA-71 section.

Another ancillary piece of equipment was the Digital Message Device Group OA-8990/P. This equipment was also known as the KY-879, and also the MEROD terminal. The terminal allows sending of text messages at high speed.

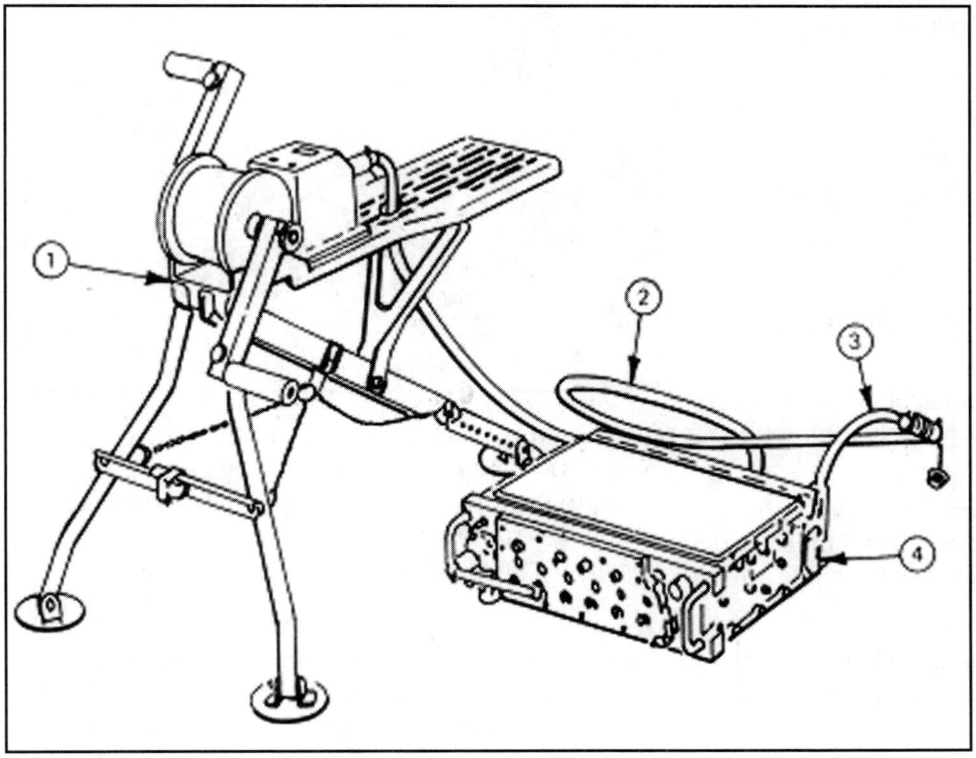

Hand Crank Generator G-76/G

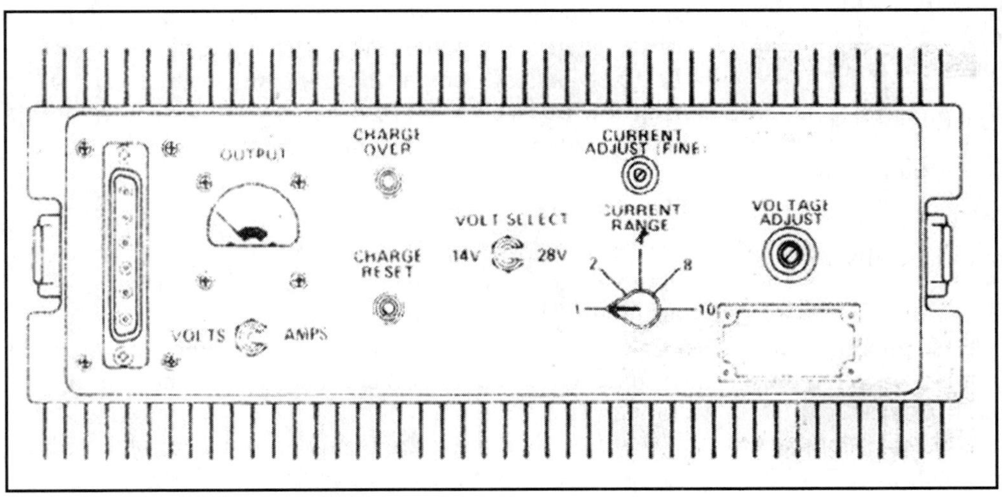

AC Power Supply PP-6148/U

PRC-2000

Application:

The PRC-2000 is a roughly mid-80s vintage, Philips-design HF manpack set. It reputedly saw service in the Falkland Islands war, a strange affair in which Britain found herself fighting what will probably prove to be the planet's last colonial war.

Tech Specs:

Frequency range: 1.6 – 30mHz in 100 Hz steps

Type of modulation and power out: USB, LSB, or CW, Data, 4 watts low-power or 20 watts high-power, PEP

Transmit/Receive bandwidth: 2.8 kHz (no special receive filter for CW)

Size: 11" x 3.5" x 14"

Weight: 18 pounds, with nicad battery (not including audio accessories, pack, or antenna)

Active devices: ECL and 4000-series chips, some of European variety; most components are commonly available

Power requirements: 12VDC nominal but radio really needs 14-15VDC to operate well (12 nicads). Battery compartment with a bottom cover is integral with the outer case.

Audio input/output devices: Supplied with a handset with a round 7-pin push-and-lock male connector (see text)

Antenna connection: Whip antenna connector for connection of the 9-foot whip directly to the radio panel for manpack operation. Also, 50-ohm BNC connector, which is not connected through the tuner.

Whip antenna: 9 foot, including the flexible base

Other antennas: 50-ohm coaxial-fed antennas (see text)

Antenna tuner: Internal, tunes automatically after a frequency switch, or manually

Tech Manuals: An operator's guide is generally supplied with the surplus sets the author has encountered, called "PHILIPS Callpac PRC/VRC 2000 Tactical HF Radio Operator's Handbook," about 1985. Members of the author's group have obtained tech manuals but there is a caveat. (see text).

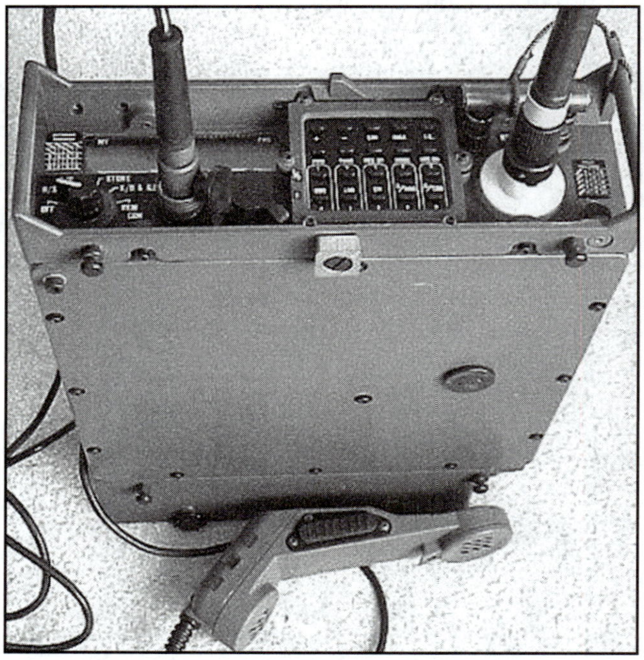

Construction Style: Heavy-duty die-cast case and front panel. Circuitry on several separate boards, connected to the top and bottom covers, and in the center section. Boards are easily accessible for maintenance, no extender cards are required.

Description:

Those accustomed to earlier manpack designs with decade switch frequency entry, knobs for AF GAIN adjustment, etc., may be taken a little off guard by the PRC-2000 which is completely microprocessor controlled. There is a main function switch only, and everything else is handled by multifunction pushbuttons – even the volume level. The dimensions inside the fixed battery compartment are sufficient to allow insertion of twelve D-sized battery cells. The radio has the capability for 10 preset memory channels.

Operation:

Power was originally supplied by a conformal-wrapped nicad battery pack that fit in the battery compartment. There's enough room to put in twelve D-size cells if they are arranged in a staggered fashion so as to leave room for connections to the battery compartment pins.

There is also an external power connector located on the bottom of the set to the side of the battery compartment. The mating connector to this one is a

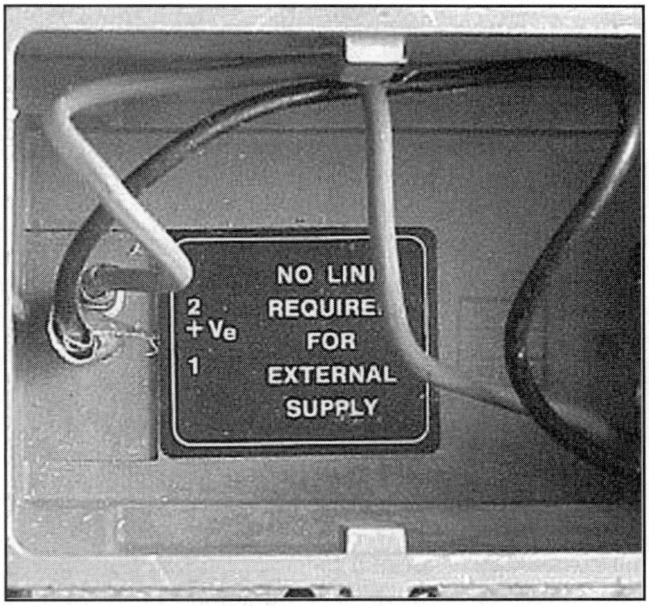

PRC-2000 battery pin-outs

External PRC-2000 power connector

female counterpart to the mike connector. It's made by Thorn EMI Electrical Components, a Bendix licensee, and is their part number: 05 1505 10 7SE

The pin-outs are:
Pin A: External supply +
Pin B: Ground
Pin C: External LP Select
Pin D: External Supply + Trickle

Tip: Since the center pin of the seven pins is not used, it is possible to use the far more common 6-pin push and turn connectors of U.S. manufacture. It's necessary to file off one of the locating pins.

Pin-outs for the audio connectors are as follows:
AUDIO CONNECTOR # 1 (left)
A MIC 1 (audio in)
B AUDIO COMMON
C SWITCHED +15 VDC
D AUDIO OUT
E GROUND
F PRESSEL 1 (push to talk, fed to channelisation)
G STANDBY (used with special handset)

AUDIO CONNECTOR # 2 (right)
A MIC 2 (audio in)
B AUDIO COMMON
C FR/PLUG (used with remote)
D AUDIO OUT
E GROUND
F PRESSEL 2 (push to talk, fed to channelisation)
G MORSE KEY

Attachment of the flexible base is somewhat different than the AB-series U.S. bases. You screw the base into the PRC-2000 until the square teeth on the bottom lock ring engage the antenna connector, then turn down the knurled lock ring above. The 8-foot sectionalized whip is a press-fit into the top of the base.

The author had some time to use a friend's PRC-2000, and the following condensed operating instructions and impressions are from that experience. Effort has been made to present them in the natural order one would want when trying out a new radio. For complete operating instructions, refer to the manual mentioned above.

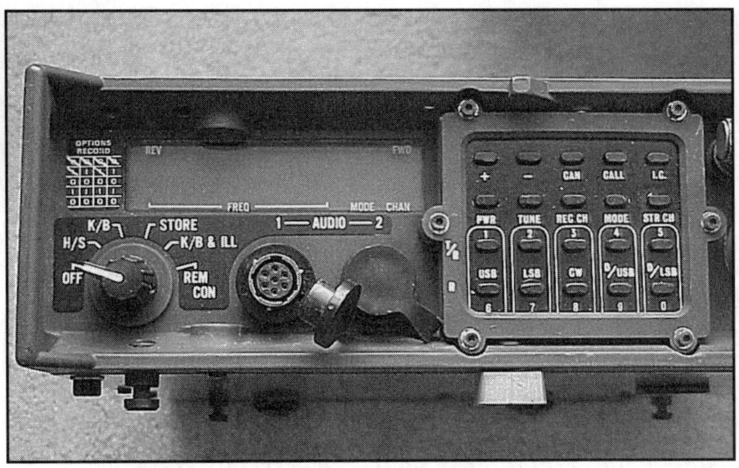

PRC-2000 front panel, showing controls

Turn the set on by turning the function switch to K/B (keyboard) if you don't have the handset with function buttons built-in. The H/S position is for when you have the special handset with channel selection controls mounted directly on it. Keyboard functions won't work in the H/S position. The set comes up with the volume set at about halfway, which is almost always way too loud.

Set the volume: Press CAN (for CANCEL) then (+) or (-). It appears as though the default keyboard function for the (+) and (-) keys is volume control, so the CANCEL button doesn't generally have to be used here as with the other functions. The usual procedure for doing just about anything on the 2000 is to clear any previous operation by pressing CANCEL, then press the rest of the buttons needed.

Select power level: CANCEL, then PWR, then (+) or (-), then CANCEL again. You can toggle between two values: four and twenty watts.

Use the following chart to select the desired mode:

USB Voice or Key, transmit or receive: 1
LSB Voice or Key, transmit or receive: 2
CW Key, transmit or receive: 3
USB Data, transmit or receive: 4
LSB Data, transmit or receive: 5
(The receive-only functions below disable transmit)
USB Voice or Key, receive only: 6
LSB Voice or Key, receive only: 7
CW Key, receive only: 8
USB Data, receive only: 9
LSB Data, receive only: 0

To enter the mode: CANCEL, then MODE, then the desired mode digit, then STR CH.

Note: Selecting the mode at this point may be important if you want to ensure against inadvertent emission. The antenna tuner operates differently depending on whether you're in a TX/RX or RX-only mode. In either case, the antenna is rough-tuned without emitting when a new channel is selected. In a TX/RX mode, the antenna is fine-tuned using RF energy the first time the PTT is depressed. In an RX-only mode, the antenna will NOT be further tuned automatically using actual emission.

To set a frequency directly into the display: press CANCEL, enter the digits, then STR CH (STORE CHANNEL). Note that the tuner will rough-tune automatically if you've entered in a new frequency.

To tune manually for max noise in the receiver without emitting: CANCEL, then TUNE, then (+) or (-) for max signal strength digit in the right side of the display. You will not emit no matter whether the mode is set to a TX/RX or RX-only mode.

To initiate a tune cycle at any time when in one of the TX/RX modes (1-5): CANCEL, STR CH, then hit the mike PTT.

Caution: Bear in mind, however, you WILL initiate a tune cycle (and emit) if you change frequency/channel and press the PTT. If radio silence is important, be sure you are in one of the receive-only modes (six through nine, and zero).

Select a frequency increment: CANCEL, then input digits. Eg: inputting 1, then 0 would set the increment to 1 kHz. A 1, then two 0s would set an increment of 10 kHz, and so on. Then press (+) or (-) to move up or down using the selected increment. Holding the (+) or (-) buttons will cause a gradually increasing scan rate that ends up being really fast after the first few iterations. This is NOT a browsing-friendly receiver!

Note: The PRC-2000 does not appear to be capable of split frequency operation. If you use the above method to increment the display, the new indicated frequency will be both the RX and TX frequency.

To store one of the ten memory channels: Enter frequency/mode as above but do not hit STR CH. Instead, turn the function switch to STORE; a beeping tone will be heard. Press STR CH, then the number of the channel to be stored, then set the function switch back to K/B or K/B&ILL (illumination).

Note: After moving to the K/B&ILL position, the display illuminates for ten seconds then turns off. After that, it illuminates again for ten seconds on any key press.

To check the channel contents: Function switch to STORE, REC CH, number of channel to check (blank display means nothing stored).

To instantly erase the memories: Function switch to STORE, press any three keys simultanously.

Tip: Use this one when the bad guys are closing in!

Additional Comments:

Note: The "data" mode is USB or LSB without the speech processor enabled. This mode is useful for many of the data modes where additional audio processing is not needed.

The maintenance manuals available for these radios in the past have been of poor quality and, in some cases, have not matched the circuitry in the radios. A member of the author's group encountered this problem while troubleshooting a set. The microprocessor board departed significantly from what was shown in the manual. He said the manual helped but a lot of hand tracing of circuitry was still required. As of Spring 2005 Combat Radio in Britain (See Appendix C for web link) is offering a shop manual on CD. This manual may have cleared up the quality and completeness issues.

Note: The PRC-2000 internal automatic antenna tuner is relatively narrow-range, meant to tune into the nine-foot whip. It appears the rough- and fine-tuning cycles are tailored to support a short whip and not much else. One might have some success experimenting with a series capacitance, a 50- or 100-pf doorknob capacitor, in series with a longwire antenna connected to the whip antenna connector, with a suitable counterpoise hooked to the case ground terminal. Always use LOW power at first.

The radio case has no legs or other means of support while sitting upright, so it can't really be used in that manner in its current form, especially with a whip antenna.

A friend summarizes the radio: "A working unit is a decent performer, they sound good, good power output, modulation, frequency stability etc. Receiver sensitivity is also good, as is the selectivity. Battery packs are also easy to make. The operator interface (man-machine interface) is poor. You need a manual to operate it. There is nothing intuitive about changing frequencies and modes. The British must have invested heavily in operator training, as the uninitiated user would probably not be able to make it function. No recent vintage U.S. military radios are nearly as difficult to turn on and tune. The radio can be learned, but operating it is a real labor of love. The radio certainly does the job, however the PRC-2000 can easily produce operator fatigue if used in conditions where frequent mode and frequency changes are required."

What To Do If It's D.O.A.:

The author has no experience troubleshooting this radio. However, here are a few "Alarms and Fault Symptoms" from the operator booklet:

1 kHz tone and display shows BA at 6 sec intervals – indicates low battery voltage

Continuous pulsing tone, display shows PH at 6 sec intervals – Synthesizer is not locked; re-select channel or enter frequency again

Continuous pulsing tone, no display message – the set is self-protecting against a high incoming signal level; check proximity of powerful HF transmitter. A second reason is that the function switch is set to STORE.

No noise in hand/handset – 1) Check function switch is set to H/S, K/B, or K/B&ILL; 2) Check that hand/headset volume is turned up (CW); 3) If totally blank display, check fuse at base of set; 4) Press and hold the (+) key for 10 sec; 5) Check charge on battery, 6) Check connections; 7) Check that the PTT is not operating (stuck) on the hand/headset.

Antenna tuning unit does not tune – 1) Check selected mode, if modes 6, 7, 8, 9, or 0 are selected, auto tuning is inhibited; 2) Check that the PTT button is operating; 3) Check connections; 4) RT to ATU link is not connected.

The above checks are obviously elementary, but do point out that there are several states the set can get into which make it appear to be defective.

Accessories:

The radio kits supplied surplus in the past have included a handset, flexible antenna base, whip antenna, wire antenna, carry bag, and manual.

Caution: The long wire antenna supplied with some PRC-2000s obtained through surplus dealers is a length of plain stranded insulated wired connected to the center pin of a BNC connector. This is not the original PRC-2000 wire antenna, but rather one half of a wire dipole that may or may not have been originally supplied with the radio. The seller has attached the BNC connector so that it appears that the wire should be connected directly into the RT antenna jack. With this arrangement, this is true *only if the wire length is very close to a resonant length.* The RT connector on the radio is a direct 50-ohm jack and expects to look into a 50-ohm load.

See the "Field Expedient HF Antennas" section for information on simple but effective antennas you can make from a couple reels of wire, and initially set up close to resonance. The following is a procedure adapted from the PRC-2000 operator's booklet for tuning such wire antennas.

Assuming a near resonant length antenna wire is set up and connected to the center pin of the RT connector (this is done by the modified BNC arrangement on the bobbin supplied with many of the surplus sets), and the counterpoise has been connected to the radio case and laid out, proceed as follows. Note this same procedure can be used for near-resonant dipoles – just add or remove a bobbin-length from both legs:

1) Select desired frequency and mode at keyboard.
2) Enter Mode 3 (CW)
3) Key the radio and note the reverse power indication in the display. If the indication is two or more, the antenna needs adjustment
4) Increment the frequency in 100 Hz steps and at each step transmit, noting the reverse power indication; if the number increases, decrement the frequency until a reverse power indication of 0 or 1 appears.
5) If the frequency is low, shorten the antenna by a bobbin length. Repeat until an optimum match is reached.
6) If the frequency is high, lengthen the antenna by a bobbin length. Repeat until an optimum match is reached.
7) Set mode back to desired mode (if not 3)

This procedure will work with any load connected to the RT jack, including regular coaxial-fed dipoles. Remember that the set expects a 50-ohm load here, and it's important to make an effort to get whatever antenna you choose as close to resonance as possible before trying the above.

PRC-1099

Application:

Synthesized HF SSB/CW paramilitary packset made by Transworld (now Datron) in the 80s and early 90s. Datron still offered the PRC-1099 as of early 2005.

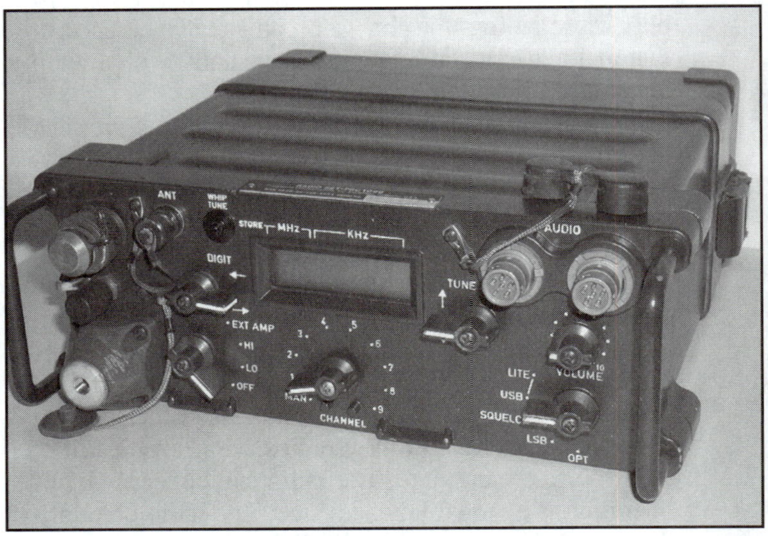

Tech Specs:

Frequency range: 1.6000 - 29.9999 mHz in 100 Hz steps. Ten preset channels.

Type of modulation and power out: USB, LSB, CW. Five watts low power, twenty watts high power. Power out during autotune cycle is reduced to 500 mW.

Size: 4" x 10.75" x 11", and same form factor as the PRC-25/PRC-77

Weight: 11 pounds (basic radio and battery case), or 18 pounds (radio, 4.5Ah NimH battery pack, harness, long whip antenna, handset)

Active devices: All solid state, microprocessor controlled. Most components and ICs are commonly available.

Power requirements: 12VDC nominal; acceptable range 10.5 - 15V
Batteries include the military BA-5598 lithium battery, the BA-4386 magnesium battery (operation restricted to low power), and Datron's own SLA battery packs.

Current drain:
Transmit—5A maximum, all modes; 1.5A typical average on voice
Receive—Non-squelched: 170mA maximum; Squelched: 120mA typical at 20deg C

Audio input/output devices: Most any accessories usable with the PRC-25/77 or PRC-74 radios, such as the H-250, H-189, H-138 handsets, H-161 headset, LS-454 speaker

Antenna connection: Same whip antenna connector as PRC-25/77. BNC connector for 50-ohm antennas (bypasses internal tuner)

Whip antenna: AB-591 base with AT-271 sectionalized mast, 10 foot total length

Other antennas: Radio also tunes the AS-1887/PRC-74 whip, and a wide variety of long wire antennas using an easily built long wire adapter.

Antenna tuner: Internal automatic latching relay type—tunes the 10 foot whip 2-30mHz. Nominal tuning time 1 second. Last tune data stored on all ten preset channels.

Tech Manuals: TW-PRC-1099-MS1, "Service and Technical Manual, RADIO SET PRC-1099"; description, theory, troubleshooting, repair, schematics, parts list at the component level.

TW-PRC-1099-TM1, "Operator's and Organizational Maintenance Manual Including Repair Parts List, RADIO SET PRC-1099"; operating instructions, no theory or schematics. "Repair Parts List" is misleading -- this list is NOT at the component level.

Note: neither of these are US military TM numbers.

Construction Style: Replaceable, repairable modules, with components easily reachable on single sided, single layer boards.

Description:

The PRC-1099 is a cost-reduced HF packset built into exactly the same form factor as a PRC-25 or PRC-77. That was presumably a way for Datron to obtain a proven, mil-spec case configuration. Many controls and antenna connections are in the same location.

The 1099 is not a standard US military radio, even though it's built in a PRC-77 case. Datron claims MIL-810D compliance for shock, vibration, humidity, immersion and rough handling, but personal experience with these radios suggests they are not up to US military standards in practice. Still, they are very useful sets after you learn to put up with (and fix) some quirks.

The 1099 uses the same accessories as the PRC-77, including the ST-138 pack harness. As with the 77, if the BA-5598 lithium battery is used there's room in the battery box for a spare. Datron's own overpriced batteries are nothing but commercial Panasonic SLAs in a plastic shell with a fuse and a socket. The battery box is exactly the same as a PRC-25 box—there's plenty of room to stuff it with nicads, NimHs, or whatever.

Operation:

Power connections inside the battery box are Pin B: +12VDC, and Pin A: NEGATIVE

This radio uses audio accessories with U-229 connectors. New family audio accessories are available from Fair Radio and others (see Appendix C), and often at swap meets. People who don't mess with mil radios usually don't want anything to do with "oddball" connectors. Anything you find with U-229 connectors is worthy of further attention.

PRC-1099 audio connector pin-outs:
Pin A: Ground
Pin B: Audio out
Pin C: Push to talk
Pin D: Mike
Pin E: CW key
Pin F: (center pin, if present) +12VDC switched

PRC-1099 controls and their function(s):
POWER Switch
OFF—Turns transceiver off.
LO—Turns transceiver on - Low Power, 5W.
HI—Turns transceiver on - High Power, 20W.
EXT AMP—Turns on the external amplifier when connected.

FUNCTION Switch
USB—Transceiver operates on the normal (upper) sideband, no squelch.
SQUELCH—Receiver squelched - Opens when signal received, USB operation.
LSB—Transceiver operates on opposite (lower) sideband.
LITE—Spring loaded position backlights Frequency display, USB operation.
OPT—Normally data mode without AME or other option installed.

CHANNEL Switch
MAN—Selects the manual tune position. Front-panel channel switch must be in this position to use optional remote handset channel switch.
1-9—Selects pre-set channel frequencies.

DIGIT Selection
LEFT ARROW—Spring Loaded–selects digit to be tuned; right to left.
RIGHT ARROW—Spring Loaded– selects digit to be tuned; left to right.

TUNE Control
UP ARROW—Spring Loaded–increments frequency higher in 100 Hz steps (larger tuning steps if selected by DIGIT switch).
DOWN ARROW—Spring Loaded–increments frequency lower in 100 Hz steps (larger tuning steps if selected by DIGIT switch).

WHIP Tune
DEPRESS—(momentary) Activates automatic antenna tuner to tune whip or long wire antennas.
STORE -- DEPRESS (Hold) Permits resetting frequency on channels 1-9.
BITE -- DEPRESS (Hold while turning tranceiver on) Activates BITE (Built-In-Test)

VOLUME Control—Varies receiver and sidetone volume.

AUDIO Connectors—Provide connection for handset or audio device.

ANT Mount—Provides connection for antenna AT-271A or Long Wire Adaptor.

ANT Connector—Provides connection for fixed base or 50ohm vehicular antenna, or external power amplifier.

POWER Connector—Provides connection for external power supply when PRC-1099 is used as part of vehicular radio set configurations. When the connector cover is removed, the PRC-1099 is disconnected from the battery pack. Rechargeable batteries may be recharged through this connector. Provides control for external accessories such as power amplifiers.

The following are the operating procedures adopted from the manual. Author's comments are within square brackets []:

Presetting Channel Frequencies

The PRC-1099 will store up to 10 different frequencies. These frequencies remain in memory even if the battery pack is disconnected. [If the radio fails to remember the preset channels, the 3V lithium coil cell backup battery is probably bad. It's located in the processor module.] Channels 1-9 are usually used to store preset channels; however, the last frequency entered in the "MAN" channel will also be retained in memory.

a. PRESETTING CHANNELS 1 TO 9
(1) Select the desired channel number.

(2) Depress switch STORE and hold down while setting the new frequency. This switch prevents accidental change of the channel frequencies during normal operation of the transceiver.

(3) Turn the DIGIT switch until the desired digit flashes.

(4) Turn the TUNE switch either "UP" or "DOWN" until the correct number appears.

(5) Repeat steps 3 and 4 until all digits are correctly selected.

(6) Release the STORE switch. The new frequency is now entered in memory. [The STORE button only has to be held down during the last toggle of the last input digit to remember the complete frequency].

b. PRESETTING MAN CHANNEL
Follow the preceding procedure except that it is not necessary to depress the STORE switch.

[MAN channel can be used for a sort of ad hoc scanning. Select any of the digits with the DIGIT switch, then hold the TUNE switch either up or down to scan. The digit change rate is the same in all cases -- you vary the scan speed via the digit you pick. The other channels 1-9 work the same way, but remember that you are "clarifying" when you move the receive frequency without using the STORE button -- if you key the mike the 1099 will transmit on whatever frequency was originally stored.]

c. MEMORY DUMP FEATURE
To erase the frequencies stored in memory, simply hold the WHIP-TUNE button in, hold the DIGIT switch up, and turn the radio from OFF to LO. This will cause the radio to reset all programmed frequencies to 00.0000 MHz.

Receiver Fine Tune
On channels 1-9 the receiver only may be fine tuned by incrementing the last digit by using the TUNE control (do not press the STORE switch). The receiver frequency offset is deleted from memory when a new channel is selected. [This "feature" can also be used to operate split-frequency. Receive can be set to an arbitrary frequency. The tuner will tune for best match on the transmit frequency.]

Operating Procedure
The audio accessories may be plugged into the following audio jacks:

Accessory	AUD-LEFT	AUD-RIGHT
RCH Remote Control Handset	X	
H-250/U Handset	X	X
H-189/U Handset	X	X
H3M Headset Boom Microphone	X	X
KYR CW Key	X	
PRC-Selcall	X	
Data Terminal	X	

[**NOTE:** The H-161 headset/boom mike with the Y-cable ending in two U-229 connectors can also be used. Just hook both U-229s to the radio -- order makes no difference. The ICS position of the toggle on the chest set will key the radio constantly, and the RAD momentary position of the toggle will act as a PTT. This is the author's favorite headset for pedestrian mobile ops.]

a. If bite is to be activated, hold down the WHIP TUNE/STORE button while performing step b.

b. Set the Power Switch to HI or LO.

c. Set the Function Switch to USB. A background noise should be heard in the handset.

d. Set the Channel Switch to the desired channel.

e. Set the VOLUME control to the desired sound level.

f. Transmit by pressing the push-to-talk switch on the handset and speaking close to the mike element.

NOTE: The H-250/U Handset has a noise canceling microphone. There will be almost no output when speaking more than 1/2 inch from the microphone element.

g. To receive, release the push-to-talk switch on the handset.

h. The background noise in the handset can be eliminated by turning the function switch to SQUELCH. This also reduces the receiver current drain.

[The SSB squelch is actually reasonably effective. HOWEVER, the current drain goes down because the dynamic range of the receiver is reduced somewhat. You may notice the squelch doesn't open on very weak signals that are otherwise readable].

Antenna Tuning Procedure
a. The antenna tuning procedure is fully automatic and is activated by a momentary depression of the WHIP TUNE switch. (Do not hold switch down as this is a dual function switch and will change to STORE operation after 1 second). A tone will be present in the handset during tune cycle. A pulsed tone at the end of the tune cycle indicates that the tuner has not been able to match the antenna correctly.

b. The tuning information is stored in memory and is automatically selected when the channel is changed.

c. The automatic antenna tuner will only operate when the AB-591 support or the LWA Long Wire Adapter is installed.

d. The antenna tuning varies considerably according to the positioning of the transceiver and the location. The tune information should be updated frequently. Always retune the antenna if the communications are unsatisfactory. [Antenna tuning is extremely important during pedestrian mobile ops. You may find that moving just slightly, or nearer a hill or tree, etc., will change the strength of incoming signals. This is a clue to retune.]

BITE Operation
BITE (Built-In-Test-Equipment) is a series of automatic "built-in" tests to allow the operator to check vital components and functions of the PRC-1099. These tests are briefly described below.

Test 1: Battery; tests battery condition
Test 2: Antenna, checks if antenna is installed
Test 3: LCD Display, tests LCD Display
Test 4: Microprocessor, tests Microprocessor
Test 5: Receiver, USB, tests receiver in USB
Test 6: Receiver, LSB, tests receiver in LSB
Tests 7-12: Transmitter, tests transmitter and Antenna Tuner antenna tuner at different frequencies

a. Set power switch to OFF.

b. Make sure the antenna and handset are installed.

c. Set the mode switch to USB and the VOLUME control to mid-position.

d. While pressing the WHIP TUNE/STORE button, set the power switch to LO.

e. Release WHIP TUNE/STORE button.

f. The BITE tests will begin, displaying "TEST:xx" then "PASS:xx" (where "xx" represents the test number) at the conclusion of each test. Test 3 requires that the operator verify that the number "8" is displayed once in each digit position of the display and that a decimal point is displayed once between each digit. If any of the above are not displayed, the display may be malfunctioning, which indicates failure of this test. Tests 5 and 6 require that the operator listen in the handset for an audible tone. If the tone is not heard, the operator must fail the test. Failure of any of the other tests are indicated on the LCD display by "Err:xx" (where "xx" represents the test number). If a failure occurs with any test, the problem should be referred to qualified personnel.

[*Not necessarily true.* The author's radio consistently fails test 11, which is the transmitter test on 4.900 mHz -- the radio indicates failure to tune. Yet the radio tunes fine on Test 12 (2.900 mHz) and works operationally with no problem otherwise around 4.900 mHz. The point is that just because the BITE fails doesn't necessarily mean you have a problem].

NOTE: The following built-in tests are available during normal operation.

g. BATTERY VOLTAGE—Low battery voltage is indicated by a flashing display in the receive mode. [This happens when the battery drops to about 10.5V.]

h. POWER OUTPUT—The power output is detected by a special peak reading power detector. The sidetone in the handset is only switched on when the correct power output is reached. The monitoring circuit adjusts automatically for low power operation.

i. RECEIVER/EXCITER PERFORMANCE - The sidetone is generated by listening to the low level exciter output in the receiver and provides a complete operational check of these components. Any excessive distortion or low level on the sidetone indicates a defect in the exciter/receiver.

j. ANTENNA TUNING - If the antenna tuner cannot achieve a correct match, this is indicated by a pulsating tone in the handset.

LSB - Lower Sideband Operation
When LSB is selected on the mode switch, the sidebands are inverted in transmit and receive modes.

CW Operation
The PRC-1099 will operate in the CW mode (telegraphy) when the key is installed in one of the audio connectors. The transceiver automatically switches to the CW mode and semi break-in keying is used. This means that the transmitter is held on when the key is depressed and will automatically return to receive when there is a short break in the transmission. A receiver sidetone is provided to monitor the keying.

[CW is generated by simply injecting an 800 Hz tone into the balanced modulator. This means that in USB, the actual carrier frequency will be 800 Hz above the display frequency, and in LSB the actual carrier frequency will be 800 Hz below the display frequency. If you find the break-in delay annoying, you can get around it by setting up a shorting switch at your key that grounds the PTT line before you start sending.]

Channel Scanning Operation
[***Note:*** When the PRC-1099 is turned on, the firmware version appears on the LCD display as "SOFt:XX"

where XX is the version. The scan function is present in firmware version 26 and above.]

PRC-1099-SCAN is a retrofittable software option which allows sequential scanning of eight channels by the transceiver. When this option is fitted, the unit will scan channels one through eight at a rate of one channel every three seconds. The internal antenna tuner, if utilized, will automatically set itself to the memorized position for each channel during the scan cycle.

a. When the internal tuner is to be used, manually tune channels one thorough eight (1-8) with the actual antenna in place before beginning the scan. This sets up the tuner memory for each channel.

b. To enable the SCAN feature, hold the DIGIT switch up until SL:xxx is displayed, where xxx can be either OFF or the digits 1-8. Enter the number of channels you want to scan by using the TUNE switch. You can also disable SCAN by selecting OFF with the tune switch.

c. Set the channel select knob to channel nine (9). The set will then begin scanning the channels.

d. To manually stop the scan, select one of the channels one through eight (1-8) or MANUAL.

[To disable the SCAN function and return channel 9 to regular operation, select OFF in step b.]

e. If the transmitter is activated, the scan will stop and the frequency of the current channel will be displayed. Sixty seconds after the end of the last transmission, the unit will begin scanning again automatically.

f. The scanning feature can be utilized with the PRC-SELCALL selective calling system. If a call is received during the scan cycle, the answerback will automatically stop the scan as the alarm is sounded.

[All of this is moot, because the scan "feature" is relatively worthless -- it isn't associated with the squelch; that is, it does not detect a signal but rather merely cycles through the selected channels.]

Additional Comments

The built-in antenna tuner is designed to tune the 10-foot whip from 2-30 mHz. Datron says a gadget called a "special long wire adapter" must be used if wire antennas are to be tuned reliably. After doing some research, the author determined Datron's hundred-dollar-plus LWA is nothing but a 50pf doorknob capacitor in series with the wire antenna to add enough capacitance so the tuner can handle it.

The author built this LWA out of a piece of sprinkler system pipe, an old push-type binding post, a 50pf doorknob cap like the one shown, and the bottom

Home-built long wire adapter

of an AB-591 antenna base. Normally he wouldn't have sacrificed the AB-591, but he also needed the top part to fix the whip on a Southcom SC-130, so it all worked out.

The threaded section is 3/8" of 5/16-24 threads followed by 5/16" inch of additional unthreaded section for actuation of the internal antenna switch. The capacitor is just series wired between the threaded section and the binding post, and everything is glued together with epoxy. Suitable doorknob capacitors were listed in the 2004 Fair Radio catalog. With this gadget, you can tune a wide variety of screwball loads from 1.6-30mHz, including (from the manual):

Long whips
16ft—need LWA above 15 mHz
24ft—need LWA above 9 mHz
32ft—need LWA above 6 mHz

Long wires
25ft, 50ft, and 100ft— need LWA

In practice the variety is much more extensive. If the LWA won't tune a given wire length, house gutter, etc., moving the lead underneath the LWA, bypassing it, often gives a match the tuner can live with. Experiment! Unlike the tuners in the PRC-74 and PRC-2000, the one in the 1099 is relatively wide-range. Remember, you're only running the tuner at 500mW -- you aren't going to burn anything up.

Tip: The importance of a counterpoise (ground) wire during manpack ops cannot be over-emphasized. The author uses the whip and an 11-foot drag-along counterpoise with great success for pedestrian mobile ops on 18157.5, the popular HFpack main calling frequency. Use Teflon-coated wire or flexible test lead to avoid kinks and snags. Strip and tin one end and connect it to the ground binding post on the 1099, then just pull it along behind you. Looks weird, works great.

Tip: Wear the radio "backwards" on the pack frame. That is, arrange it so the controls are facing the back of your head instead of toward someone behind you. The 1099 on the ST-138 harness rides high enough on the back to make the controls easily accessible.

Tip: The center pin F on the audio connector is 12VDC, present when the radio is on. This can be used to power outboard gear like a data modem, once the thermistor TH1 in series with pin F is shorted. If left in, TH1 will limit the current draw through Pin F to only a few hundred milliamps. Be sure your outboard gear is fused!

The PRC-1099 was listed as being "rated for data burst operation" in their glossy brochure, and the Operator's Manual states, "The PRC-1099 can be used for digital data (burst or RTTY) transmission and reception." This is not strictly true. The Coder-Burst Transmission Group AN/GRA-71 (or any device that keys the CW line at a high rate) will not function correctly. The author believes this is due to the 1099 keying circuits not being able to keep up with the 300 wpm rate at the Morse key input. The dots and dashes slur together into unreadable mush at the receiving end. This problem doesn't occur if the burst is handled in transmit audio only.

In the early 90s, Datron modified the PRC-1099 for a DEA contract and produced it as the PRC-1099A, which is still offered for sale. The main difference is an enhanced processor allowing remote control operation of the unit through the accessory connector. Rumor has it the radios were meant to be rack-mounted in DEA vans and used for surveillance!

Tip: Before purchasing any 1099, insist on an operational demonstration. These radios can be problem children until the bugs are worked out. They are eminently repairable, but it's always a good idea to see what you're getting into!

All that said, a working PRC-1099 is an excellent unit for pedestrian mobile ops. The author has more hours on his than any other piece of portable surplus gear. My fellow HFpackers often scoff at the 18 pound total weight for the 1099 /PM setup while holding up their FT-817s. The obvious response is that once they tack on an outboard linear, a wide range antenna tuner, a 5Ah battery, a decent antenna, and a pack or frame…they'll be pushing that weight! And the 1099 is self-contained; no interconnecting cables to worry about.

What To Do If It's D.O.A.:

When the backup battery in the processor module quits, the 1099 will no longer retain the preset channels and other erratic operation may result. The battery is a standard CR2032 3V lithium coin cell with three connection feet attached, available from Digi-Key or Mouser.

The number-one problem reported with PRC-1099s is distortion in the transmit audio. This is due to insufficient supply voltage regulation or RF getting into the synthesizer causing FMing of the VCOs or PLL chips. The result can be anything from a mild burbling all the way to audio so rotten it sounds like the operator is drowning. The main culprit is the 8V supply, derived in the M7 Processor module with an LM317T adjustable three-terminal regulator. The LM317 was a bad choice because it requires at least a 3 volt difference between the input and output voltages to regulate correctly. This is cutting it way too close if the 1099 is powered off a 12V nominal pack. Much better low drop-out regulators are available.

The varactor diodes on the VCOs and the 73mHz pulled oscillator undergo very tiny changes in capacitance over their range of operation in the circuit. A clue to this is the miniscule value of fixed capacitance across the varactors, only a few picofarads. Even the smallest voltage change across the diodes will move the circuit frequency all over the place. Fixing the FMing problem involves four steps, which should be followed in the order listed. Do a test following each, as all steps may not be necessary with your particular radio.

1) Re-seat the synthesizer chips if your 1099 is earlier production with socketed ICs. Partially lift them out and re-seat them. A small amount of De-Oxit on the pins will ensure good contact.

2) Replace 5V regulator Q3, a 2N6427, with a 78L05 regulator. This stabilizes the regulation to the PLL chips. The input pin goes to the main 8V post. The output pin goes to C30/C31 and should have an RF bead on it. The author used an Amidon FB 73-101 bead (for 50mc and below) left over from some long ago project. You may be able to get away without it. Bend the ground pin back underneath the device and solder to board ground. Add a .33uF cap from the main 8V post to ground. Do not lift or adjust any other components.

3) Replace the LM317T with an LD29150V80, available from Digi-Key (their part number 497-1498-5-ND). This will ensure good regulation on the main 8V rail. Since this is a fixed-voltage regulator with a different pin-out, you'll need to modify the board slightly. Cut the trace to R21 to isolate the voltage-setting parts originally used with the adjustable regulator. The GROUND terminal of the LD29150 is con-

nected via an insulated wire passing through the hole that was originally used for the ADJUST on the LM317, to the negative terminal of C17 on the solder side of the board. Tack a .33 uF cap across the INPUT and GROUND terminals.

4) Add copper shielding in the synthesizer. Most of the FMing may be gone after the previous three steps but shielding should get rid of the last bit. Hobby Lobby has thin copper sheet in the form of an "ArtEmboss" copper embossing kit made by American Art Clay Company. Cost is about $7 and you get a 9 X 12in sheet of thin copper rolled up in a clear plastic tube with a wooden embossing tool and tracing patterns for you artiste types. I bet they didn't figure on their product ending up in an HF packset!

Be sure to add access holes for the 5120kHz reference adjust (if you have the early discrete crystal reference oscillator), and for the SYN2 adjust C104. The author didn't have to mess with C104, which adjusts the SYN2 VCXO for loop locking across its range. Before adding the shield, attach a short pigtail between the right pin (ground) of crystal Y101 and its case. Cut a small tongue in the shield and solder to Y101 after the shield is installed. If you have problems with microphonics due to the shields, put small pieces of cloth between the shields and module cover to damp them out.

The number-two problem reported with PRC-1099s is sidetone cut-out. This is very often due to insufficient gain in the mike audio circuits. Try replacing R70, a 1500 ohm resistor at the VOGAD chip with a 4.7K ohm. This will raise the gain of the VOGAD chip and often solves the problem.

Almost everyone who has owned one of these radios has had problems at some point with intermodule connectors inside the radio coming loose. They are simple DIN-type press-together types with no other method of retention. If your radio quits, that's the first thing to check. Other connectors that can become intermittent are the five brown ones connecting to the M7 Processor Module. Bending the pins on the module very slightly so as to better press against the contacts inside the brown connectors often fixes intermittent paths to the processor.

The 1099 is a fairly complicated set, and you'll need the TM and troubleshooting skills to work on it. Murphy's Surplus may still have the manuals. See Appendix C.

Accessories

Tip: You can extend the length of the standard AT-271 by removing the screw-on end cap, adding a short double-threaded extension, and screwing on the top section of a second AT-271. This extends the antenna to almost exactly a quarter wave on 17 meters. Of course, if you don't have a spare AT-271 you can always braze a stiff wire or small diameter brass pipe onto the extension and get the same result. Don't forget to paint it green!

In fact, you can really start to get serious about antennas. The author was mulling over how to get longer length for pedestrian mobile ops on 80m when a friend mentioned trying the old MS-116, -117, and -118 mast sections used with the GRC-9 and innumerable Korean War era tank and vehicular installations. It turns out there is a gimmick called an M-422 Adapter which screws into the top of an AB-591 base and accepts an MS-116. The problem is that after any more than a couple of mast sections, the whole thing flops over due to the torque on the AB-591. The base was never meant to take that much weight. The same friend suggested using a short length of PVC pipe to stiffen the base. The author eventually found something even better.

The reinforcing tube came, literally, from the junk box. It's a section from an old telescoping shower curtain rod! It even had a nylon cap at the far end that could be drilled out to accept the MS-116. With this arrangement, you can run 3 MS-116s, an MS-117, and an MS-118 for a total length of 16 1/2 feet. With a 44-foot dragging counterpoise, the tuner matches this antenna well on 80 meters. The author, who lives near Colorado Springs, used this setup to set the (current as of Spring 2005) 5 watt SSB world record for 80m pedestrian mobile-to-pedestrian mobile with Ken Chong, WB6MLC, in San Francisco.

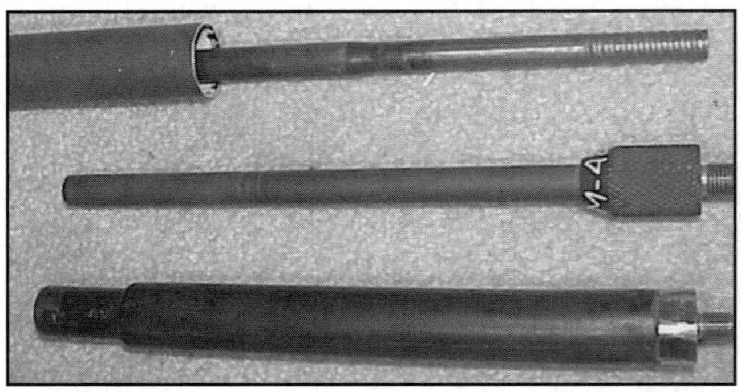

AB-591 base (bottom), M-422 Adapter (center), MS-116 with reinforcing tube (top)

GN-58, G-43

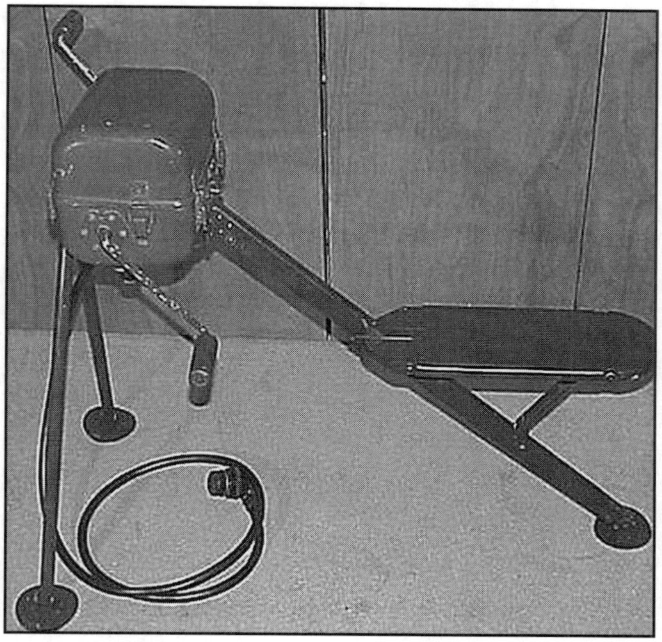

Application:

The GN-58/G is the hand-crank generator for the GRC-9 radios. It and the G-43/G appear to be equivalents. The author's GN-58 is pictured above.

Tech Specs:

Power input requirements: One (1) cranker

Output:
1.4VDC at 465 mA, 105VDC at 320 mA, 6.3VDC at 2.5A, 425VDC at 115 mA
Total output power: about 85 watts

Size: 27" x 38" x 19" (assembled)

Weight: 32 pounds, with cable

Active devices: None. Inductively and electromechanically regulated.

Tech Manuals: TM 11-5122, "DIRECT CURRENT GENERATORS G-43/G AND G-77/G"

Construction Style: Frame—lightweight aluminum with wooden seat. Generator—clamshell enclosure with gear drive, generator motor, and electromechanical regulator module.

Description:

The GN-58 and G-43 are collapsible, human-powered generators that produce DC input power directly for the GRC-9 tranceiver. They will also power the GRC-109 transceiver if the correct outboard regulator (CN-690) or correct power supply (PP-2684) is used. (See the GRC-109 section).

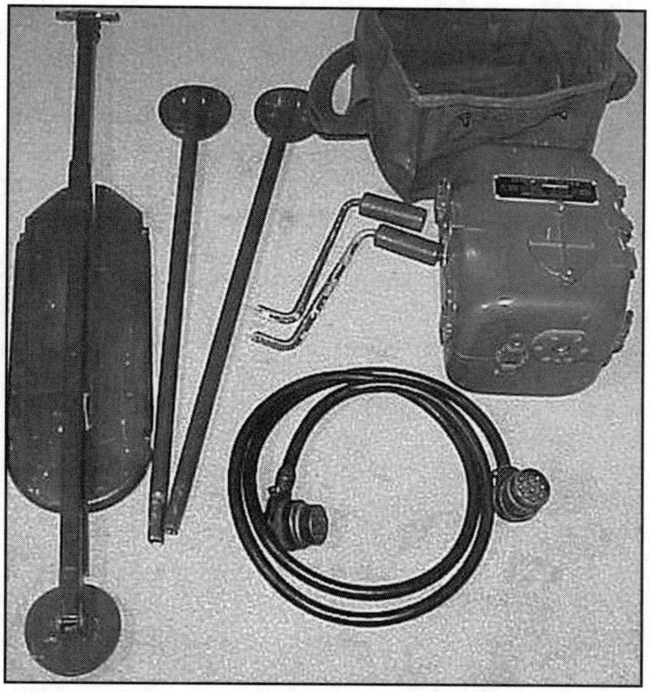

Generator GN-58, collapsed, with Bag BG-175. Some assembly required; batteries definitely not included

Unlike the G-77 used with the PRC-74, the GN-58 and G-43 have no electronics in the regulators at all. Regulation is accomplished by a "REG-OHM" Electric Regulator module:

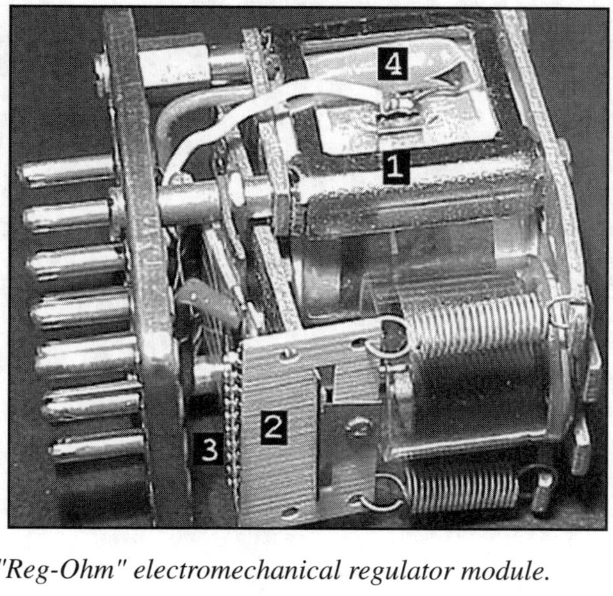

"Reg-Ohm" electromechanical regulator module.

This is a fascinating gimmick and worth some explanation in itself. It's about 2" x 2" x 1.5" in size. The coil (1) is hooked in series with the field winding of the generator. As the current through the field winding increases, the armature (2) of the regulator begins to engage the contacts (3) of a tapped resistor, which is also in series with the field coil. The contact surface of armature (2) is inclined, so that the contacts (3) are engaged one at a time, starting from the bottom of the picture. The armature gradually engages each resistance in turn by removing the short across that particular tapped section of the resistor. This is all arranged so that at 50-75rpm, and combined with load variations, the regulation is sufficient for good transceiver operation.

Two paralleled resistors, one positive- and one negative-temperature compensated (4) are also included to make the regulation "almost independent of temperature changes."

This is an interesting, if rather quaint, method of voltage regulation from an era when semiconductors were not yet in general use.

Operation:

Setup is straightforward for the most part. The author's GN-58 requires the legs to be separately installed onto the generator body. The G-43 appears to be more like the G-77 with an integral seat-legs assembly that folds open.

Resistor R2 is initially adjusted to obtain 6.3VDC between pins 20 and 24, or 34 and 24 of the generator output connector, with the generator properly loaded and cranked at normal speed. You shouldn't mess with this adjustment unless the output is obviously erroneous.

Location of output adjustment resistor R2

The pin-outs at the generator output socket are: Pin 24–GROUND; Pin 32–450VDC; Pins 20, 34–6VDC The connector that engages the generator is the same one (PL-294, male) as on the GRC-9 cable. Once you're wired up, crank clockwise (forward over the top with your right hand) at about 1 rev/sec.

Connections at GN-58 or G-43 connector

Additional Comments:

The GN-58 or G-43 will work with the GRC-9 or the GRC-109 (see GRC-109 section for special instructions).

You want aerobic exercise? Try cranking one of these babies under load for 30 minutes! Your arms will end up like wet noodles.

What To Do If It's D.O.A.:

There's even less to go wrong in the GN-58 or G-43 than the G-77 because they contain no semiconductors at all. If possible, it's best to use the info in this section to check the unit before you purchase -- if there's no output, there is probably something seriously wrong with the generator itself. The author's GN-58 has a small bag clamped inside the cover with spare brushes. Removing and replacing the original ones might help.

Accessories:

The original cable that connects the GN-58 or G-43 to the GRC-9 is the CD-1086. See GRC-9 section for details.

G-77

Generator G-77/G, collapsed, with Bag CW-420/G

Application:

The G-77/G is the hand crank generator for the PRC-74 series radios. It was also intended to charge nicad battery packs. Once its output voltage is adjusted, it can be used for any 12VDC nominal radios in the 15-20 watt range.

Tech Specs:

Size: 33" x 24" x 28" (assembled)

Weight: 17 pounds

Active devices: 2N5302, 2N3739, 2N2905A, 2N3506, 2 op-amps of unknown origin

Power requirements: One (1) cranker

Output:
10.5V at 5.0 Amps
17.5V at 2.0 Amps
17.5V at 100 mA
(see text)

Tech Manuals: TM 11-5122, "DIRECT CURRENT GENERATORS G-43/G AND G-77/G"

Construction Style: Frame—lightweight aluminum. Generator— clamshell enclosure with gear drive, generator motor, and regulator electronics. Electronics cannot be reached without removing the generator/gearbox.

Description:

The G-77 is a collapsible, human-powered generator that produces DC input power for the PRC-74 series of transceivers and also charges MA6982-3A nicad battery packs. A lamp on the generator lights when 2 amps of charging current is reached.

Operation:

Setup is straightforward for the most part. Unlike the GN-58 for the GRC-9, the legs on the G-77 are integral with the seat and just unfold. The generator housing latch can be a real pain though -- literally, if you let it snap your fingers. Use your foot to set the latch during assembly, and a crank handle end to pop it free during teardown. Your fingers will thank you!

WARNING: As supplied, the G-77 puts out nearly 18VDC at a load current of 100ma. This is fine for the PRC-74, but it may over-gee other 12V nominal radios. The author's unit has an output voltage adjust pot that can be adjusted for the radio to be used.

Start by setting it for 13.5VDC no load. Then experiment to see how your particular radio loads down the regulator. The author's TM states that both the output adjust pot and the pot that controls the current at which the charge light illuminates are located on the circuit card. This card is inaccessible without pulling the whole generator apart. The bracket holding the pot shown above looks like a later tack-on. In fact, the G-77 shown has a sticker showing that MWO-11-5820-

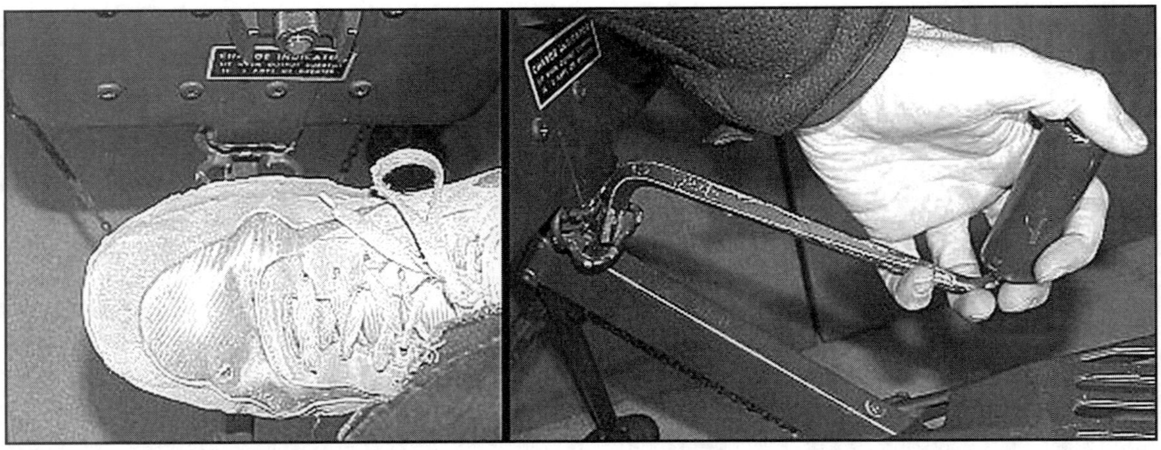

Dealing with the generator housing latch

GN-77 internals, showing output adjust pot

GN-77 Power Connector pin-outs

590-15 has been added. The techs probably yelled bloody murder about having to yank so much of the mechanical parts.

Note: Electrical ground is not case ground. It floats.

Author's modified metal battery box

The generator end of the cable uses the same connector (PL-294, male) as the one on the GRC-9. Once you're wired up, crank clockwise (forward over the top with your right hand) at about 1 rev/sec.

Additional Comments:

The G-77 should work with any 12VDC radios in the 15-20 watt class, once you adjust the voltage output. It's much easier to crank under load than the GN-58!

What To Do If It's D.O.A.:

It's best to use the info in this section to check the unit before you purchase -- it is a pain to get to the electronics if the regulators go bad. (Remember to take that cheapie VOM to the hamfest…)

Accessories:

The original cable that connects the G-77 to the PRC-74 is the CX-13103/PRC-74. The radio end connects to the CY-7773 plastic battery box POWER connector to run the radio. To charge the battery, you separate the PRC-74 from the battery box and plug the cable into the CHARGE connector on the CY-7773. The author has chosen to modify one of the better-quality metal PRC-74 battery boxes, adding a screw-type connector and a battery cutout toggle switch.

Caution: It is definitely not a good idea to charge SLAs, nicads, etc. in enclosed spaces. They can produce flammable gas during the charging process. Here's an object lesson: a friend had just finished charging the nicads in his Kachina MP-25 HF packset when he took it outside and turned it on. The radio exploded, cloaking him in a mist of ice crystals!

$H_2 + O_2 = H_2O +$ *ENERGY*

The original MP-25s had battery boxes that were open to the rest of the radio circuitry. When he hit the power switch, the spark set off the volatile hydrogen-air mixture which, once it had blown the radio apart and was free to expand, cooled so rapidly that the resultant water vapor crystallized into an instant snowstorm! The explosion damaged toroids inside the set, and he was never able to fix the radio. That episode *definitely* got his attention. We spend too much time coddling this stuff and getting it to work in the first place to risk blowing it up!

PP-4514

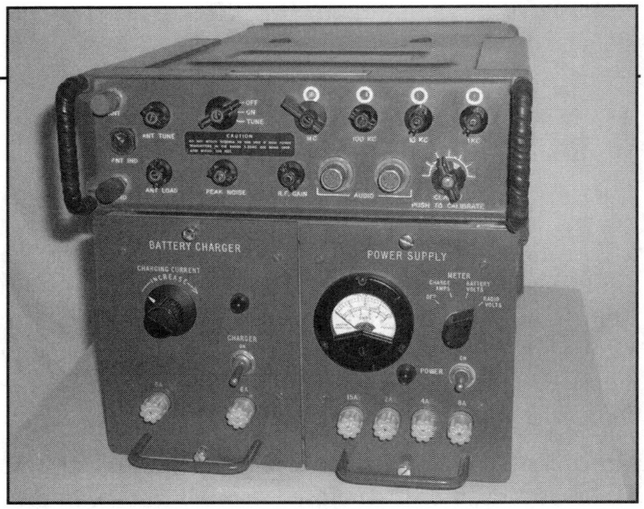

Application:

AC/DC power supply for the PRC-74. Also included is a battery charging function used to charge the original wet cell pack (10 BB-418/U nicads filled with caustic sodium hydroxide—something to avoid).

Tech Specs:

Size: 11.5" x 10" x 19"

Weight: 55 pounds

Active devices: 60s-vintage semiconductors, no chips

Power requirements:
INPUT
28VDC +10% to -25%
110VAC or 220VAC single phase +15% to -25%, 47 to 400 Hz

OUTPUT (power supply)
14VDC +/- 3VDC at 0 to 6.5A

OUTPUT (battery charger)
Variable current, 1 to 5A

Tech Manuals: The PP-4514 is covered in the following PRC-74 TMs:

TM 11-5820-590-35-1 "Direct Support, General Support, and Depot Maintenance Manual, Radio Sets AN/PRC-74B and AN/PRC-74C, Power Supplies PP-4514/PRC-74 and PP-4514A/PRC-74 and Battery Boxes CY-6121/PRC-74 and CY-6314/PRC-74 and CY-6314A/PRC-74"—detailed theory, troubleshooting, and schematics.

TM 11-5820-590-12-1 "Operator's and Organizational Maintenance Manual, Radio Sets AN/PRC-74B and AN/PRC-74C; and Power Supplies PP-4514/PRC-74 and PP-4514A/PRC-74; and Battery Boxes CY-6121/PRC-74 and CY-6314/PRC-74 and CY-6314A/PRC-74"—operating info and some specs on the components. NO schematics.

Construction Style: Heavy stamped metal outer case with two large pull-out modules, one for the DC power supply itself (right, in the picture), and one for the battery charger. Repairable.

Description:

The PP-4514 is the companion AC/DC power supply that allows use of the PRC-74 series radios off 117VAC, or 28VDC vehicular systems. It's pretty big and heavy for what it does, but you may find it handy for powering other 12V radios within its current range.

Operation:

The supply should come with five (5) different cables. For those who are short these cables, pin-outs are included here. For ALL power input, use the large connector inside the hinged lid. The matching female connecter is a Cannon 22-5.

W1: 21-31VDC power input cable (with DC vehicular system plug)
W2: 110VAC power input cable–Pins A and C (neutral)
W3: 220VAC power input cable–Pins F and C (neutral)
W4: 24VDC power input cable (with alligator clips)
For both W1 and W4–Pin B: +21 to 31VDC,
Pin A: Ground

For the charger output, use the small 2-pin connector inside the hinged lid. The matching male connecter is a Cannon 12S-3.

W5: Battery charger OUTPUT cable—
Pin B: POSITIVE (+), Pin A: NEGATIVE(-)

Once suitable power is hooked up, set METER switch on the power supply to RADIO VOLTS and turn ON the supply. You should get 13 or 14VDC on the meter. During AC operation, you should hear the DC-DC converter squealing inside.

The battery charger side simply produces a steady charging voltage and the ability to vary the current from 1 to 5A or so.

Additional Comments:

The author hasn't run into any trouble running the supply with no load for short periods, but he would recommend keeping no-load operation to a minimum.

What To Do If It's D.O.A.:

If you don't get a reading on the RADIO VOLTS switch setting, and you don't hear the DC-DC converter, check the fuses. The biggest problem the author has had with his supply is the pilot lights (327 type) get kicked too hard and don't last.

GRA-71

Application:

CW Burst Coder equipment for the GRC-109, TRC-77, PRC-74; adaptable to any other radios capable of being keyed at high speed.

Tech Specs:

Keying Speed: 300 words per minute.

Size: 7" x 9" x 5.5", units in case

Weight: 10 pounds, case with all units.

Active devices: Solid state, early 60s semis; Keyer Adapter has a sub-mini vacuum tube regulator (6542).

Power requirements (KY-468 only): 12VDC, supplied by MX-4498 Keyer Adapter when GRC-109 is used, or by CX-10239 cable directly from a PRC-74 or other radio with 12V on Pin E of the audio connector, or by CX-11389 cable for a TRC-77 (see TRC-77 section for details).

Current drain (KY-468 only): 40mA max at 12VDC (KY-468)

Length and type of magnetic tape used: 12.5 feet, 3M Type 428 1/4" Instrument Grade

Tech Manuals:
TM 11-5835-224-12, "OPERATOR AND ORGANIZATIONAL MAINTENANCE MANUAL, CODER-BURST TRANSMISSION GROUP AN/GRA-71", operating instructions, basic theory and maintenance; no detailed theory, troubleshooting, or schematics.

TM 11-5835-224-35, "DS, GS AND DEPOT MAINTENANCE MANUAL INCLUDING REPAIR PARTS AND SPECIAL TOOL LISTS, CODER-BURST TRANSMISSION GROUP AN/GRA-71", theory, troubleshooting, repair, schematics.

Construction Style:

The electromechanical units have ingeniously-designed, clockwork-style construction. Electronics in the Keyer Adapter is repairable/replaceable and easily reached on the circuit card. However, the keying circuitry in the KY-468 Keyer is built on four PCBs, sandwiched cordwood-style and looks to be more difficult to work on.

Description:

The GRA-71 is a fascinating piece of 1960s technology, both in construction and function. A lot of good general descriptive info and history is available on the Web already, so the presentation here will focus on some additional interesting details.

The tape is 1/4" (see Tech Specs for exact type). The right reel is driven by the KY-468 keyer during operation; the left reel provides tension on the system.

MA-9 Tape Magazine mechanism, covers removed.

The maximum number of turns possible is governed by the placement of the stop nubs on the two gears that run off the driven reel. The mechanism is shown positioned at the beginning of the tape. The stop nubs are engaged, and prevent the tape from being pulled

Tape Magazine mechanism, bottom

all the way to the end of the driven reel (and pulled off altogether). Since the top gear has one less tooth than the bottom one, after the first revolution, its stop nub will be one tooth behind the nub on the other gear. Thirty-five turns later, the nubs synch up and hit again just before the physical end of the tape. Pretty slick.

The reeled ribbon rewind spring keeps the tension on the system substantially constant throughout the tape's travel. The stop nubs on the right set of gears do not engage at any time on this particular Tape Magazine.

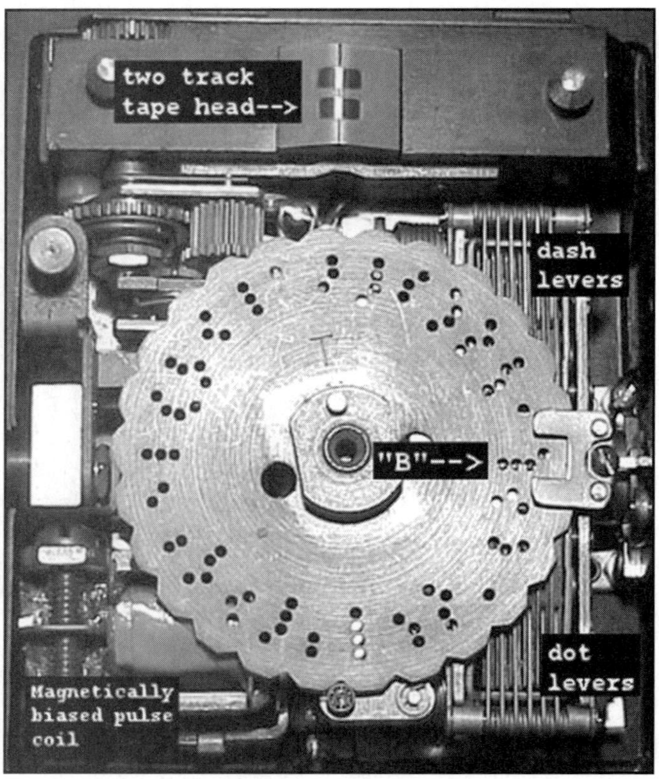

MX-4496 Tape Coder, cover removed

The tape coder has no active devices. The information to be written onto the tape is generated by moving an armature in a magnetically biased coil (lower left). The designers made some assumptions -- they left out the capability to code numbers to reduce the character set to those requiring only four elements (A through Z). In the illustration, the letter "B" is staged to be coded when the lever is operated. The top row of holes is dashes and the bottom row is dots. Thus, the letter "B" reads dash-dot-dot-dot from right to left. You can read the rest of the letters in order, clockwise around the wheel. The first dash in the C is obscured by the bracket.

The pulse coil always generates four pulses when the coding lever is operated. The dash and dot levers determine which track, if any, the pulse is written to. If there's no hole in a dash or dot position, no element is written to either track. During transmission, the KY-468 keyer generates a dash or dot depending on which track contains a pulse.

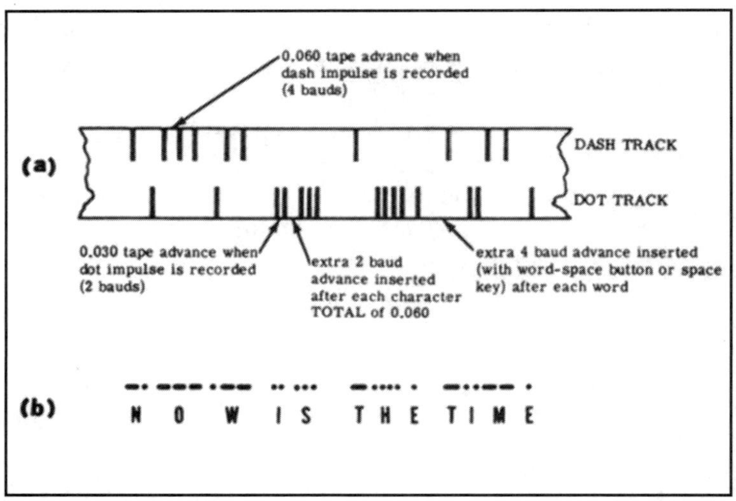

Two track dot and dash recording method, showing a sample message

The sample message shows how the dot and dash marks are laid down on the tape (dashes above, dots below). Part a) shows the recorded impulses corresponding to the message in Part b).

The MX-4495, also supplied with the GRA-71, provides the ability to directly code dots, spaces, and dashes. It's a simpler version of the MX-4496, with a lot less mechanical stuff to go wrong. So in addition adding the capability to code special characters, it's also a mechanical backup.

The Keyer Adapter is needed for use with the GRC-109, but not for any other sets with new family connectors (which accept U-229s) with the correct pinouts and the center pin set up with switched 12VDC (see Operation section).

MX-4495 Tape Coder, cover removed

KY-468 Keyer, cover removed

Once the message is coded on a tape with either coder, the KY-468 is used to actually send the message. The clockwork mechanism utilizes a centrifugal governor to keep the reel speed constant similar to those used on old-time steam engines!

KY-468 electronics, showing nasty cordwood construction

The traces on the boards are easily reached with the top and bottom covers removed, but replacement of components isn't much fun.

Operation:

Setup for the GRA-71 varies depending on the radio you're using, but operation is pretty much the same.

The KY-468 sending unit is used with the MX-4498 Keyer Adapter when the radio is the GRC-109, and with special purpose cables direct to the radio when the TRC-77 or PRC-74 are used.

Caution: The Keyer Adapter is necessary for operation with the GRC-109. Do not attempt to wire the KY-468 directly to the KEYER jack on the -109 transmitter! (The plugs don't match, and for a good reason).

Special Purpose Cable CX-10239 (KY-468 to PRC-74) detailed wiring:

7-pin Amphenol KY-468 end	6-pin U-229 PRC-74 end	Function
A	A	Ground
B	no connection	
C	A	Ground
D	E	Keying Closure From KY-468
E	no connection	
F	A	Ground
H	F	+12V to KY-468

Special Purpose Cable CX-11389 (KY-468 to TRC-77) detailed wiring:

7-pin Amphenol KY-468 end	PL-55 TRC-77 end	Wire leads TRC-77 end	Function
A	PL-55 sleeve	-12V	Ground
B	no connection		
C	PL-55 sleeve	-12V	Ground
D	PL-55 tip		Keying Closure From KY-468
E	no connection		
F	PL-55 sleeve	-12V	Ground
H		+12V	+12V to KY-468

The 7-pin Amphenol connector at the KY-468 end on either cable is generally available surplus. (See connector source in Appendix).

GRA-71 Operation. The following text is from the Operator's Manual. See author's comments within the "[]".

2.2 Message Recording

2.2.1 MX-4496 Coder Message Recording. The MA-9 Cartridge will hold at least 125 five-letter words plus word spaces. The maximum word capacity varies, depending upon the number of characters per word and the message length. Because of the automatic rewind feature, the MA-9 must not be removed from the MX-4496 until the entire message has been recorded. If it is accidentally disconnected from the MX-4496 before message completion, erase the tape and start over.

The following procedure describes the steps necessary to record a message in the MA-9 Cartridge with a MX-4496 Coder.

Step 1. Open and remove the MX-4496 Coder and MA-9 Cartridge lids by sliding them off their hinge pins. Attach a fully erased MA-9 Cartridge onto the MX-4496 Coder by lining up the locking pins and cartridge drive gear on the coder with the sockets (in locking plates) and tape drive gear on the MA-9 Cartridge. Gently snap the two units together.

Step 2. Place the MX-4496 Coder on a smooth, flat surface. Depress the WORD-SPACE BUTTON about 10 times to move exposed tape past the recording head.

[The author suggests 25 to 50 times].

Step 3. Rotate the CHARACTER DIAL until the first character of the message clicks into position opposite the index mark. The character dial may be rotated in either direction before or after the operating handle is pulled down.

WARNING: The character dial must not be rotated during the downstroke of the operating handle. If the dial becomes jammed, DO NOT FORCE IT. Operate the handle again, more slowly. When the jam clears, insert two word spaces and repeat the entire group in which the jam occurred. Forcing a jammed dial can seriously damage the MX-4496 or MX-4495 mechanisms.

Step 4. Depress the EXTENSION HANDLE RELEASE BUTTON and lift up the handle. Raise the HANDLE to the "starting" position. [all the way up and back].

Step 5. Using the thumb and three fingers, grasp the HANDLE and pull it down with a firm, even stroke until it snaps against the stop at the end of its travel. DO NOT HESITATE DURING THE DOWNSTROKE. The character has now been recorded in the MA-9 Cartridge.

[The author has found that operating the handle too rapidly will also cause problems such as slurring of the characters. It takes a little practice…]

Step 6. Select the next character of the word and repeat step 5.

Step 7. After the word has been recorded, insert a word space by firmly depressing the WORD-SPACE BUTTON once and releasing it, allowing it to return to normal position.

Step 8. Continue the procedure of steps 5, 6, and 7 until the entire message has been recorded.

Step 9. After the entire message has been recorded, detach the MA-9 Cartridge from the MX-4496 Coder by pressing the CARTRIDGE RELEASE BUTTON inward. Lift the cartridge away from the Coder. As soon as the cartridge drive and coder drive gears separate, the cartridge begins rewinding the tape. Upon completion of rewind (about 5 seconds), the message recorded in the cartridge may be transmitted. Replace the coder LID immediately and fold the EXTENSION HANDLE back down by depressing the release button. Replace the cartridge LID as soon as possible.

2.2.2 MX-4495 Coder Message Recording. The MA-9 Cartridge will hold at least 125 five-letter words plus word spaces. The maximum word capacity varies, depending upon the number of characters per word and the message length. Because of the automatic rewind feature, the MA-9 must not be removed from the MX-4495 until the entire message has been recorded. If it is accidentally disconnected from the MX-4495 before message completion, erase the tape and start over.

The following procedure describes the steps necessary to record a message in the MA-9 Cartridge with a MX-4495 Coder.

Step 1. Depress the LID RELEASE BUTTON and open the MX-4495 Coder LID. Open and remove MA-9 Cartridge LID by sliding it off the hinge pins. Attach a fully erased MA-9 Cartridge onto the MX-4495 Coder by lining up the locking pins and cartridge drive gear on the MX-4495 Coder with the sockets (in locking plates) and tape drive gear on the MA-9 Cartridge. Gently snap the two units together.

Step 2. Place the MX-4495 Coder on a smooth, flat surface. Place fingers on the keyboard as shown in Figure 2-2. Depress the SPACE key about 10 times to move exposed tape past the recording head.

[Again, the author suggests 25 to 50 times….]

Step 3. Depress the DOT (.) and DASH (-) keys accordingly to record a character. For example to record the letter Q (--.-), depress the DASH key twice, depress the DOT key once, and depress the DASH key again once. The letter Q is now recorded on the tape as DASH DASH DOT DASH.

Step 4. Depress the SPACE key once to create a space at the end of the character.

Step 5. Record the next character in the word by repeating steps 3 and 4.

Step 6. After a complete word has been recorded, insert a word space by depressing
the SPACE key twice.

Step 7. Continue the procedure of steps 3, 4, 5, and 6 until the entire message has been recorded.

Step 8. After the entire message has bcen recorded, detach the MA-9 from the MX-4495 by pressing the CARTRIDGE RELEASE BUTTON inward. Lift the cartridge away from the coder. As soon as the cartridge drive and coder gears separate, the MA-9 Cartridge begins rewinding the tape. Upon completion of rewind (about 5 seconds), the message recorded in the MA-9 Cartridge may be transmitted. Close the MX-4495 Coder LID immediately. Replace the MA-9 Cartridge LID as soon as possible.

2.3 Message Transmission

The following procedure describes the steps necessary to transmit a recorded message with the KY-468 Keyer.

KY-468 Keyer and MA-9 Tape Cartridge shown connected through MX-4498 Keyer Adapter to T-784/GRC-109

Step 1. Grasp the latches at opposite corners of the GRA-71 case and lift them up. (When replacing the LID, close the trunk latches at opposite corners. This prevents the rubber gasket from becoming distorted, maintaining the water-tight seal.) Open and remove the KY-468 and MA-9 LIDS by sliding them off their hinge pins. Insert the plug (P2) at the end of the keyer connector connector cable on the MX-4498 into the CONNECTOR RECEPTACLE on the KY-468 control panel.

Insert the plug (P1) at the end of the transmitter connector cable into the appropriate receptacle on the T-784/GRC-109 Transmitter.

Step 2. The KY-468 Keyer drive motor must be fully wound before attaching the MA-9 Cartridge and Keyer connector cable. With the MOTOR ON-OFF switch in OFF position, fold out the WIND-UP CRANK and wind it clockwise until the TAPE DRIVE GEAR begins to rotate. A definite increase in winding resistance occurs when the motor is fully wound. About 30 turns of the crank will wind a fully run down motor.

[The author disagrees with this method of winding. He believes this just puts undue strain on the mechanism brake. Twenty turns gives over 30 seconds of good operation, corresponding to about 150 groups. This is more than the stated capacity of the tape. Winding any more than 20 turns is unnecessary].

Step 3. Attach the MA-9 Cartridge to the KY-468 by lining up the locking pins with their respective sockets on the MA-9 so that the MA-9 and the KY-468 gears will mesh. Gently snap the two units together. The KY-468 is now ready for operation. Be sure that the T-784/GRC-109 Transmitter is switched ON and is properly adjusted.

Step 4. To transmit IDY, press the spring--loaded IDY SWITCH upward and hold it all the way up for the required IDY interval. The transmitter output indicator will be activated during IDY transmission.

["IDY" is just dits at 300 wpm. This can be used to allow the distant station to make last minute adjustments before the burst comes in, or for other purposes].

Step 5. To transmit the recorded message, slide the MOTOR ON-OFF switch to ON position. The output indicator on the transmitter should show that the message is being transmitted. When message transmission is completed, slide the MOTOR ON-OFF switch back to OFF and immediately disconnect the keyer connector cable. When the MA-9 is detached, the tape will rewind automatically. The MA-9 may now be reattached and the message transmitted again, if necessary.

[Leaving the keyer connected can result in additional (spurious) transmissions as the MA-9 is detached and starts rewinding the tape over the KY-468 pickup head].

2.4 Message Erasing

The following procedure describes the steps necessary to erase an MA-9 Cartridge with the KY-468 Keyer.

Step 1. Be certain that the KY-468 Keyer drive motor is fully wound. With the MOTOR ON-OFF switch in OFF position, fold out the WIND-UP CRANK and wind it clockwise until the TAPE DRIVE GEAR begins to rotate.

[See author's comments on excessive winding, above].

Step 2. Connect the KY-468 Keyer to the MX-4498 Keyer Adapter. Connect the MX-4498 Keyer Adapter to the T-784/GRC-109 Transmitter and be certain that the transmitter is in operating condition.

Step 3. Attach the MA-9 to the KY-468 by lining up the locking pins with their respective sockets on the MA-9 so that the MA-9 and KY-468 gears will mesh. Gently snap the two units together. The MA-9 is now ready to be erased.

Step 4. Slide the ERASE switch up, in the direction of the arrow, and hold it in the upward position. Then slide the MOTOR ON-OFF to full ON position. This will lock the ERASE switch in position. Hold the MOTOR ON-OFF switch firmly against its stop until the tape has run completely through and stops.

[Running the tape completely through is unnecessary; running it until it is obviously farther along than the end of the recording is sufficient.

Step 5. Return the MOTOR ON-OFF switch to OFF position, allowing the ERASE switch to return automatically to its original position.

Step 6. Disconnect the keyer adapter connector cable and then detach the MA-9 Cartridge to allow automatic rewinding. The tape is now ready for another recording.

Operation with the PRC-74 is substantially the same, except the KY-468 is connected directly to the radio with the CX-10239 cable (or your fabricated equivalent). Similarly, operation with a TRC-77 is the same, except a direct connection via cable CX-11389 is used.

Additional Comments:

Here are the directions for MA-9 tape replacement from the Operator's Manual:

2.5 Tape Replacement

When necessary to replace the recording tape in the MA-9 cartridge, proceed as follows:

[Replacing the tape may become necessary if you start to notice dropped or distorted characters, especially toward the beginning of the tape, which is used more heavily].

Step 1. Remove the hinge pin and remove the lid of the cartridge.

Step 2. Remove the three attaching screws on the spool cover and remove the cover.

Step 3. Remove the end of the old tape from the core of the takeup spool and allow it to dangle free.

Step 4. Remove (pull) the tape from the storage spool. Apply finger pressure to the spool to prevent automatic rewind action, then disconnect the end of the tape from the storage spool. Discard the old tape. Allow the storage spool to rewind slowly.

Step 5. Inspect the cores of both spools. Remove any residue remaining from the old tape with the camel's-hair brush provided.

Step 6. Rotate the storage spool clockwise (to increase spring pressure) until it encounters the takeup auto-stops, and hold it in this position.

Step 7. Place the shiny side of the new tape towards the core of the storage spool. Firmly attach one end of the tape to the backside of the storage spool core with a small piece of pressure-sensitive adhesive tape.

Step 8. Allow the storage spool to rewind (the tape) slowly. Control the rewind speed with finger pressure on the storage spool while maintaining sufficient tension on the tape to prevent it from wrinkling or winding up loosely. There should be about 1 inch of loose tape remaining after a complete rewind.

Step 9. Rotate the takeup spool counterclockwise until it encounters the rewind auto-stops.

Step 10. Thread the free end of the tape over the tension idler, and over the tape guides. Be sure that the dull surface of the tape is on the outside.

Step 11. Pull out enough tape to make a complete extra turn around the core of the takeup spool. Attach the tape to the core with a short length of pressure-sensitive tape. The purpose of this extra turn is to provide a full turn of tape completely around the core of the spool when it is at rest. This insures a lasting attachment and prevents the adhesive tape from contaminating or touching the overlying layers of recording tape.

Step 12. Check the rewind action by rotating the take-up spool fully clockwise and then allowing the storage spool to rewind completely.

What To Do If It's D.O.A.:

As with any electromechanical device, it seems to be the mechanical half that produces the most headaches. The GRA-71 is no exception. The author has found that dirty or corroded contacts in the two tape coders are most likely to cause problems.

Tip: You may notice a fine white powder on the inside of the rubber stowage compartments in the MX-4498. The author has read that this is sulfur or related material from the rubber and will eventually tend to affect certain internals of the units. Switch contacts were mentioned specifically. The first thing to try if you notice dropped dashes or dots is to carefully clean the cammed switch contacts inside the MX-4496 and/or MX-4495.

If you get dropped characters or character elements when using both coders, it may be a bad section on the tape itself. To detail-troubleshoot these units, you really need the dash-35 TM, with theory, schematics and details on adjusting the mechanical intricacies.

DY-88

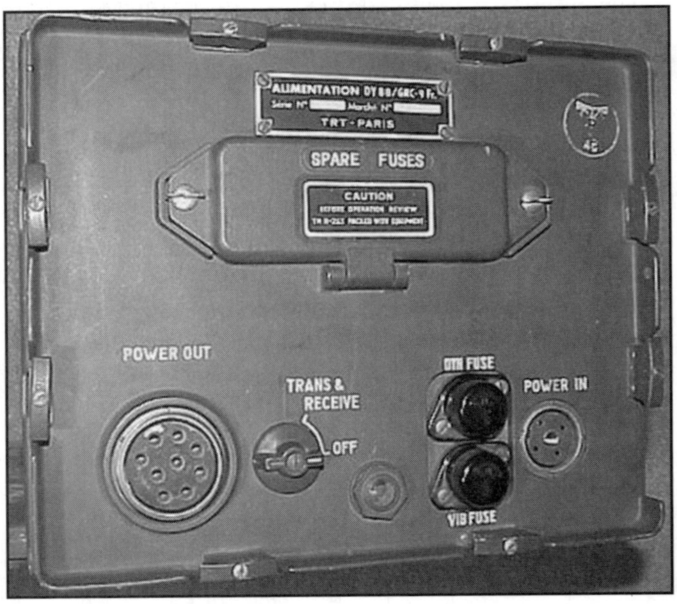

Application:

Dynamotor DC power supply for the GRC-9 transceiver.

Tech Specs:

Size: 14" x 9" x 13"

Weight: 37 pounds

Active devices: Vibrators and ballast tubes (see text)

Power requirements: 6VDC (22.2A), 12VDC (12.2A), or 24VDC (6.7A); with heavy start-up current—see text

Tech Manuals: TM 11-263; there is a schematic card inside the case.

Construction Style: Heavy die-cast front panel connected to a metal chassis, which slides into an outer case retained with Dzus fasteners. Point-to-point wiring, with all components easily reachable.

Description:

The DY-88 supply will power the GRC-9 from a variety of vehicular systems and/or storage batteries -- a great relief in portable setups that frees you from cranking that blasted generator. The author's unit is labeled "ALIMENTATION," the French word for nourishment or feeding. This unit was manufactured under license by TRT in Paris for NATO. Incredibly, these units derive all the voltages with no vacuum tubes (regulators or otherwise) -- it's all done with resistive ballast regulation.

Operation:

Caution: Be sure to inspect the CD-1086 power cable per the suggestions in the GRC-9 section before using it the first time. Some cables may be defective.

An internal switch must be set to the input voltage being used. The window at bottom center in the front panel image will show the current setting. If it doesn't match your input voltage, remove the unit from case by undoing the Dzus fasteners, and set the switch to the correct input voltage.

Underside of chassis; pen points to voltage selector switch lever

The CD-1086 connects the POWER OUT jack on the DY-88 to the POWER jack on the GRC-9. The CX-2031 is used for the vehicular DC power connection to the POWER IN jack on the DY-88. The detailed pin-outs for the jacks may be found in the GRC-9 section if you don't have these cables.

Once the supply is hooked up, the only operational action is to move the switch on the DY-88 from OFF to TRANS & RECEIVE. This only enables the supply and will not turn on the radio until control E on the GRC-9 is moved from OFF to either SEND or STANDBY. SEND will start the dynamotor and ready the set for PTT operation. STANDBY will run the receiver only and leave the dynamotor off to conserve the battery.

Additional Comments:

Note: While the steady-state DC input current in high-power transmit at 24VDC is listed as 6.7A, the dynamotor startup current goes far over that for a short period. The author powers his supplies with a big Perkin magnetically regulated supply; when he moves the GRC-9 power switch to SEND, the current meter pegs at 30 amps for a half second or so until the dynamotor gets up to speed. This won't be a problem with most vehicular systems or storage batteries, but it's something to be aware of. The usual 20 to 25-amp ham equipment supply won't handle the start-up current.

What To Do If It's D.O.A.:

Check fuses first. Make sure the correct value fuse is installed for the input voltage you are using! If there are no fuses behind the SPARE FUSES door on the front panel, do not despair -- there should be numerous spares on the chassis inside in the SPARES box. You'll also find a complete set of dynamotor brushes below the fuses.

In STANDBY or SEND, you should hear vibrators buzzing. If not, try pulling them and tapping them a few times on a hard surface. You may end up having to cut them open to unstick the armatures. One or more ballast tubes may have opened up -- use the schematic packed inside the unit to determine the pins on the various ballasts and ohm them out.

If the dynamotor doesn't turn at all, check the control circuits. The full input current doesn't flow through control E on the GRC-9; that control actuates a relay in the DY-88 with heavier contacts.

Caution: After sitting for many years, the lube in the dynamotor and hand generator bearings will dry out and cause them to wear rapidly. It's best to check them before using the either for the first time. This is a lot of work and involves extensive disassembly; it depends on how much you plan on using the units. (Hey, we're all fanatics, right?) The procedure is the same for both, once you have the bearings exposed:

Top view of chassis, showing schematic card and spares box (cover removed)

a. Clean old grease from the bearings with cleaning compound. Do not let the cleaning compound flow into the commutator or armature windings.

b. Relubricate with grease [type "GL" is called out] during re-assembly. Spread the grease evenly and avoid an excess of grease. Never apply grease to the brushes, commutator, or brush holders.

ME-61

Application:

HF field strength meter.

Tech Specs:

Frequency range: 1.5 - 24 MHz

Size: 5" x 5" x 4.25"

Weight: 2 pounds

Active device: 1N125A rectifier

Power requirements: None (other than received RF)

Audio input/output devices: Phone jack (PL-55 plug/standard 1/4") provided for monitoring.

Whip antenna: Built-in 18" whip

Tech Manuals: unknown

Construction Style: Panel/chassis fits into a slip-on case. Hinged lid retained by single latch. Components are repairable/replaceable.

Description:

The ME-61 is useful for general field strength measurements. Unlike the cheap imports you often see, it has a tuned antenna circuit to limit unwanted responses.

Operation:

Extend the antenna. Set the BAND SWITCH to the range covering the frequency you wish to measure, and rough-tune the TUNING knob to that frequency. As you transmit, peak the tuning, increasing or decreasing the METER SENSITIVITY to get a useful reading on the meter. If desired, you can monitor the signal at the PHONES jack.

The "0—5" scale is relative only and not meant to be an indication of any specific power level.

Additional Comments:

Modern HF sets with automatic antenna tuners vary considerably in their ability to accurately tune antennas -- especially into screwball loads. They often over- or undershoot the correct match and have to be retuned a couple of times before they narrow in (if ever). The only way to really tell if you are radiating is with a field strength meter.

What To Do If It's D.O.A.:

If the unit appears totally inoperative, the IN125A may be smoked -- pull the chassis by unscrewing the four corner captive screws and check the diode. Also, never discount the contacts on those rotary switches!

GRR-5

Application:

Mobile HF receiving set. The date on the author's manual suggests the GRR-5 was designed in the late 40s or very early 50s. It was a contemporary to the RT-66/-67-68 series of VHF FM transceivers and, interestingly, is shown as an "antiaircraft-warning receiving set in conjunction with Radio Set AN/VRC-8, 9, or 10 and Intercommunication Set AN/UIC-1."

The VRC-8/-9/-10 are the RT-66/-67/-68 with PP-109 or -110 power supplies. The UIC-1 appears to be an AM-65 audio amplifier/power supply away from its usual role as a supply for the RT-70. The C-375 is a remote junction box for the control and audio lines to two radio sets.

The author has learned that "antiaircraft warning" as applied to this system did not refer to distant stations informing the AAA sites of incoming aircraft, but rather the opposite. It was most likely used to warn our own people about falling flak from the AAA batteries. A personal friend served in Germany in the late 1960s and was issued an M37 truck with two RT-524s and a GRR-5. The system was called the AAIS or Anti-Aircraft Information System. The GRR-5's role was to receive AM voice signals informing the mechanized infantry when AA batteries were going to open up, and where the spent flak might be expected to fall! Apparently, similar systems were put together with T-195 transmitters and (as above) with the older FM gear.

My friend listened to the BBC and Radio Luxembourg (which played rock music). Later in Vietnam he managed to scrounge a GRR-5 and haywire it into another M-37.

Tech Specs:

Frequency range: 1.5 to 18mHz, continuous tuning in four bands, 1.5 - 2.7, 2.7 - 5.0, 5.0 - 9.5, and 9.5 to

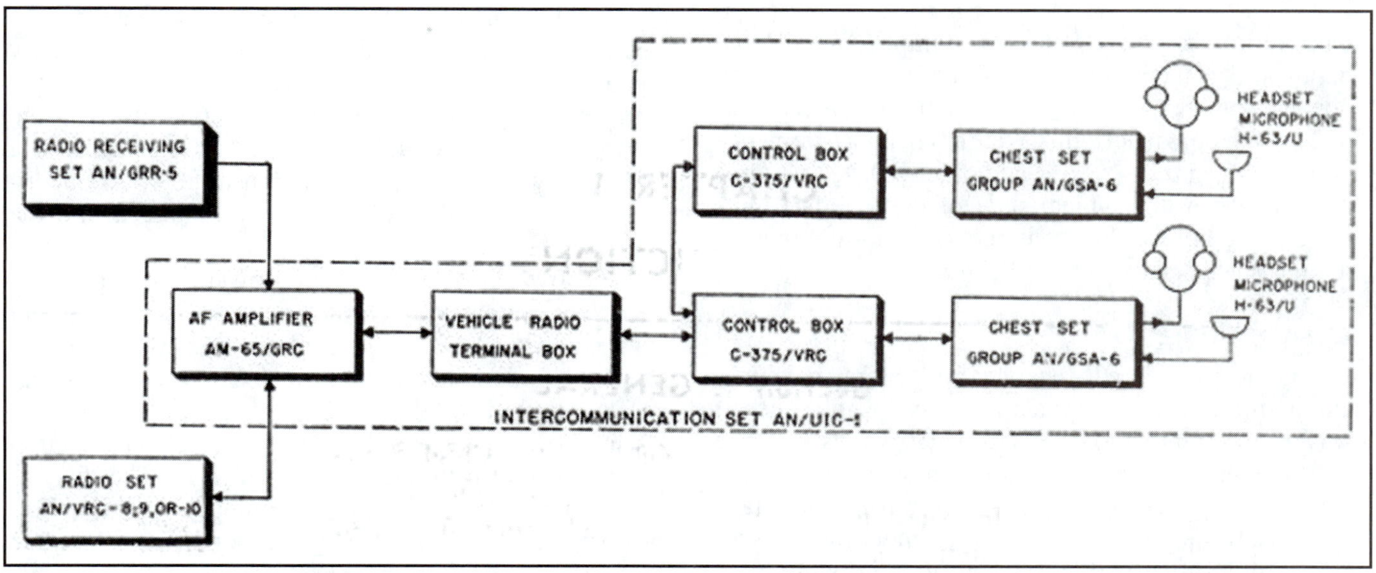

GRR-5 used in the antiaircraft-warning role

18mHz, with some overlap at the ends on each band. Ten channels may be mechanically preset.

Modes received: AM, CW and SSB (see text).

Size: 14" x 14.5" x 11"

Weight: 60 pounds

Active devices: Vacuum tubes. Receiver R-174/URR: 1L4(3), 1R5(3), 1U5, 3V4. Power Supply PP-308/URR: 1007, 6AG7(2), 0B2

Power requirements: When receiver is used with power supply: 6VDC at 6.9A; 12VDC at 3.22A; 24VDC at 2.55A; 115VAC/50 or 60Hz, @455ma, 52.5 watts

In field operation directly from battery power: two batteries BA-419/U, 90VDC @ 27mA, and one battery BA-405/U, 1.5VDC at 350mA

Audio input/output devices: Most any audio accessories using the U-77 connector, including the LS-166, HS-63, H-33, etc. Also, built-in speaker.

Antenna connection: Screw-in provision for MS-116 mast section, or antenna and ground binding posts.

Whip antenna: MS-116, MS-117, MS-118 mast sections (see text).

Other antennas: Virtually any other long wire, wire reel, house gutter, you name it

Antenna tuner: Integral antenna trim control

Tech Manuals: TM 11-295, "RADIO RECEIVING SET AN/GRR-5", description, theory, troubleshooting, repair, alignment, schematics.

TO 16-30-GRR5-5 -- Air Force version of the manual

Construction Style: Separate stamped metal front panel and chassis assemblies for the receiver and power supplies slide into a common outer metal case of similar construction. Circuit wiring is point-to-point between tube sockets and other chassis-mounted components in the typical style of the era. The units are both reparable, but construction is quite compact in places, especially in the receiver.

Description:

The GRR-5 is a great receiver for general listening, although tuning SSB signals is pretty touchy. The set was never intended to do that anyway, so it's sort of a bonus. The audio quality is good enough for international BC reception (a speaker with more bass response helps). The receiver has so much excess sensitivity, one often ends up turning the RF gain down even when using a short antenna. In fact, if reasonably well aligned, the receiver will pick up all sorts of stuff with nothing but your finger near the antenna post!

The flexibility of being able to run off 6, 12, 24, or 115AC power makes this set a conversation piece at camp-outs. The author has set his up on a picnic table, connected it to his pickup battery, and let participants tune around. A simple whip antenna is more than enough to pick up gobs of stuff.

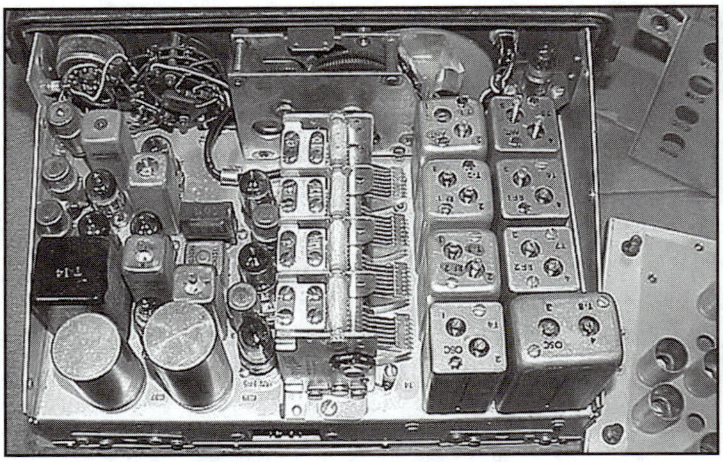

Receiver R-174/URR, top

As shown in the receiver top view, the internal chassis construction is not quite as heavy as a lot of the GRC gear, showing an obvious goal of reducing the tonnage. The covers are all lightweight aluminum. Somehow, the thing still ended up at 60 pounds. Note the integral tube shields on the cover at lower right -- that's one way to keep the techs from leaving them out! Everything is splash proof (note enclosed MONITORING INPUT jack at upper left), and MFP'd (moisture and fungus-proofed).

Visible at upper right is the neon bulb antenna circuit protector. These things are necessary, especially in a mobile environment -- you never knew when you might come near a high-power transmitter. If the receiver antenna circuits are tuned to resonance on a frequency where high power transmitters are operating, significant RF voltages can be induced into the front end.

The front of the receiver is toward the top in the receiver bottom view. You can see the point-to-point wiring, and the ganged ceramic rotary switch (an expensive component when this thing was designed, to be sure). The antenna trim variable cap is visible just above the right side of the switch, and the BFO variable behind the panel at upper right. The relay at top center mutes the receiver —among other functions (more on that later). There is a complete schematic glued inside the bottom cover, which is only 6" x 3.5". A real eye chart but useable in a pinch.

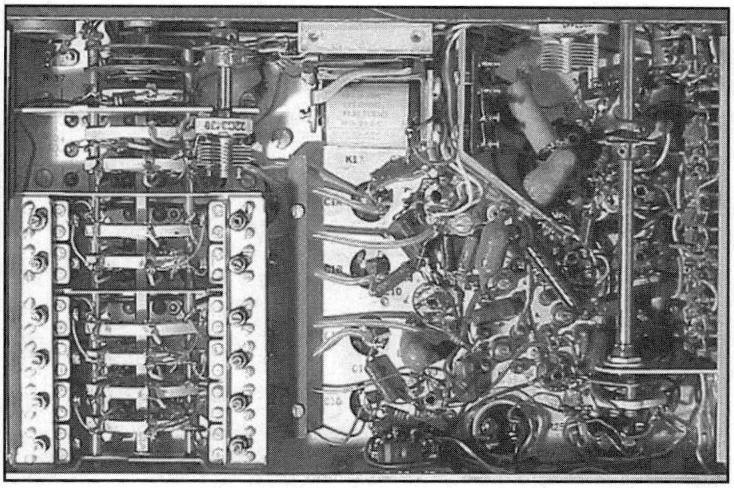

Receiver R-174/URR, bottom

Power Supply PP-308, top

Power Supply PP-308, bottom

High-quality hermetically sealed transformers and chokes are used in the power supply as shown in the supply top view.

The filament voltage adjust pot (and test point, both accessible from the chassis side without removing the cover) is visible at upper right. Note shield cans over vibrator sockets to suppress hash -- a consideration that's long since disappeared from the comm gear design field! A complete power supply schematic (another eye chart) is included inside the bottom lid.

Operation:

IMPORTANT: Before attempting to operate the set for the first time, pull the power supply partially out of the case so the filament adjust test point and adjustment pot are visible on the right side of the chassis. Connect a VOM between the test point and the chassis. At start-up the first time, adjust the filament voltage for 1.4VDC.

Other than an antenna and one of several possible power sources, the GRR-5 is a self-contained receiving system. Power connections vary depending on the input.

115VAC operation. Connect the AC to pins A and D of the PP-308 POWER IN jack.

Direct operation from external batteries. Set the POWER SELECTOR screwdriver switch to DRY BAT. Connect voltages as follows:

+90VDC to Pin A of the PP-308 DRY BATTERY jack

-90VDC to pin D of DRY BATTERY jack

Note: -90V is not on chassis ground—it floats. See section on MONITORING INPUT jack below on how to run the R-174 if you have no PP-308 at all.

+1.5VDC to pin B of DRY BATTERY

-1.5VDC to pin C of DRY BATTERY (chassis ground)

6, 12, or 24VDC operation. Connect POSITIVE to pin B of POWER IN, and NEGATIVE to pin C of POWER IN. Set POWER SELECTOR screwdriver switch to the correct voltage setting "6V", "12V", or "24V".

Operation of the GRR-5 is the same from this point, and is reproduced from the TM. Refer to the line drawings for control location. The author's comments are in the [].

Section III. OPERATION

Caution: Reception on the loudspeaker of early models of the receiving set has a tendency to become distorted after short periods of operation, because of air pressure which builds up within the power supply and has no outlet. To prevent this condition, loosen the cap of the SPARE FUSES compartment before operating

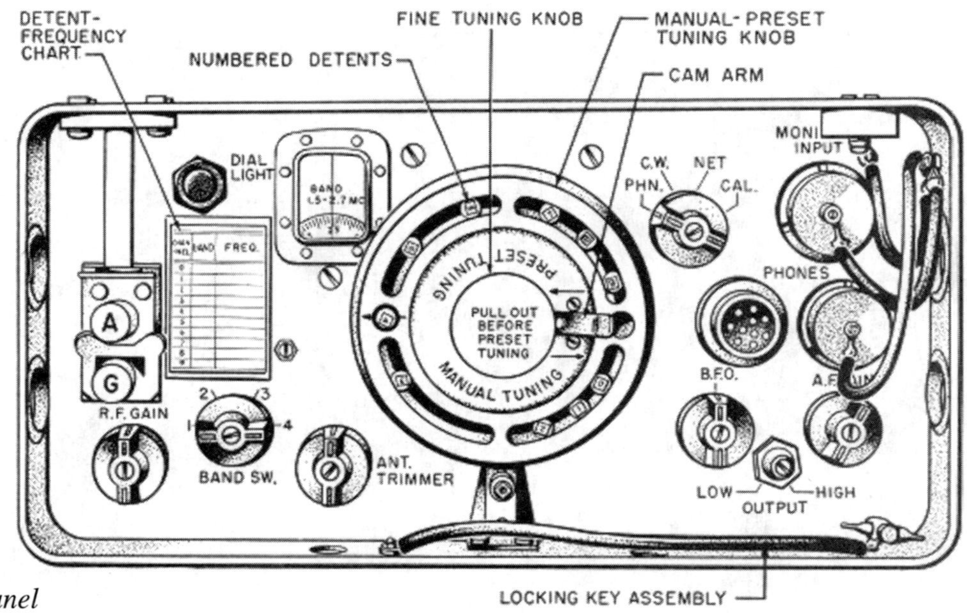

R-174 front panel

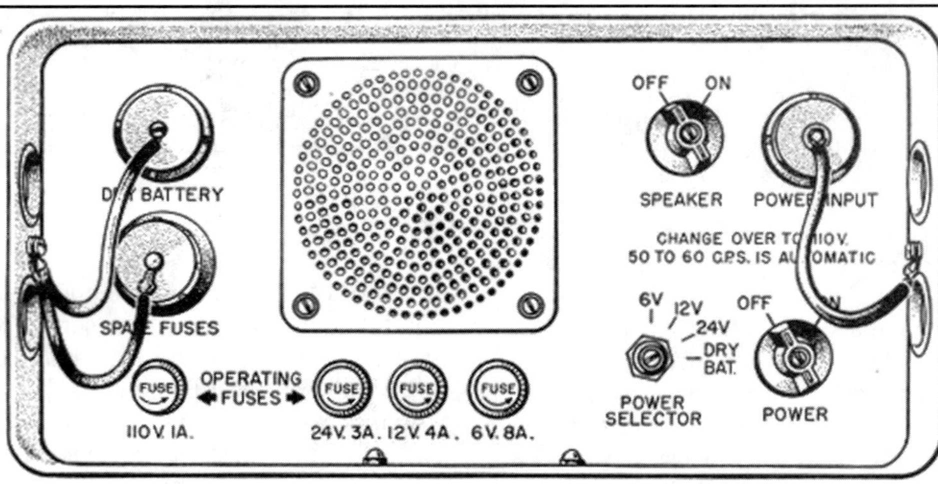

PP-308 front panel

the set. A hole has been drilled in the rear of the SPARE FUSES compartment to allow reduction of air pressure. All units which incorporate this expedient have a caution notice affixed to the lower flange of the receiver front panel.

20. Preliminary Control Setting
Before starting the equipment (and during shut-down periods) set the front panel controls as follows:

POWER SELECTOR: 6V, 12V, 24V, or DRY BAT. switch.
POWER switch: OFF.
SPEAKER switch: ON (OFF, if headset is used alone).
OUTPUT switch: HIGH (LOW when dry batteries are used).
A.F. GAIN control: Halfway clockwise.
R.F. GAIN control: Halfway clockwise.

NOTES
1. For 115-volt a-c operation, the POWER SELECTOR switch may be in any position except when the receiver is used with a transmitter. The switch must then be in either the 6V or DRY BAT. position.

2. Although the OUTPUT switch normally is set at LOW, when power source is dry batteries, it may be set at HIGH for emergency. The normal LOW setting conserves the batteries.

21. Reception of Modulated Signals [AM signals, that is…]
With the equipment connected properly for the available power source (par.15), and controls set as directed in paragraph 20, operate the equipment in the following manner.

a. Turn the POWER switch to ON.

b. If the equipment has not been used for one week or longer, connect a 20,000-ohm-per volt meter to J102 on the power supply and adjust R116 for a reading of 1.4 volts.

[This is the filament voltage adjustment mentioned earlier, and is important -- an incorrect setting will shorten the life of the tube filaments. Once set, it won't have to be checked as often as mentioned, especially if your radio isn't getting jounced around in the back of a halftrack!]

c. Set the BAND SW. switch (S1) to the band which includes the desired frequency.

d. Turn PHN.-C.W.-NET-CAL. switch S2 to PHN.

e. Tune the receiver to the desired frequency. Depress the DIAL LIGHT switch while tuning, only if the receiver is being operated in an area that is poorly lit.

(1) If the desired frequency has not been preset, tune to the signal by rotating the outer ring of the MANUAL-PRESET TUNING control.

[Directions for presetting channels are presented shortly. The center (small) dial can be used for finer tuning].

(2) If the desired frequency has been preset, pull out the fine-tuning control, lift the cam arm, and rotate the MANUAL-PRESET TUNING CONTROL until the arrowed hole is over the desired numbered detent.

f. Adjust the ANT. TRIMMER control for maximum output signal.

g. Adjust the R.F. GAIN and A.F. GAIN controls to the desired level.

22. Code Reception
With the equipment connected properly for the available power source, and controls set as directed in paragraph 20, operate the equipment in the following manner.

a. Turn the POWER switch to ON.

b. If the equipment has not been used for a week or longer, connect a 20,000 ohms per-volt meter to J102 on the power supply and adjust R116 for a reading of 1.4 volts.

[Does this sound like the Army, or what? They tell ya what they're gonna tell ya, they tell ya, then they *tell* ya what they *told* ya.]

c. Set PHN.-C.W.-NET-CAL. switch to C.W.

d. Set the BAND SW. switch (S1) to the band that includes the desired frequency.

e. Tune in a coded signal and adjust the B.F.O. control to obtain the desired pitch of the beat note. The desired tone may be obtained by setting the B.F.O. control to either side of zero beat. Depress the DIAL LIGHT switch for illumination, if necessary.

(1) If the desired frequency has not been preset, tune to the signal by rotating the outer ring of the MANUAL-PRESET TUNING control.

[Again, use the center dial for fine tuning].

(2) If the desired frequency has been preset, pull out the fine-tuning control, lift the cam arm, and rotate the MANUAL-PRESET TUNING CONTROL until the arrowed hole is over the desired numbered detent.

f. Adjust the ANT. TRIMMER for maximum audio output.

g. Keep the R.F. GAIN control low for strong signals to prevent overloading; the volume can be regulated by the R.F. GAIN control.

[*Caution:* Watch your ears! This receiver has massive excess sensitivity].

[*Tip:* For SSB reception, move the BFO slightly to the left of center (about 1/8 turn) for USB, and slightly to the right of center for LSB. Typically, the audio will wander around some in frequency -- this is NOT a receiver with the inherent stability necessary for SSB reception].

23. Net Operation
a. Repeat the procedure outlined in paragraph 21 or 22, using the frequency of the net control station.

b. Set a numbered detent (if available and not already set) to this frequency channel.

c. Set the PHN.-C.W.-NET-CAL. switch to the NET position.

d. Without disturbing the receiver, tune the exciter of the local transmitter to the net-control-station frequency by tuning (the exciter) for a zero-beat indication on the receiver.

e. Set the PHN.-C.W.-NET-CAL. switch to the PHN. or C.W. position, as required.

f. Complete the tuning of the transmitter.

24. Calibration
To calibrate the receiver, perform the following steps:

a. Turn the POWER switch to ON.

b. Adjust the filament voltage.

[Telling us what they told us again!]

c. Turn PHN.-C.W.-NET-CAL. to CAL.

d. Short the antenna with the shorting switch and check the calibration of the MANUAL PRESET TUNING control over the complete range of tuning for all four bands. A beat note should be heard at every integral multiple of 200 kc. If necessary, tune ANT. TRIMMER control for maximum beat signal. When the beat note does not occur at integral multiples of 200 kc, it is necessary to realine the receiver.

[This is a pretty strange way to put it. They didn't really mention how to actually use the calibrator. Just go to the 200kc point closest to your desired listening frequency. While in CAL, zero-beat the receiver and note the dial setting, which will give you a good idea as to the required offset needed. Remember, the GRR-5 is not really a precision receiver. You're lucky if you can get within 10kHz of a desired frequency].

25. Stopping Procedure
a. Turn the POWER switch to OFF.

b. Turn the MANUAL-PRESET TUNING control to the l-f (low-frequency) end of the tuning range of any band. This protects tuning-capacitor plates.

[This is the first instance the author can recall in which a TM actually specifies rotating the rotor of a tuning cap all the way into the stator to protect the plates].

Additional Comments:

The whip antenna originally specified for the GRR-5 consists of Mast Sections MS-116 (2ea.), MS-117 (1ea.), and MS-118 (1ea.), for a total length of about 12.5 feet. In reality, the set will work with just about anything you connect to the binding posts, including your finger!

The unit can also be used with a companion transmitter by using the control circuits accessible via the MONITORING INPUT jack on the R-174. Here's the section on that jack from the TM:

68. Monitoring Operation

a. For purposes of monitoring sidetone in system applications, connection is made from an external transmitter to MONITORING INPUT receptacle J1. In this mode of operation, a 12-volt d-c or a 24-volt d-c input may be used to energize relay K1. This opens the plate and screen voltage for a-f amplifier tube V6, disabling the tube, and shorts the antenna to ground. The low and high a-f amplifiers are still operative.

(1) For 12- and 24-volt vehicular operation, the disabling voltage is obtained directly from the power supply. In 24-volt operation, R23 drops the 24-volt supply to 12 volts to activate relay K1.

The coil of disabling relay K1 is completed to ground at the transmitter through pin C of J1, and normally will be connected in such a manner as to cause operation of the relay when the push-to-talk circuit of the transmitter is closed.

(2) For 6-volt vehicular, dry-battery, or 115-volt a-c operation, external disabling voltage of 12 or 24 volts dc must be supplied by the transmitter. When a 12-volt d-c source for disabling is used, the +12 volts is applied to pin A of J1, and the ground return circuit is connected to pin C of J1. When a 24-volt d-c source is used, the +24 volts is applied to pin B of J1, and the ground return is connected to pin C of J1.

b. The sidetone output of the transmitter should have an impedance of 600 ohms, and one side free to be grounded. Sidetone input is connected to pin D of J1 and pin E of J1 (ground).

(1) Sidetone input voltage is impressed across R36 and is fed into V7 (LOW AF-amplifier) through fixed attenuator R12 and coupling ca-pacitors C27 and C47.

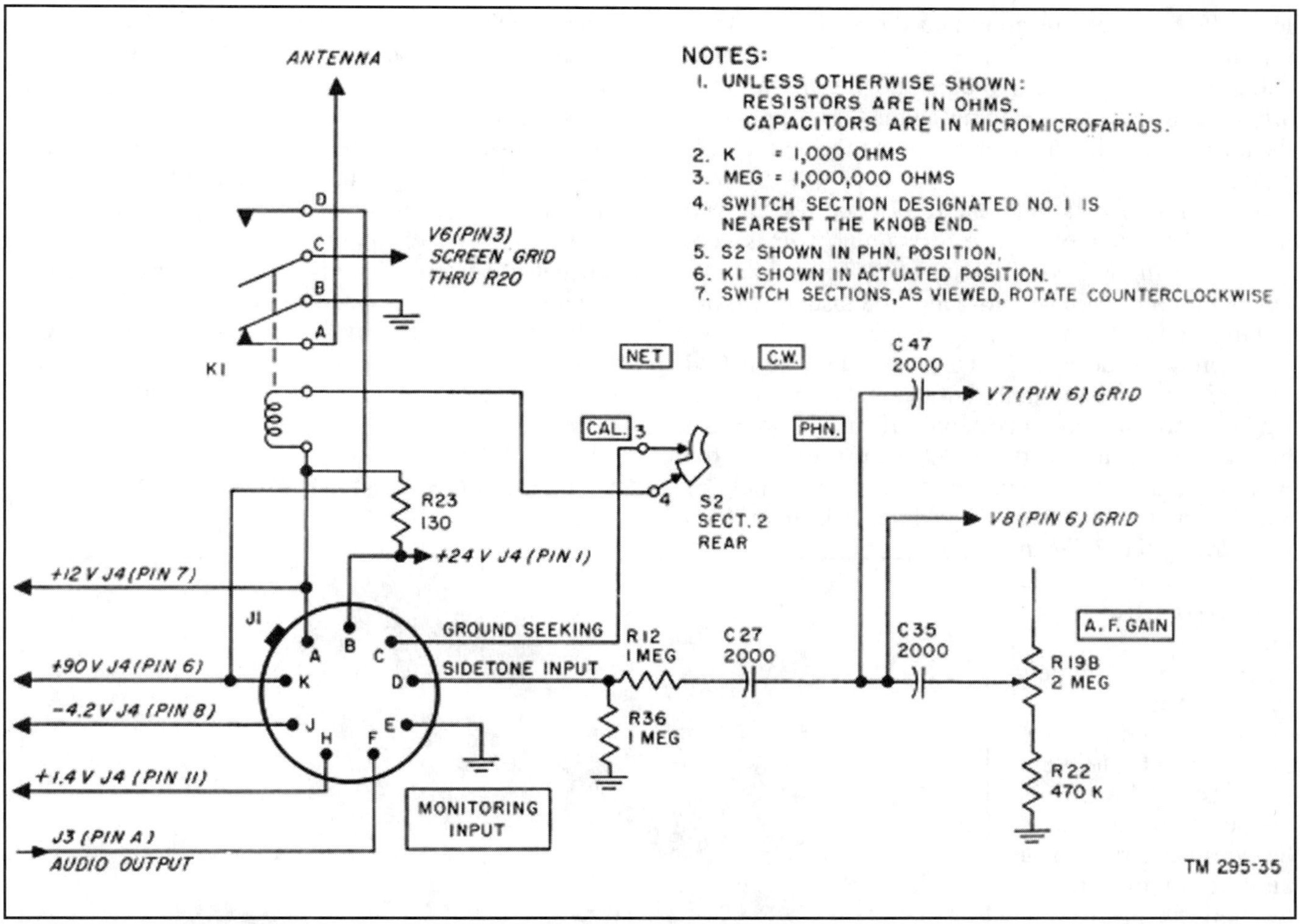

MONITORING INPUT jack pin-outs and wiring

(2) Sidetone input voltage is fed into V8 (HIGH AF-amplifier) through R12 and C27.

c. The audio output connections, pin F of J1 and pin E of J1 (ground), are used to supply the receiver output to the transmitter for local monitoring. The audio output of the receiver is taken from the 600-ohm secondary terminal of T14 (pin A of J2 and J3).

d. It is possible to use Radio Receiver R-174/URR without Power Supply PP-308/URR.

For this purpose, A+, B+, B-, and ground connections were brought out to the MONITORING INPUT jack J1.

(1) For such operation, the positive terminal of the 1.5-volt d-c source (A+) should connect to pin H and the negative terminal (A-) to pin E (ground). The positive terminal of the 90-volt d-c source (B+) should connect to pin K and the negative terminal (B-) to pin J. Also, it will be necessary to connect a 160-ohm resistor across pins 8 and 10 of J4.

[***Tip:*** J4 is the power jack on the back of the receiver. If you wish, you can also put the resistor across pins J (-4.2V) and E (ground) on the MONITORING INPUT jack. They are grounding the audio amp bias].

(2) A suitable cabinet should be used that will fit properly against the waterproof gasket around the rim of the receiver.

[The author has never seen any such thing for use with the R-174 in a standalone role].

The speaker can be turned on and off with the SPEAKER switch on the PP-308. Both the speaker and audio accessories can be operated simultaneously

What To Do If It's D.O.A.:

Vibrators run in the PP-308 no matter what source of input power is used for the GRR-5. If you don't hear vibrators with power applied and the set turned ON, check the fuses. If they're bad, you might get lucky and find some spares under the SPARE FUSES cap. Of course, blown fuses often suggest more serious problems. If the fuses are good and the vibrators aren't starting, they may be defective or "stuck." Sometimes actually opening up a vibrator and freeing a stuck armature will get them going again. Try tapping on them first, or pulling them and tapping them on a hard surface.

If the vibrators start and the R-174 doesn't run, you need to check supply voltages *first thing*. This is easily done at the MONITORING INPUT jack per the above pin-outs. Voltages missing or substantially out of range indicate power supply problems.

If there's no audio at all, the speaker may be turned off. This may sound dumb, but the author has forgotten on more than one occasion! If there's no audio hiss at full volume with the RF gain turned all the way down, a good place to look to start with would be V6, V7, and V8. If there's no increase in hiss when you turn up the RF gain all the way, the trouble is likely earlier in the receiver.

"Dead" or poorly performing receivers can often be traced to acute misalignment. There are all sorts of adjustments inside, associated with the various bands. You may be tempted to try tweaking them to bring the radio back to life. A suggestion from the author: **DON'T.**

To align this radio, you MUST have the manual. The alignment is fairly involved, and requires stepping through the bands adjusting each in turn in a specific order. The author's unit arrived in such a horrendous condition of misalignment that he concluded the previous owner must have screwdrivered the entire set. Last time he checked, Fair Radio had repro manuals for twelve bucks.

Equipment needed for alignment is uncomplicated—a calibrated sig-gen, (or an uncalibrated sig-gen and a frequency counter) is all you need. The author used a Wavetek frequency synthesizer and a step attenuator. The procedure is tedious, but well worth it. You'll end up with a neat general-purpose HF receiver with sensitivity that'll knock yer socks off. And there's a certain charm to a radio that has the entire 40m band in three-quarters of an inch of dial space!

Accessories:

The complete Radio Receiving Set AN/GRR-5 looks like this:

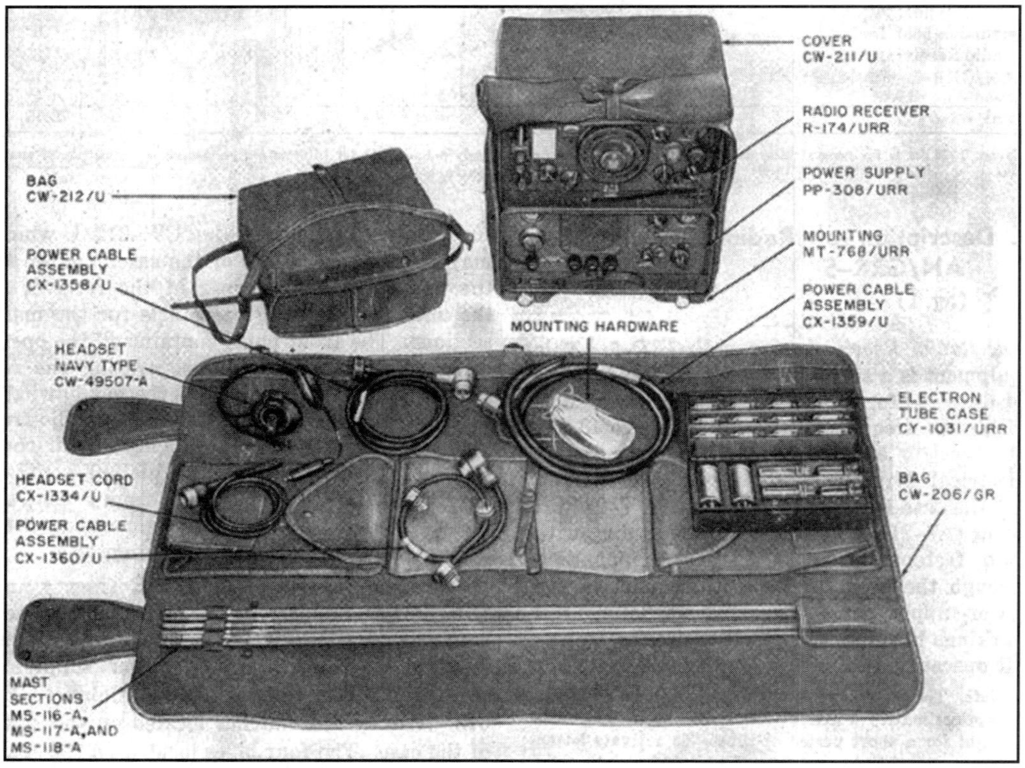

Radio Set AN/GRR-5, complete

R-392

Application:

General-purpose HF receiver for mobile applications. Was generally used in conjunction with a T-195 transmitter in Radio Set GRC-19.

Tech Specs:

Frequency range: 500kHz to 32mHz in 32 bands

Calibrator: built-in, every 100kHz

Signals received: CW, MCW, voice (AM), RTTY, SSB

Size: 11.5" x 14.5" x 11.5"

Weight: 52 pounds

Active devices: Vacuum tubes; 6AJ5, 26A7, 12AU7, 26A6, 26FZ6, 26C6, 26D6, (about 26 tubes total)

Power requirements: 28VDC 3A nominal; will function from 25 - 30VDC

Audio input/output devices: Any of the old family audio accessories with U-77 plugs, such as LS-166, H-33, etc.

Antenna connection: push-type wire binding post, 50-ohm BNC jack

Whip antenna: 15-foot mobile whip

Other antennas: long wire, doublet

Antenna tuner: integral antenna trim control

Tech Manuals:
TM 11-5820-334-10 "OPERATOR'S MANUAL, RADIO RECEIVER R-392/URR"; operating instructions only, no theory or schematics

TM 11-5820-334-35 "FIELD AND DEPOT MAINTENANCE MANUAL, RADIO RECEIVER R-392/URR"; description, theory, troubleshooting, repair, alignment, schematics

Construction Style: Heavy-duty die-cast front panel and chassis assembly slips into a very rugged ribbed outer metal case. Internal construction is removable, repairable subchassis. Radio is certainly reparable, however work on the RF subchassis often requires extensive disassembly of the radio.

Description:

Question: How do you cram an R-390 into about a cubic foot?

Answer: The R-392.

The 392 is basically an R-390 repackaged into a unit suitable for the rigors of mobile use. That's actually an oversimplification, because there are quite a few other differences. Much has been written about the Collins receiver designs, so the author will limit himself to the one big difference -- the 392 uses the vehicular supply voltage (28V) as the tube plate voltage. This is unique, insofar as the author knows. All the major R-390 functions were retained, and many of the minor ones. The R-392 is a remarkable engineering achievement. In fact, the entire GRC-19 set was amazing for its time, as anyone who's ever looked inside a T-195 transmitter will agree!

R-392 chassis, top

Does that top view show a Collins design or what? Slug racks on the RF subchassis are at front, crystal oscillator subchassis at rear.

R-392 chassis, bottom

The Calibration Oscillator Subchassis is at top center in the bottom view, Permeability Tuned Oscillator at center-center, Audio Subchassis at bottom center, I.F. Subchassis at right. And all them thar gears behind the front panel. Yep, it's a Collins.

Operation:

Setup is simple. Power connections at the POWER INPUT - TRANS CONT front panel jack are: Pins A,D: +28VDC; Pin E: GROUND. The author has found that if an AC power supply is used, the input voltage should be very well filtered, and preferably regulated. AC hum tends to get into the circuits otherwise. Connect an LS-166 speaker, or other audio accessory to one of the AUDIO connectors. Connect an antenna to the binding post or BNC jack.

Mechanically-related cautions from the TM:
Caution: Be careful when operating the MEGACYCLES control, because on some receivers the stop mechanism fails to provide a positive stop. The limits between which the control can be rotated without damage are 00 and 31 megacycles.

Caution: To prevent any possible damage to the tuning mechanism, do not tune past 000 kilocycles counterclockwise or 999 kilocycles clockwise.

Operation instructions from the TM; author's comments in []:

AM VOICE OPERATION

1) AF GAIN to mid-position.

2) DIAL DIM to ON, or as required.

3) Function Switch to NORMAL; 5min warm-up recommended.

4) AGC to ON.

5) BAND WIDTH to 8KC.

6) ANT TRIM to 0.

7) BFO to OFF.

8) RF GAIN SQUELCH THRESH fully CW.

9) DIAL LOCK fully CCW (unlocked).

10) Set the MEGACYCLES and KILOCYCLES control to the desired frequency.

11) Tune the KILOCYCLES control for maximum on the CARRIER LEVEL METER.
[Narrowing the bandwidth helps to find the center (carrier) frequency].

12) Adjust ANT TRIM control for max on the CARRIER LEVEL METER.

13) Lock the KILOCYCLES control if desired with the DIAL LOCK.

14) Adjust AF GAIN for comfortable level.

15) If objectionable noise or static is heard, set the function switch to LIMITER. [The Collins 392/390/390A/389/391 noise limiters are notorious non-performers. Most of the time they introduce more garbage than they eliminate].

16) Narrow the BAND WIDTH to avoid nearby interference.

17) To eliminate noise during periods of no signal reception, move the Function Switch to SQ (squelch) and turn the RF GAIN SQUELCH THRESH control CCW only until the noise is eliminated.

Note: Do not use the squelch if the signal is weak or subject to fading, or if any more than a small CCW movement of the RF GAIN is necessary to eliminate the noise. [What they're saying here is that the carrier operated squelch is really pretty ineffective and shouldn't be used except when the channel is clear or sigs are barreling in bigtime].

CW OPERATION: Same as AM operation, except narrow down the BAND WIDTH to 2kHz, turn ON the BFO, and zero beat the signal with the KILOCYCLES control. Then use the BFO PITCH to tune set the beat note to a comfortable frequency.

SSB OPERATION: Same as CW operation, except set the BFO PITCH slightly to the left of zero for USB reception and slightly to the right of zero for LSB reception, before calibrating the receiver as follows:

1) BAND WIDTH to 2KC.

2) AGC switch to CAL.

3) KILOCYCLES control to 100kHz point nearest operating frequency.

4) Adjust ANT TRIM for max on the CARRIER LEVEL METER.

5) Turn DIAL ZERO to clockwise limit.

6) Adjust KILOCYCLES CONTROL for zero beat.

7) Release DIAL ZERO.

Note: Calibrate for CW or AM operations with the BFO centered.

If the receiver isn't to be used for short periods, turn the Function Switch to STAND BY. It's probably not a good idea to leave the receiver in this state for long periods for the same reason there are warnings not to do so with the R-390...shortened tube life from running the filaments with no B+.

Additional Comments

AUDIO connector pin-outs:

Pin A	A-F OUTPUT
Pin B	GROUND
Pin C	MICROPHONE
Pin D	NO CONNECTION
Pin E	GROUND
Pin F	PUSH TO TALK SWITCH ON MICROPHONE OR KEY
Pin H	GROUND
Pin J	NO CONNECTION
Pin K	CARRIER RELAY CONTROL
Pin L	A-F OUTPUT

A few may need some explanation. The MICROPHONE and PTT are just straight pass throughs to the POWER INPUT - TRANS CONTROL connector for use when the 392 is used in a system with the T-195, etc.

Pin K is on GROUND when the receiver is unsquelched, and OPEN when the receiver is squelched. This can be used for retransmission control.

POWER INPUT - TRANS CONT connector pin-outs:

Pin A	+28VDC POWER IN
Pin B	BREAK-IN RELAY CONTROL
Pin C	MICROPHONE
Pin D	+28VDC FILAMENT
Pin E	GROUND
Pin F	PUSH TO TALK SWITCH ON MICROPHONE OR KEY
Pin H	A-F OUTPUT
Pin J	NO CONNECTION
Pin K	CARRIER RELAY CONTROL

The +28V on Pin D of the POWER INPUT - TRANS CONTROL connector is used to actuate the break-in and antenna relays when the Function Switch is set to STAND BY. This grounds the antenna and the input to the first audio amp, muting the receiver. In the NORMAL, LIMITER, or SQ positions of the Function Switch, break-in operation is in effect. A ground applied (usually by the associated transmitter) to Pin B will short the antenna and mute the receiver.

What To Do If It's D.O.A.:

Check fuses and tubes to start with, and *don't trust that tube tester -- substitute if there's any doubt!*

The R-392 is triple conversion below 8mHz, and double conversion above that. Use the block diagram in the TM to zero in on particular stages that show fault symptoms during operation. For instance, if the squelch doesn't work, you might look at V602B, the Squelch Rectifier, or V605, the Squelch Control. If the receiver works above 8mc but not below, the First Mixer V203 may be bad. And so on.

Detailed troubleshooting will require the dash-35 TM.

GRC-106

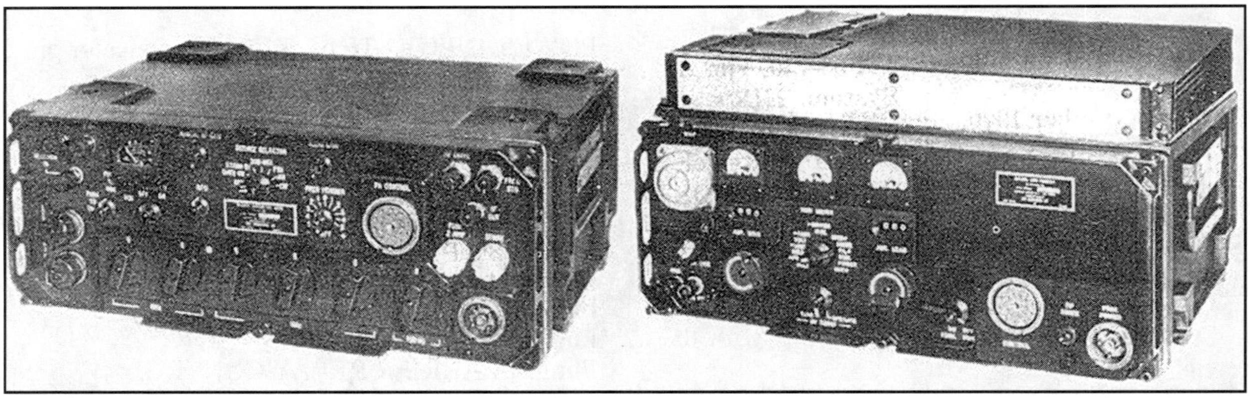

Application:

Synthesized medium-power HF set used in a wide variety of mobile and fixed applications, including enclosed radio shelters and open vehicular mountings.

Tech Specs:

Frequency Range: 2.000 to 29.999mHz in 1kHz steps with a +/-600Hz vernier (when RT-662 Receiver/Exciter is used); 2.0000 to 29.9999mHz in 100Hz steps with a +/-600Hz vernier (when RT-834 Receiver/Exciter is used).

Type of modulation and power out:
AM-3349—USB voice, 400 watts PEP (two-tone signal); CW, 200 watts; Compatible AM (AME), 200 watts (two-tone signal); FSK, 200 watts (with appropriate ancillary converter)

The RT-662 or RT-834 Receiver/Exciters are rated at 100mW output nominal, 125mW output maximum

Operating bandwidth (transmit/receive): 3.2kHz

Size: RT-662 and RT-834, 7" x 13" x 18"; AM-3349, 9" x 13" x 18"

Weight: RT-662, 51lbs; RT-834, 46.5lbs, AM-3349 71.5lbs

Active devices: 1960s discrete transistors (140-plus in the RT unit alone!) No chips.
AM-3349 Linear Amplifier, 3 vacuum tubes, 10 transistors
RT-662 Receiver/Exciter, 2 vacuum tubes, 140 transistors
RT-834 Receiver/Exciter, 2 vacuum tubes, 151 transistors

Power requirements: RT-662 or RT-834, about 50 watts during transmit. AM-3349, about 1000 watts, two-tone transmit (about 37 amps at 27VDC) during transmit. In STANDBY mode, 9 amps (!)

Audio input/output and keying devices: H-33 handset, M-29 microphone, H-227 phones, LS-166 speaker, KY-116 key and numerous others with old family connectors (U-77/U).

Antenna connection: Special screw-type connector on AM-3349 for heavy wire lead to whip antenna. The RT units use BNC connectors.

Whip antenna: A 15-foot whip was used in the mobile application

Antenna tuner: The tuning network inside the AM-3349 supports tuning of a whip as short as 15 feet, or a doublet (nominal 50 ohms)

Tech Manuals: TM11-5820-520-12 (operating info, no schematics or theory)

TM11-5820-520-35 (theory, troubleshooting, schematics)

Construction Style: Removable, repairable modules. Extender cables necessary in many cases.

Description:

The AN/GRC-106 consists of two main units, the receiver-exciter which can be either the RT-662 or RT-834, and the AM-3349 power amplifier which sits atop the receiver-exciter. This is a heavy beast, at around 125 pounds total, and an electromechanical marvel (or nightmare to some) with large rotating tuned circuit turrets in both units.

The differences between the GRC-106 and GRC-106A are as follows:

The GRC-106 uses the RT-662 receiver-exciter; the GRC-106A uses the RT-834. The GRC-106 tunes in 1kHz increments using five controls; the GRC-106A tunes in 100Hz increments using six controls. Early model RT-662s had a noise blanker, but the space used by this control is unused after serial number 200.

The RT-834 has a 100Hz control where the FREQ VERNIER is located on the RT-662. The FREQ VERNIER on the RT-834 is located where the BFO control is placed on the RT-662. The BFO control on the RT-834 is located in the space previously used by the NOISE BLANKER switch on early RT-662s.

The easiest way to tell the difference between the two sets is that the 106 has five main tuning knobs on the RT, and the 106A has six main knobs. The RT-834 is the more desirable of the two exciters since it tunes both receive and transmit in 100Hz steps. The RT-662 also has a receive slider, but you are restricted to 1kHz steps on transmit.

Note: The AM-3349 will operate with either receiver-exciter.

Operation:

Physical mounting of the set will vary depending on what mounting hardware you have. A typical cabling setup is illustrated.

If your set came without cabling, Fair Radio still

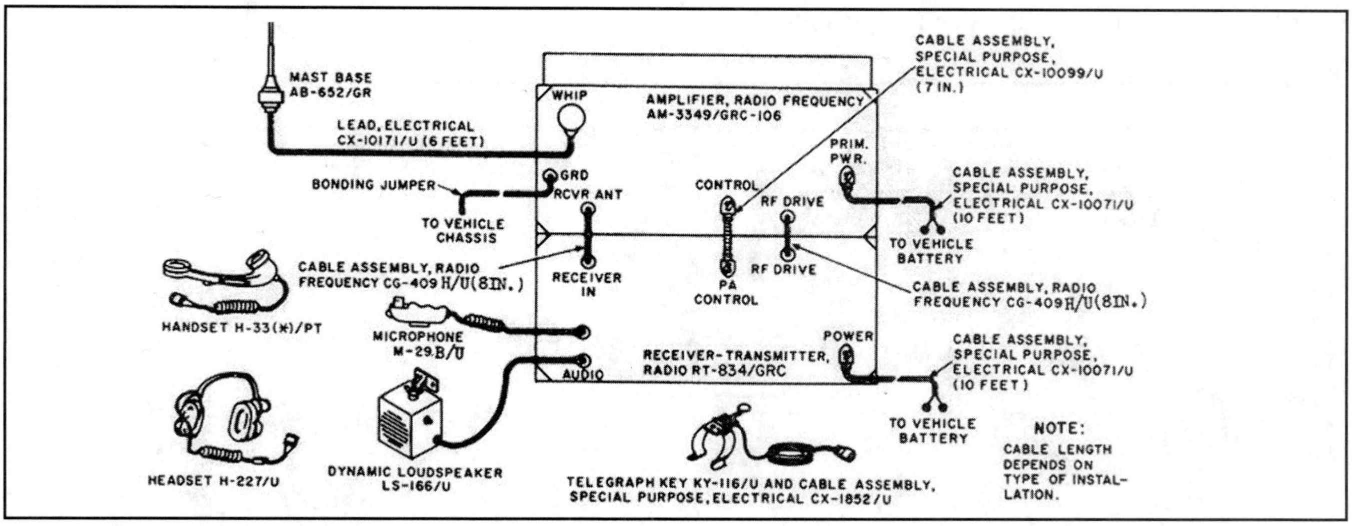

Cabling for the AN/GRC-106 Set

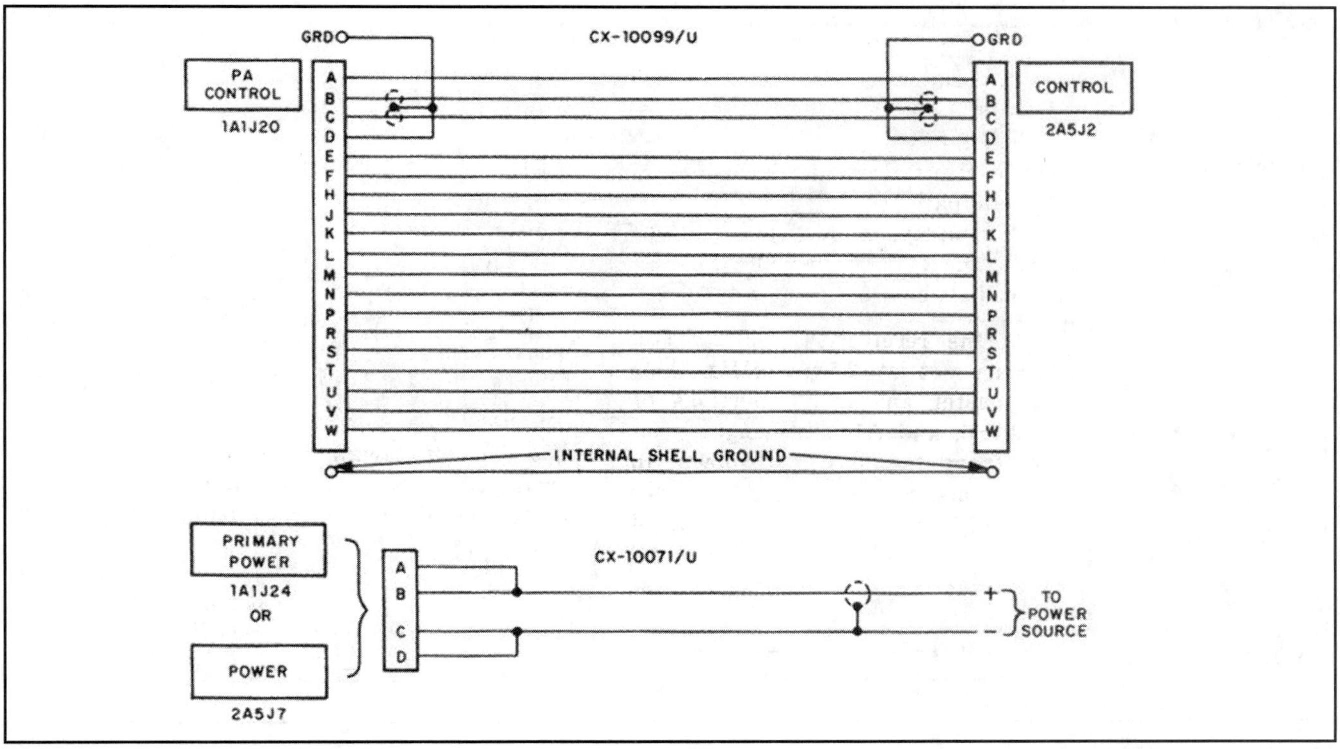

CX-10099 control and CX-10071 power cable pin-outs

lists the CX-10099 cable, which is the main interconnect between the RT unit and the amplifier. The twenty dollar price tag is well worth it. For those of you very short on funding and LONG on time, the wiring data appears in the pin-out illustration.

The following instructions are adapted from the operator's manual. Before you fire up the set, review the power requirements. This set draws nine amps just sitting there idling!

Caution: Review this section in its entirety before operating the GRC-106 for the first time. This is a radio that can be readily damaged in a number of ways by improper use. Author's comments in the []:

Operator controls, indicators, and connectors:

=RT-662 or RT-834=

SERVICE SELECTOR switch—Selects mode of operation
- OFF No primary power applied
- OVEN-ON Primary power is applied, oven in frequency standard is energized
- STAND BY Primary power is applied. GRC-106 initially inoperative, but ready to operate after 60-second delay
- SSB NSK All STAND BY functions repeated. Provides squelch and vox capability. USB voice transmission with squelch and VOX capability

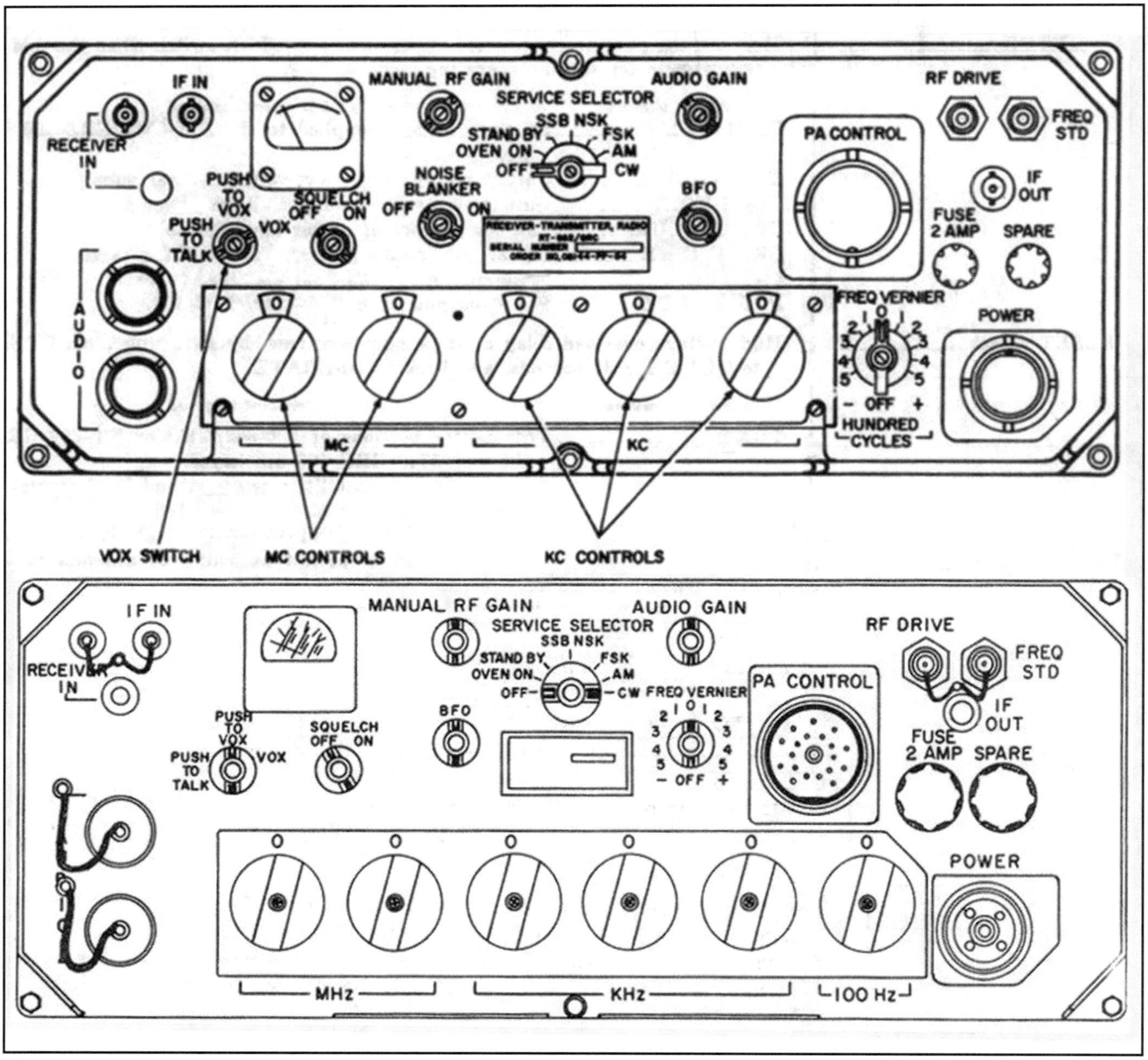

RT-662 (top) and RT-834, front panel views

enabled. NSK (narrow frequency shift keyed) transmission/reception enabled with suitable RTTY gear.

FSK All STAND BY functions repeated. Squelch and VOX capability disabled. FSK transmission/ reception enabled with suitable RTTY gear.

AM All STAND BY functions repeated. Transmission of USB with carrier enabled. Conventional double sideband signals may also be received.

CW All STAND BY functions repeated. BFO circuit is enabled. External modulation capability disabled; squelch and VOX disabled. CW transmission by keying KY-16/U.

RECEIVER IN connector – Coaxial connector used for connecting the received signals routed through AM-3349.

RECEIVER IN binding post – Spring-loaded post used for connecting a long wire antenna when the RT unit is used only as a receiver.

IF IN connector – Coaxial connector used for connecting an external 1.75mHz IF during transmit mode of operation (requires internal jumpering).

IF OUT connector – Coaxial connector used for connecting the internal third IF signal to external equipment (receive operation only).

RF DRIVE connector – Coaxial connector used for connecting the RF output from the RT unit to RF input of AM03349.

PA CONTROL connector – Used for connecting all control functions to and from AM-3349.

FREQ STD connector – Coaxial connector used for connecting an external 5mHz frequency standard or for connecting the internal 5mHz standard to external equipment.

(INT-EXT switch in the frequency standard module must be set to EXT if external standard is used.)

POWER connector – Used for connecting the +27VDC primary power from the vehicle generating system.

AUDIO GAIN control – adjusts the audio output levels.

BFO control – Varies BFO circuit output frequency

SQUELCH switch – In the ON position, eliminates receiver background noise in the absence of received signals.

FREQ. VERNIER control – Provides for continuous frequency tuning of +/- 600Hz about any 1kHz increment in receive mode in the RT-662. In the RT-834, it affords the same function about any 100Hz increment.

MANUAL RF GAIN control – Varies the sensitivity of the receiver circuits.

Signal level meter – Provides a relative indication of RF output during transmission; Provides an indication of RF input signal level in dB above the AGC threshold in receive mode.

Vox switch – Selects the method by which the GRC-106 is keyed in SSB or AME mode.

VOX GRC-106 is keyed by voice

PUSH TO VOX GRC-106 is keyed by voice when the M-29 or H-33 PTT switch is depressed

PUSH TO TALK GRC-106 is keyed directly by the PTT switch on the M-29 or H-33

MC controls – Provide selection of the 10mHz and 1mHz digits of the operating frequency.

KC controls – Provide selection of the 100kHz, 10kHz, and 1kHz digits of the operating frequency.

100Hz control – Provides selection of the 100Hz digits of the operating frequency (RT-834 only).

NOISE BLANKER – [Apparently, the noise blanker was not very effective. From the wording in the TM it appears that in the ON position, it actually desensed the receiver. It was present only on early model RT-662 RT units below serial #221.]

AUDIO connectors – Parallel-connected, used for connecting various audio and keying accessories.

FUSE 2 AMP – Protection for primary +27VDC power input line.

SPARE – Spare 2A fuse for above.

=AM-3349=

ANT. TUNE control – Used in conjunction with ANT. LOAD control to match antenna to 50 ohm output of the AM-3349 final amplifier stage.

ANT. TUNE counter – Provides an indication of ANT. TUNE control position. Initially set by ANT. TUNE control to setting indicated on antenna tuning and loading chart or LOGGING CHART according to desired operating frequency and type of antenna being used.

ANT. TUNE meter – Indicates relative degree of mistuning between antenna and AM-3349. The unit is correctly tuned when indication is at center zero.

ANT. LOAD control—Used in conjunction with ANT. TUNE control to match antenna to 50 ohm output of the AM-3349 final amplifier stage.

ANT. LOAD counter—Provides an indication of ANT. LOAD control position. Initially set by ANT. LOAD control to setting indicated on antenna tuning and loading chart or LOGGING CHART according to desired operating frequency and type of antenna being used.

ANT. LOAD meter – Indicates relative degree of mistuning between antenna and AM-3349. The unit is correctly tuned when indication is at center zero.

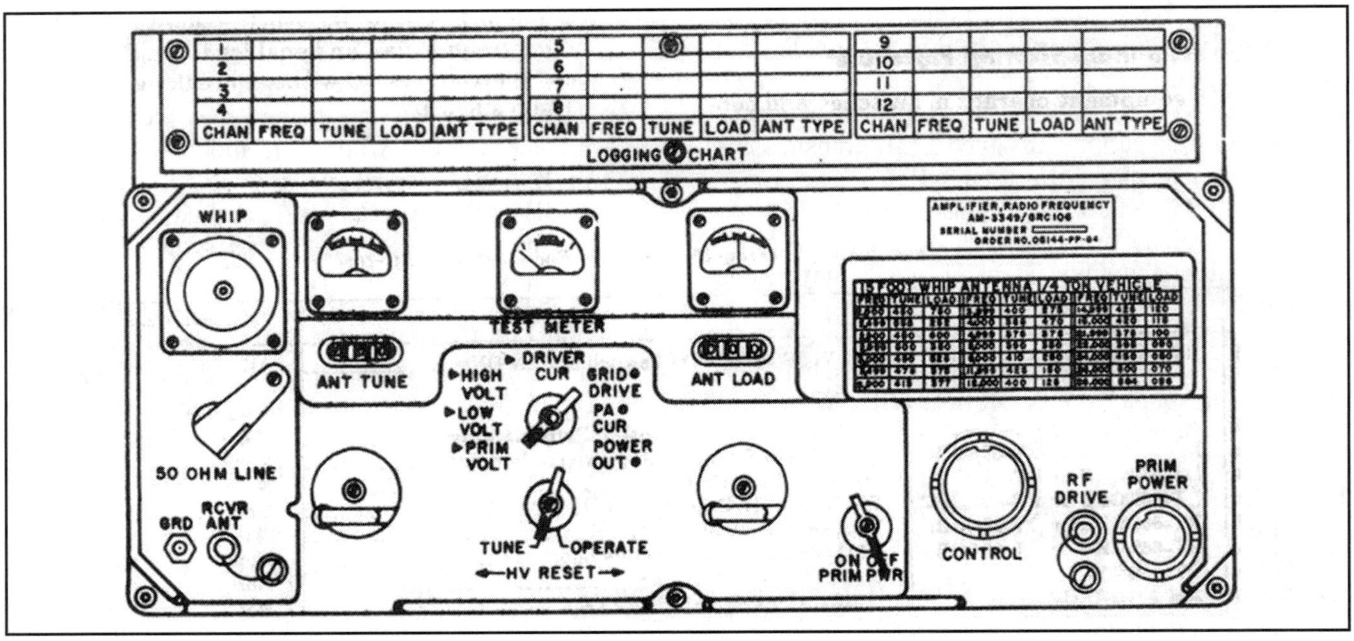

AM-3349, front panel view

TEST METER – Monitors the parameters selected by the TEST METER switch when the AM-3349 is keyed in an operating mode.

Switch Position	Equipment Response
PRIM. VOLT	Primary voltage supplied to the AM-3349
LOW VOLT	Output voltage from low voltage power supply
HIGH VOLT	Output voltage from high voltage power supply
DRIVER CUR.	Cathode current of driver amplifier tube
GRID DRIVE	Signal level at the grid of final amplifier tubes
PA. CUR	Plate current of the final amplifier tubes
POWER OUT	Power output from the AM-3349

HV RESET switch—High voltage overload relay must be manually reset by switching from OPERATE to TUNE for 30 seconds, and back to OPERATE.

Switch Position	Equipment Response
TUNE	Transmitter portions of RT and amplifier are keyed All modulating capabilities removed and replaced with local carrier insertion for tuneup Sensitivities of tuning meters are increased
OPERATE	All circuits are connected for transmitting Keying of the GRC-106 is accomplished at the RT unit

PRIM. PWR switch—Applies or removes primary voltage when the RT front panel SERVICE SELECTOR switch is in any position other than OFF or OVEN ON. Also acts as a low voltage circuit breaker – will automatically remove primary power when overloaded. To reset, set switch at OFF, then at ON.

Caution: The ON OFF PRIM PWR switch shaft is easily broken. Use a minimum turning force when setting this switch. [This caution was added in a later TM version].

Antenna Tuning and loading chart – Lists approximate initial settings of ANT. TUNE counter for various operating frequencies and types of antenna. Chart listings vary according to type of antenna. Insure that the correct chart for the installation and type of antenna is being used. ANT. LOAD counter settings are given also.

Logging chart – Used to log ANT. TUNE and ANT LOAD counter settings for specific operating frequencies. Provides quick reference for future tuning.

WHIP connector – Used for connecting 15-foot whip antenna

50 OHM LINE connector – Used for connecting doublet connector

CONTROL connector – Used for connecting control functions to and from the RT unit.

RF DRIVE connector – Used for connecting output RF signals from the RT portion to the input of the AM-3349

PRIM. POWER connector – Used for connecting +27VDC primary power

Operating procedures, adopted from the TM; author's comments are in []:

=Startup=
Determine that the GRC-106 is in a shutdown status (PRIM. PWR switch to OFF and SERVICE SELECTOR switch to OFF) and proceed as follows, assuming a suitable source of power has been connected:

a. Place the RT unit SERVICE SELECTOR switch to OVEN ON and allow several minutes for warmup.

[***Tip:*** The above step is not in the TM, and is not strictly necessary, but is highly recommended. The current draw through the SERVICE SELECTOR switch can be quite high during startup, and the author's group has determined it's best to power up the GRC-106 in steps. The frequency standard takes about 10 minutes or so to move roughly onto frequency anyway. Many GRC-106 owners who are operating their sets off DC supplies like the PP-4763 leave the switch in the OVEN ON position constantly.]

b. Place the RT unit SERVICE SELECTOR switch to STANDBY and the AM-3349 PRIM. PWR switch to ON, and allow 90 seconds for warmup. Signal level meter will move to the extreme right, and the blowers will start.

[***CAUTION:*** Do ***not*** turn on the equipment by simply moving the SERVICE SELECTOR switch directly to an operating mode, such as SSB NSK. This is because high startup currents are pulled through the switch when the blowers, ovens, and circuitry are all energized at once. The author's group has experienced failures in the SERVICE SELECTOR switch in the past. Use the warmup periods!]

c. Set the RT unit SERVICE SELECTOR switch to SSB NSK, or other desired operating mode. Signal level meter will return to extreme left portion of meter scale.

d. Set the AM-3349 TEST METER switch to PRIM. VOLT. Observe that the test meter indicates within the area of the two dark green wedges in SSB, NSK, FSK, AM, and CW positions.

=Tuning=

a. Set the RT unit MHz and kHz to the desired operating frequency.

b. Note the AM-3349 ANT TUNE and ANT LOAD predetermined setting on the antenna tuning and loading chart, or the logging chart.

c. Adjust the AM-3349 ANT TUNE and ANT LOAD controls to the numbers on the chart used.

CAUTION: The AM-3349 HV RESET switch should not stay in the TUNE position for more than two minutes. If this time is exceeded, move the HV RESET switch to OPERATE and the RT unit SERVICE SELECTOR switch to STANDBY for 5 minutes cooling.

[***Comment:*** The above should not happen. If you can't get your GRC-106 to tune in a minute or less, there's almost certainly something wrong with your procedure, the antenna, or the radio itself.]

CAUTION: ANT TUNE and ANT LOAD controls will interact with each other. To center their respective meter pointers, rotate them slowly in the direction opposite to that of their pointer deflection.

CAUTION: Be sure the antenna is attached for proper loading to prevent damage to the equipment while performing the tuning procedure below.

d. Set the AM-3349 HV RESET switch to TUNE. Wait for a deflection on the ANT TUNE and ANT LOAD meters.

e. Adjust the AM-3349 ANT LOAD control for a center scale reading on the ANT LOAD meter.
1. Rotate the control in the direction that the meter pointer is to move. Adjust the ANT TUNE control for a center scale reading on the ANT TUNE meter.
2. Rotate control in the direction that the meter pointer is to move, keeping the ANT LOAD meter as close to center scale as possible.
2.1 Place the HV RESET switch on the AM-3349 to OPERATE.
2.2 Place the TEST METER function switch on the AM-3349 to PRIM VOLT position.
2.3 Place HV RESET switch of AM-3349 to TUNE position and observe that the test meter pointer indicates within the area of the two dark green wedges (top scale) of the test meter.

[2.1-2.3 appear to be only testing the DC supply voltage under load].

3) Tuning of the AM-3349 is complete when simultaneous center scale readings are obtained on the ANT TUNE and ANT LOAD meters.

[*Note:* In practice, the tuning procedure can be very touchy. It should be performed as rapidly as possible. Gyrations of the needles, inability to get a stable center or near-center reading, or other obvious amplifier instability indicates antenna problems or possibly a radio failure].

f. Major functions can be checked by rotating the TEST METER switch to the various positions:

TEST METER position	Nominal TEST METER indication
LOW VOLT.	Within green area of top scale
HIGH VOLT.	Within green area of top scale
DRIVER CUR.	Within two dark green wedges of top scale
GRID DRIVE	Just to left of gray area of bottom scale
PA CUR.	Just to left of gray area of bottom scale
POWER OUT	Just to left of gray area of bottom scale

g. Set the AM-3349 HV RESET switch to OPERATE

CAUTION: If the operating frequency of the GRC-106 is changed by more than 100kHz, the unit will automatically reprogram to the new frequency and prevent keying the AM-3349. The HV RESET switch must be set to TUNE and the amplifier retuned to the new frequency. Failure to retune may result in damage to the amplifier.

=Operating Procedure=
To receive in any mode of operation, proceed as follows:

1) Set the HV RESET switch to OPERATE.

Note: When receiving only, to conserve power, set the PRIM. PWR switch at OFF and the HV RESET switch at OPERATE.

2) Set the RT unit SERVICE SELECTOR switch at the desired mode: SSB NSK, AM, FSK, or CW.

3) Adjust the AUDIO GAIN control for a comfortable listening level.

4) If the noise level is undesirable in the absence of received signals, set the SQUELCH switch at ON.

Note: In CW and FSK mode, the squelch circuit is automatically disabled. In NSK mode, the SQUELCH switch should be set to OFF. In AM or SSB, the audio output is not completely squelched with high noise levels. A low level of audio is always present to indicate the receiver portion of the RT unit is operating.

Note: The RT unit MANUAL RF GAIN control setting should not be altered unless the GRC-106 is being operated in close proximity (10 miles or less) with only one other radio set. [In other words, if the received signals are VERY strong. This squelch is like all of the SSB or AM squelches of the era – totally miserable]. The MANUAL RF GAIN control can then be rotated to some counterclockwise position that provides an indication on the signal level meter that is roughly one or two divisions below the indication of the received signal strength. This will reduce background noise and minimize adjacent channel interference when not receiving a signal.

[These original TM notes are quite revealing. The GRC-106 has a known (and notorious) problem. Running the unit with the RF GAIN wide open with reasonably strong stations nearby (but well outside the passband) often results in some horrendous IMD and adjacent channel interference. The front ends on these radios are pretty soft and often act up in this manner, especially when antennas with more gain than the specified whip or doublet are used. Often, backing down slightly on the RF gain will alleviate the IMD problem somewhat].

Rotating the MANUAL RF GAIN control counterclockwise desensitizes the RT unit. If the operating frequency or location is changed, the MANUAL RF GAIN control should be returned to its maximum clockwise position to insure that the signals are not lost. In net operation the MANUAL RF GAIN control is left in the maximum clockwise position.

5) When using the MANUAL RF GAIN control to reduce noise and improve reception, note the signal level meter indication in the presence of a signal. Rotate the MANUAL RF GAIN control slowly counterclockwise until the signal level meter indication is slightly lower (one or two divisions). This should improve reception.

[Again, the designers don't admit it here, but this is necessary because of the tendency for this radio to exhibit IMD and adjacent channel problems].

6) When receiving CW signals, adjust the BFO control for a comfortable tone.

7) When receiving AM, FSK, or NSK signals from radio sets other than the GRC-106, adjust the FREQ. VERNIER control for the best reception obtainable.

[The FREQ. VERNIER moves the received frequency only].

=Transmit=
To transmit in any mode of operation, proceed as follows:

Note: The AM-3349 must be keyed in order to check the low voltage power supply, high voltage power supply, driver cathode current, or final amplifier current.

1) Set the AM-3349 HV RESET switch at OPERATE.

2) Set the RT unit SERVICE SELECTOR switch at the desired operating position (SSB NSK, FSK, AM, or CW). Ensure that the PRIM. PWR switch is ON.

3) If using SSB NSK or AM, check that the VOX switch is in the right position (see above).

4) The set is now ready for keying by handset, mike, key, or ancillary RTTY equipment depending on the mode setting. In the VOX settings, you need to speak with a moderately loud voice to ensure keying the VOX.

Note: If a cable is connected to the AM-3349 50 OHM LINE connector, the rotating switch breaks the RF output path to the WHIP connector.

=Stopping Procedure=

Radio set AN/GRC-106 may be placed in standby or completely shut down. Normally, the complete shutdown takes approximately 3 minutes. In an emergency, the GRC-106 may be stopped immediately.

a. Standby. Keep HV RESET switch at OPERATE. Set the RT unit SERVICE SELECTOR switch at STANDBY.

b. Complete shutdown. Use the complete shutdown procedure for the GRC-106 only when the equipment is to be off for long periods. For periods of 1 hour or less, place the equipment in standby (a) above. To shut the equipment down completely, proceed as follows:

1) Set the RT unit SERVICE SELECTOR switch to STAND BY. Allow 2 minutes for the GRC-106 to cool before proceeding. Keep the HV RESET switch at OPERATE, when the GRC-106 is turned ON or OFF.

2) Set the AM-3349 PRIM. PWR switch and the RT unit SERVICE SELECTOR switch to OFF.

[The delay for cooling is for the 4CX350s in the AM-3349. External anode tubes require forced air cooling even when running filaments only. During operation with B+ applied they can run very hot indeed. The GRC-106 has a very elaborate cooling scheme using several blowers (internal and external) to get the heat away from the tubes into the radio chassis and then from there to surrounding air. The radio internals must remain sealed, so the transfer is accomplished by conduction between the internal cooling system and external cooling system. The cool down time with the blowers running is to bleed off the residual heat. The author believes this cooling system is right on the hairy edge of the needed cooling efficiency for 300-400 watts worth of heat dissipation. Radio shelters employing the GRC-106 such as the GRC-142 have auxiliary blowers that force extra air through the AM-3349. This is almost certainly mandatory for continuous duty transmitting, as in RTTY].

c. Emergency stopping. To turn the GRC-106 off in an emergency, set the RT unit SERVICE SELCTOR switch at OFF.

Note: When operating the GRC-106 in cold climates and shutdown is to be for 10 hours or less, set the RT unit SERVICE SELECTOR switch to OVEN ON.

Additional Comments:

GRC-106 examples the author's group has acquired have been all over the map on power output. Most were on the high side: 350 to 400watts PEP or "whistle power." While the sets are certainly capable of this power level, and for extended periods, it's hard on the final tubes and doesn't add all that much to the signal effectiveness. But most importantly, running the amp at the 400 watt level leaves very little margin of error during operation. The TUNE and LOAD controls are very touchy and it's easy to move quite a ways off resonance while tuning. They're easy to bump during operation, too, especially if the handles are extended. The military had trouble with personnel burning up the finals and/or tank circuits in the AM-3349s from hogging them around too much.

The author's group typically dials back their GRC-106s to 200-250watts PEP output.

Use the following procedure to adjust the SSB output on the GRC-106:

Tune up the GRC-106 as usual on 29.500mc into a 50 ohm load. Then in SSB/NSK mode, adjust 1A1R13 for 250watts out on a good whistle. 1A1R13 is located just behind the RT-662 or RT-834 front panel, at top center. You access this adjustment by pulling the RT chassis out a few inches. You may need to swivel aside an access plate.

In mobile systems using whip antennas such as Jeeps, where the radio is out in the open and exposed to the antenna, RF pickup problems often occur. Sometimes they can be severe, with the symptoms being a "frying eggs" sound on the transmitted audio, white noise, extremely distorted audio, etc. This is usually due to RF pickup on the H-33 cord.

Tip: To alleviate RF pickup in mobile installations, try changing the orientation of the mike or handset cord in relation to the antenna. Grounding in mobile installations is absolutely critical – when in doubt add more grounding! Also, laying down a few radials attached to the chassis of your vehicle may also help. More is better, but try three to start with. Use the formula

$$\text{Length of ground radial in feet} = \frac{234}{\text{GRC-106 operating frequency in MHz}}$$

Example for an operating frequency of 14.320 MHz:

234/14.320 = 16.3ft each radial

If you're getting zapped off the mike or handset, ground radials can be a big help. The idea is to get the RF off the radio case and onto the antenna where it can do some good!

What To Do If It's D.O.A.:

A totally dead set is indicative of blown fuse(s) or possibly problems with the circuit breakers. The author's group has had problems with the SERVICE SELECTOR switch in the past. It's a multi-layer rotary switch that can fail in a way that will render the RT units inoperative. We think this comes from moving the switch too rapidly from OFF to an operating mode like SSB NSK without spending the recommended warmup time in OVEN ON and STANDBY.

One member of the group found the contacts on his SERVICE SELECTOR switch had become charred/dirty and was able to clean them with Caig De-Oxit and restore operation. It was necessary to pull the front panel, remove the switch, desolder switch wires, etc. Best to save yourself some trouble in the future by adhering to the warmup procedure!

Other faults that have been encountered:
1) Failures on the input protection board related to the vacuum relay.
2) 100kHz synthesizer module unlocking – usually can be traced to unlocked regenerative dividers; fixable via re-alignment.
3) Frequency standard module unlocks – also a regenerative divider alignment problem.

Alignment or serious troubleshooting will require the dash-35 TM and a good understanding of this moderately complex radio system.

Accessories:

The most useful accessory is probably the PP-4763A/GRC-106 supply, which provides primary DC power for the set from 117VAC. It's a BIG, heavy chunk of iron that cranks out the 40amps or so necessary to run the radio.

Note: While the PP-4763 supplies can often be acquired at reasonable cost, the internal regulator board contains a unijunction-type transistor that has become very difficult to find. If that device fails, you're basically dead in the water unless you can find another board. Fair Radio sells the supplies – less the control board. That's probably because of the rarity of the unijunction device.

RF-5020

Application:

HF DSP-based receiver/exciter used in a variety of suites. Early 1990s vintage. The RF-5022, a slightly updated 5020, is in current use.

Tech Specs:

The RF-5020 is an amazingly capable piece of equipment. Here are the specs adapted from the RF-5000 System User's Manual. Author's additional comments in [].

=GENERAL=
Frequency Range: 1.6 to 29.99999 MHz in 10-Hz steps [Tunable in 1-Hz steps via RS-232 interface]

Channels: 2.84 x 1E6 in 10-Hz increments [2.84 x 1E7 in 1Hz increments]

Frequency Stability: +/-0.25 x 10-6 [OCXO; ovenized crystal oscillator]

Modes of Operation: LSB, USB, AME, CW

Preset Channels: Up to 100

Self-Test: Full built-in-test to the module level

RF/Input/Output Impedance: 50 ohms nominal, unbalanced

Power Input: +26.4 VDC per MIL-STD-1275

Remote Control: Detachable Front Panel; RS-232 Asynchronous interface

Data Interfaces: Asynchronous [RS-232 for controlling radio]; Synchronous

Temperature Range: -40degC to +70degC

Environmental: MIL-STD-810D

Size: 7.5W x 12.1D x 5.5H in. (19.0W x 30.8D x 14.0H cm)

Weight: 6.8 kg, (15 lbs)

=RECEIVER=
Sensitivity: -113 dBm for 10 dB SINAD

Audio Output: 3 watts to internal speaker

Squelch: Front-panel adjustable

Selectivity: 3 kHz at -3 dB; 3.8 kHz at -60 dB

[*Note:* this doesn't really tell the whole story, since it's the specs for only USB/LSB in the 3.0 kHz band-

width setting. Selectable bandwidths for all modes are as follows:

USB/LSB: 1.5, 2.0, 2.4, *2.7, 3.0 kHz
AME: 3.0, 4.0, 5.0. *6.0 kHz
CW: 0.35, 0.68, *1.0, 1.5 kHz
MODEM: *3.0 kHz
LPC VOICE: *3.0 kHz

"*" indicates default bandwidth settings.]

Image and IF Rejection: Greater than 80 dB

Spurious Responses: Below -120 dBm on 99.9% of channels

AGC: Mode dependent, automatically selected with manual override [SLOW, MEDIUM, FAST]

[**BFO RANGE:** -4000Hz to +4000 in 10Hz steps or 1Hz steps when controlled via RS-232]

=TRANSMITTER=
Power Output: Nominal 100 mW; Peak or Average

Carrier Suppression: Greater than 60 dB

Undesired Sideband Rejection: Greater than 60 dB

Intermodulation Distortion: -45 dB

Spurious Outputs: Below -120 dBm on 99.9% of channels

Audio Input: Dynamic microphone (1 mV) or 600 ohm @ 0dBm for full rated output

***

Active devices: Virtually all ICs

Audio input/output devices: Any of the new family headsets and handsets, including the H-250, H-189, H-138, H-161, etc. Audio output at AUDIO connectors is low-level and receiver will not drive an external speaker such as the LS-454 directly. (See text).

Antenna connection: BNC rear panel jack, 50-ohm

Antennas: This is a 100mW receiver/exciter and is not meant to drive an antenna directly

Tech Manuals:
Service Manual—RF-5020 R/T MAINTENANCE MANUAL, (Harris) PUBLICATION NUMBER 10181-0033, OCTOBER 1991

Operator's Manual—ARF-5000 TACTICAL COMMUNICATIONS SYSTEM
(Harris) PUBLICATION NUMBER 1081-0032F, MAY 1991

Construction Style: Rugged cast aluminum outer case. Detachable front panel. Circuit cards plug into an internal mother board. There are no moving parts save for the three front panel controls. Extender cards required for servicing.

Description:

This early '90s production radio is being replaced by the RF-5022, so RF-5020s are just beginning to show up in small numbers. Unless you get lucky, they are not inexpensive pieces of equipment. They are, however, outrageously capable radios. The basic R/T unit is a receiver-*exciter* with an output power rating of 100mW. You will need an outboard amplifier of some sort to make a practical system. The 5020 "talks" with any of several companion RF amplifiers via a serial link, which includes the receive/transmit commands in serial data format. There is no T/R line directly accessible from the 5020. This would seem at first to be a problem, but there are ways to get around it that will be presented shortly.

The 5020 has a digital I.F., with DSP processing done at 30kHz. The capability Harris has crammed into a box about the size and weight of an RT-70 is just astonishing. That capability is somewhat marred by a command list that's totally menu driven off a 16-button keypad. Even the simplest commands require several button pushes to get into the various menus.

The standard RF-5020 includes the following modules:

Mod01	Synthesizer
Mod02A	Receiver/Exciter
Mod02B	Digital IF
Mod06	Main Controller
Mod08	Interface Module
Mod10	Motherboard Assembly
Mod11	Remote/rear Panel Interface Module
Mod20	Front Panel

Optional modules are:

Mod03	ALE Controller (This is the Harris proprietary RF-5120 ALE)
Mod04A	High-Speed Data Modem
Mod04B	High-Speed Data Modem
Mod04C	High-Speed Data Modem
Mod04D	ECCM Data Modem
Mod05A	ALE Controller (MIL-STD-141A, what you usually hear)
Mod06B	ECCM Controller
Mod08B	Data Encryption Controller
Mod09A	LPC Voice Digitizer
Mod09B	LPC Voice Digitizer

The author's unit has digital voice, RF-5120 ALE (This is a Harris proprietary ALE, not the more common MIL-STD-141A ALE), and high-speed 39-tone modem installed.

Operation:

CAUTION: This is a totally menu-driven unit. Operating this radio without a users' manual will be hazardous to your sanity.

The 5020 J6 connector connects to a companion amplifier. Power to the 5020 can be input here directly.

J6 Connector on 5020		**J8 Connector on P.A.**
Pin	Function	Signal Level
A	Hop Clock	RS-232C Output
B	PA Control Data-	5VDC or 0VDC Differential Data
C	PA Control Data+	5VDC or 0VDC Differential Data
D	ON/OFF	Ground Closure
E	Primary Power Return	Ground
F	Primary Power Return	Ground
G	PA Feedback-	Feedback Retrurn
H	PA Feedback+	0-25VDC (4 VDC at rated PEP)
J	+26.4 VDC	R/T Supply Voltage
K	+26.4 VDC	R/T Supply Voltage

Basic power connection at J6: Pins J,K +26.4VDC; Pins E,F Ground.

The removable front panel connects directly to the R/T unit via matching connectors. The panel can be remoted from the R/T by using a cable. The matching connector to the front panel is an MS3116J12-10S. The matching connector to the R/T unit is an

MS3116E12-10P. The cable is a pin-to-pin affair, using A, B, C, D, E, F, G, H, J, and K.

For those interested in setting up RS-232 control of the RF-5020, this is done at the back panel connector J5. For some reason, the manual doesn't give the matching connector type for J5. The author was able to make one up from scrap pieces of other connectors he had in the junk box, so there is no way to know the actual nomenclature.

Mating connector for J6

If you send a screen capture of this image to Bill Perry (see Surplus Dealers section in Appendix), he may be able to fix you up. The image is larger than actual size. The insert is about 5/8" in diameter. Connectors are a constant pain when dealing with surplus, but they don't need to be showstoppers. For quick tests, the author simply makes short pigtails out of larger 14 or 12 ga solid insulated wire and pushes it right into the female connector holes. If connectors absolutely cannot be found, a strain relief can be made to keep the wires from pulling out and the setup graduates from "haywired" to "temporary." (Eg: FOREVER).

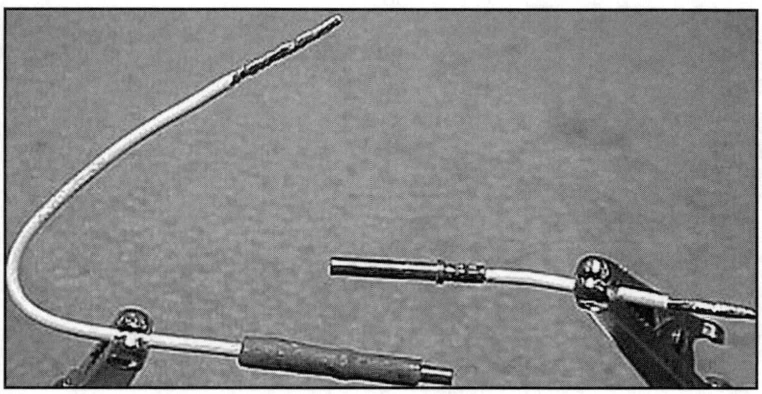

Use of female connector inserts

The best way to temporarily hook up to male connector pins is with the female inserts normally found in complete connectors. The author has broken apart a couple of oddball junk connectors to get a supply of female pin inserts. Pigtails are soldered to them, if there isn't a length of wire connected already. If there's any chance of the inserts touching during use, put on some heat shrink as shown.

If you're ambitious, a more permanent connector can be fabricated—see the end of the GRC-109 section for an example.

J5 AUXILIARY connector pin-outs:

Remote Operations Audio/Control Input/Output Signals:

Pin	Function	Interface
U	Audio Keyline Input	Ground closure
L	TX Audio (-)	-10 or 0dBm
N	TX Audio (+)	-10 or 0dBm
T	Retransmit Keyline Output	Ground closure
C	RX Audio (+)	-10 or 0dBm
P	RX Audio (-)	-10 or 0dBm
D	Remote Off/On Input	Ground closure
A	Analog Ground	

Asynchronous/Remote Control Input/Output Signals:

Pin	Function	Interface
R	TX Data Async	RS-232
V	RX Data Async	RS-232
H	Request to Send Async	RS-232
B	Clear to Send Async	RS-232
M	Digital Ground	Signal Ground
K	Receive Level Sense Detect	RS-232

Synchronous Input/Output Signals:

Pin	Function	Interface
F	TX Data Sync	RS-232
E	RX Data Sync	RS-232
G	Request to Send Sync	RS-232
S	Clear to Send Sync	RS-232
M	Digital Ground	Signal Ground
K	Receive Level Sense Detect	RS-232
J	Sync Clock	RS-232

The minimum connections needed to talk to the 5020 are TX Data Async, RX Data Async, and Signal Ground— R, V, and M.

Note: The author has written some fairly capable

proof-of-concept software to control the 5020 via the serial interface. The 5020 certainly isn't a browsing radio; just moving the frequency and operating basic functions like mode, bandwidth, and AGC hang time require various sequences of non-obvious keystrokes. The current software runs in a simple DOS window and gives quick access to mode, bandwidth, RX and TX frequency, standard squelch, BFO, and AGC. ON/OFF toggles are also included to control clipping, compression, digital voice, and digital squelch. More capability will be added from time to time.

There are a *mess* of commands used to operate this radio. The command section in the User's Guide is 26 pages long just by itself. A short operation overview is presented here. This must be considered a "quick start" guide, suitable for checking out the basic functions of the radio only.

1) Connect a suitable source of power (26VDC at about 2A max) to J6 per the above.

2) Connect an antenna to the BNC jack J7

3) Turn the Function Control (3) to SSB . This is the main frequency display.

4) Pull the SPKR ON PULL switch out to turn on the speaker and advance it a 1/4 turn or so.

Note: Depending on what state the unit is in, you may not need to press CLR in all cases. Pressing CLR ensures any parts of previous commands are flushed. Generally, pressing CLR while entering a value will back up one digit, etc.

5) TO CHANGE FREQUENCY. Press CLR button. Press FREQ button. Enter the RX FREQ as XX.XXXX with a leading zero if less than 10mHz. When finished entering RX FREQ, press ENTR. Enter TX FREQ in the same way, and press ENTR again when done. Note that TX FREQ is set to RX FREQ by default unless you deliberately change it.

6) TO SET AN INCREMENT OR DECREMENT STEP. From the main frequency display press CLR, then 6, then PRGM. The INCR can then be set for 100kHz, 10kHz, 1kHz, or 100Hz. (Press CLR to move left one digit). Press ENTR when done.

7) TO INCREMENT OR DECREMENT the frequency. From the main frequency display, press CLR, then FREQ. You must be in the FREQ change state to increment/decrement. Pressing the UP ARROW or DOWN ARROW will increment or decrement the RX FREQ by the step set earlier. Press ENTR to move to the TX FREQ, which can be changed in the same way. Press ENTR again when done.

8) TO CHANGE THE MODE. From the main frequency display, press CLR, then MODE repeatedly until RADIO MODULATION display appears. Use ARROWS to pick mode, then press ENTR.

9) TO CHANGE BANDWIDTH. From the main frequency display, press CLR, then 2, then PRGM. Use the ARROW keys to toggle between the legal bandwidths for the current mode. Press ENTR when done.

10) TO CHANGE THE AGC SETTING. From the main frequency display, press CLR, then 3, then PRGM. Use the ARROW keys to toggle between the five settings.

11) TO ENABLE/DISABLE THE BFO. From the main frequency display, press CLR, then 9, then PRGM. Use the ARROW keys to turn the BFO ON or OFF. Press ENTR when done.

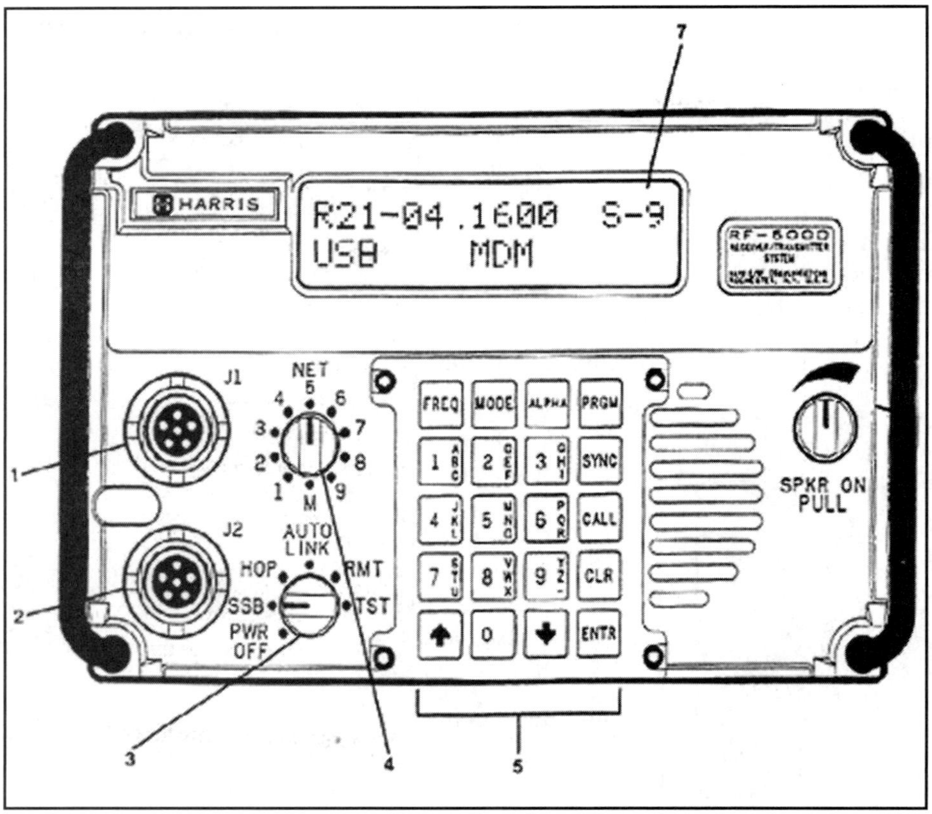

RF-5020 front panel

12) TO MOVE THE BFO. While in the main frequency display, press the ARROW keys to move the BFO up or down in 10Hz steps. (-4000Hz to +4000Hz).

13) TO TOGGLE AUDIO COMPRESSION. From the main frequency display, press CLR, then 7, then PRGM. Use the ARROW keys to toggle compression ON or OFF. Press ENTR when done.

14) TO TOGGLE AUDIO CLIPPING. From the main frequency display, press CLR, then 8, then PRGM. Use the ARROW keys to toggle clipping ON or OFF. Press ENTR when done.

15) TO SET THE SQUELCH. From the main frequency display, press CLR, then MODE repeatedly until the SQUELCH display appears. Set the level with the ARROW keys, increasing (5 unit increments) until the squelch just closes on no-signal conditions.

16) TO TOGGLE DIGITAL VOICE (if installed). From the main frequency display, press CLR, then MODE repeatedly until the DIGITAL VOICE display appears. Use the ARROW keys to toggle digital voice ON or OFF. Press ENTR when done.

17) TO TOGGLE DIGITAL SQUELCH. From the main frequency display, press CLR, then MODE repeatedly until the SQUELCH display appears. Use the ARROW keys to toggle digital squelch ON or OFF. Press ENTR when done.

Note: In regular single-channel mode the SQUELCH display has the three digit settable level. In DIGITAL VOICE mode, the SQUELCH display changes to ON/OFF only.

18) Before messing with the RS-232 serial control feature, you must make sure your computer and the 5020 are set up with the same serial parameters. To set them on the 5020 proceed as follows. From the main frequency display, press CLR, then 17, then PRGM. You are now in the DATA PORT group of menus.

a) TO SET ASYNC RATE. From the DATA PORT main display, press ENTR. Use the ARROW keys to choose the desired rate. When done, press PRGM to move to the next parameter, BITS. (step b, below)

b) TO SET NUMBER OF BITS. Use the ARROW keys to set the number of bits. When done, press PRGM to move to the next parameter, STOP BITS. (step c, below)

c) TO SET NUMBER OF STOP BITS. Use the ARROW keys to set the number of bits. When done, press PRGM to move to the next parameter, PARITY. (step d, below)

d) TO SET TYPE OF PARITY. Use the ARROW keys to set the parity. When done, press PRGM to move to the next parameter, ASYNC FLOW. (step e, below)

e) TO SET TYPE OF FLOW CONTROL. Use the ARROW keys to set the type of flow control. When done, press PRGM to move to the next parameter, ASYNC ECHO. (step f, below)

f) TO SET ASYNC ECHO. Use the ARROW keys to set echo ON or OFF. When done, press PRGM to move to the next parameter, DATA LEVEL. (step g, below)

g) TO SET DATA LEVEL. Use the ARROW keys to choose the desired data level (RS-232 or MIL-188). When done, press ENTR to exit back the the main frequency display.

19) TO SET THE AUDIO OUTPUT LEVEL. This function sets the level at the AUXILIARY transmit audio interface (J5, pins L, N). From the main frequency display, press CLR, then 27, then PRGM. Use the ARROW keys to pick the level, then press ENTR when done.

20) TO SET THE TIME OF DAY. From the main frequency display, press CLR, then 28, then PRGM. Enter the TOD in 24-hour time, and press ENTR when done.

***

The above is only a fraction of the total commands possible with the radio, depending on the options you have installed.

RS-232 SERIAL CONTROL COMMANDS

After hooking up a serial cable and ensuring the parameters are set the same at the computer end and in the RF-5020, turn the Function Switch (3) to the RMT position. When this is done, the radio will send a summary of the current settings and you should see it show up on your computer screen, and you'll get a prompt. If you were in SSB mode, it will be:

SSB>

The 5020 is now ready to accept commands from your remote keyboard. The command

FR 05

Will change the RX FREQ and TX FREQ to 5 meg WWV.

The command

MODE AM

Will change the mode to AM.

Some general commands:

The three major Function Operating Modes are:
SSB—Single Channel Mode
HOP— Hopping Mode
ALE AUTOLINK mode
(The last two work only if you have those options. You will be in SSB mode most of the time)

Other commands:
ASYNC SHOW: shows all the current parameters for the asynchronous interface
ASYNC HELP: shows a summary of all the parameter setting commands
SHOW: gives a general snapshot of the current status and settings in the RF-5020
HELP: shows a chart of all the commands for any Function Operating Mode.

Note: Only the capital letters of the commands need to be entered. "AG slow" programs AGC to slow setting.

More examples. The command

NET 02

switches the net (same as channel) setting to NET 02.

The command

FR 11175

sets the operating frequency to 11.175000 MHz. (Leading zero is required below 10mHz)

To program a channel by remote control, a text string can be sent:

Net 07
Freq 07240
Mode LSB
Ba 2.4
AGC med

In fact, as many as 100 "nets" or channels can be completely specified in a text file and uploaded into the 5020 via the serial interface. This includes HOP frequency sets, encryption keys, ALE station calls, the whole smash. That is, IF you have the required options. Things you can do in the RMT mode you can't do from the front panel:

1) Frequency can be entered in 10Hz or 1Hz steps. (great for PSK31!)
2) The BFO can be moved in 1Hz steps.
3) The SQUELCH can be moved in increments of 1 unit.

Additional Comments:

Several members of the author's immediate group have acquired RF-5020s. We're still learning about the radio. Here are a few random facts we've determined/discovered:

The RF-5020 has some truly excellent operating characteristics. In heavy QRM conditions, going to the 2.0 kHz bandwidth setting really makes the incoming sig jump out of the garbage. You'll also be transmitting with 2.0 kHz bandwidth so you can really wedge in between stations, assuming the other op has the same capability. If things really get rough, you can pack all your power into to 1.5 kHz. The audio sounds a little restricted at the other end but it will really punch through.

Using digital voice requires a good signal to noise ratio equivalent to an S8 to S9 signal. Otherwise the two units will have difficulty synching up. But if the S/N is good, the results are startling. With the digital squelch on, you hear absolutely nothing until the radio synchs with the remote unit – then you hear audio with no HF-related noise at all.

The built-in speaker often sounds terrible. We've traced this to the way the speaker is mounted on the front panel. When the speaker mounting screws are torqued, the speaker mashes down against the rubber gasket and the entire speaker frame distorts. This brings the voice coil in contact with the magnet and produces a large amount of distortion in the audio. The degree of torque applied to the speaker mounting screws has a large effect on the level of distortion. In one case, relieving the screws helped. In another case turning them down tighter lowered the distortion. You'll have to experiment.

There is no speaker-level audio available at the front panel audio connectors J1 or J2. The best you get is 0dBm into 1000 ohms, so you're stuck with the (often) distorted internal speaker, or phones.

There are several ways to key the 5020, however the unit does not have extra contact closures or ground closures available externally for operating a T/R relay. The T/R indication is sent as data over the serial line connected to a companion Harris RF amplifier.

There are several ways to key the RF-5020:

MIC: J1 or J2 handset connectors
AUX: Remote Keyline AUXILIARY (AUX) connector J5
DATA: Request To Send control line (RTS) AUX connector
DATA: Control-B data character (key); Control-C (unkey) AUX connector
REMOTE: Key/Unkey remote control commands

The transmit/receive state can be determined remotely through the RS-232 serial interface. Each time the 5020 is keyed and unkeyed, a corresponding message is sent over the serial interface. When in the SSB mode, the messages are "Key Mic" (for TX), and "Key Off" (for return to RX). These messages can be trapped by the remote software and used to actuate a T/R relay.

The ALE option that fits in Slot A3 (Harris RF-5120AFC AUTOLINK® Adaptive Frequency Controller) is not compatible in any way with the more generally encountered MIL-STD-188-141A ALE. Slot A5 is for the MIL-STD-188-141A ALE most commonly encountered on the ham bands.

Warning: Do not attempt to remove the bottom cover without first pulling all the modules and removing the motherboard. You should never have to do this.

What To Do If It's D.O.A.:

There is a BITE test that isolates faults down to the module level. It will be small comfort though if a fault is found and listed, because replacing any one of the modules will involve a sizable monetary outlay. The BITE may help you get started on component-level troubleshooting, however. Setting the FUNCTION control to the TST position initiates a system self-test.

If a fault condition occurs, its source is displayed on the LCD display as a fault code. The list of fault codes is extensive and beyond the scope of this book. If your 5020 is exhibiting a fault and you get stuck, contact the author for explanation of the fault code. He's interested in others who are working with this radio.

If your unit does not retain channel settings during power off, one (or more) of three memory batteries is probably dead. The main memory battery is located on the A6 board and is a cylindrical 3.9V lithium type. This BCX72 3B50 battery is a $26+ item from Allied if the exact replacement is ordered. The author discovered that a jumper connecting the battery to the memory circuits in his 5020 had been deliberately cut, removing the battery from the circuit and disabling the memory retention feature.

Battery jumper on A6 Module, front board

After replacing this jumper, the radio retained settings. This was probably why the author had a good battery; other members of his group had dead ones. For those opposed to paying that much for a battery, a more common 3.6V battery seems to work okay. The holder can be located in the space between the A6 and A4 modules (if you don't have the A5 option. The author has also used two AAA batteries in a plastic holder attached to the A6 module which seems to retain the memories just fine.

The other two batteries are both "Keeper" brand LTC-7PN 3.5V 750mAH types available from Mouser Electronics for about $6. One is located in the A6 module on the back board. The other is inside the control head. The batteries seem to be much longer-lived that the one on the A6 module.

One member of the author's group had a catastrophic failure in his control head. He quickly traced this to a multiple short that occurred between the metal back of the LCD display and the traces on the board to which it is mounted. The author has worked out a preventative fix for this problem. The clearance between the metal back of the display assembly and the chip pins protruding up through the board is extremely small. The metal display back spans many pins/chips and probably ended up putting supply voltage on many of them. Since the going rate for the control head alone is in the $350 range, it seems like prevention is the best ticket. Near-complete disassembly of the control head is necessary. Use this opportunity to check the backup battery, clean the LCD display, and adjust the speaker screw torque if your audio is distorted.

1) Remove the function switch and NET knobs. (Two left-hand knobs)

2) Remove the nuts and star washers from the two controls.

3) Remove the four screws on the back cover and lift the cover away from the head carefully. Unplug the connector on the cover ribbon cable.

4) Unplug the ribbon cable connector going to the SPKR control.

5) Remove the six screws holding the main PCB, four at the corners, and two reached through the PCB.

6) Lift out the PCB, taking care not to bend the connector pins going to the keypad. The keypad can be removed beforehand, but it's just as easy to leave it in.

7) Loosen the two screws holding what looks to be regulators to the display assembly.

8) Remove the display assembly and set aside.

9) Cut an index card approximately as shown.

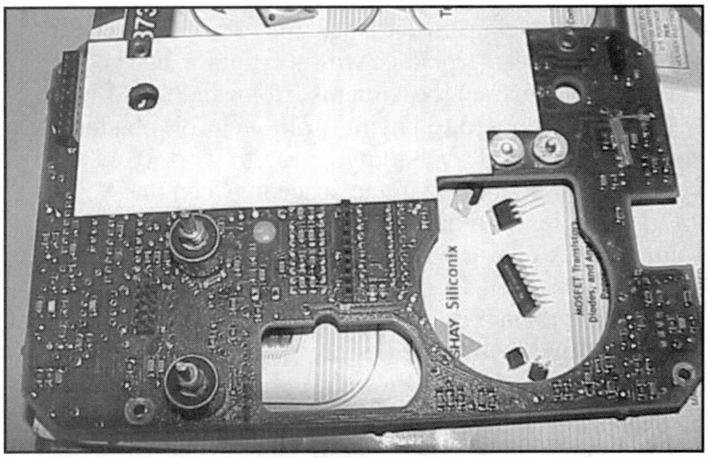

Insulator made from an index card

Don't forget the hole at upper left for the screw through to the display assembly. A little trial and error works here. Index cards are cheap.

10) Reattach the display assembly. At this point, you might want to clean the display with a soft cloth moistened in alcohol.

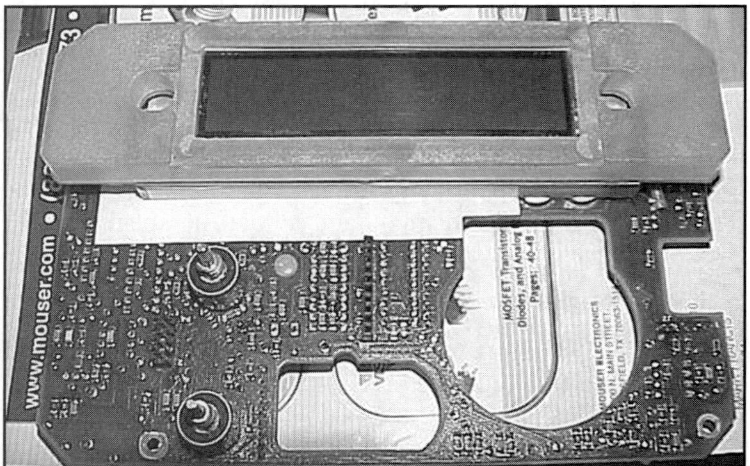

Display reattached, with insulator beneath

11) Reassemble control head, reversing the disassembly process.

WARNING! Make *absolutely sure* the connector from the back cover is connected correctly to the PC board! Eg: not offset one way or the other.

The backlight on every 5020 in the author's group is either extremely weak, or inoperative. The light function can be reached by pressing CLR, then 26, then PRGM. The three choices are OFF, ON, and MOMENTARY. The MOMENTARY setting turns on the backlight for 15sec after the NET or FUNCTION key is pressed. When the backlight is ON, press the ENTR key then use the ARROW keys to vary the intensity of the LCD display. To store the intensity level, press ENTR. The preceeding reflects a correctly functioning display. But these displays hardly ever work. Your symptoms will vary from barely discernable backlight to none at all, rapidly flickering LCD lettering when backlight is on, etc. If you backlight appears inoperative, you should make sure the 26 PRGM function is set to OFF.

The author has had two A7 Power Supply module failures in 5020s he's worked on that resulted in a dead unit. The toroid coils on the power supply are mounted with a dab of brittle cement that tends to break, allowing the coil to flop around, often breaking a lead. 5020 owners should check these coils as a matter of course. Also, the four adjacent rectifiers located at the center of the A7 run hot enough to discolor the PC board. Often one of these will be charred, cracked, or bulged.

Newly acquired balky radios often just need some Caig De-Oxit sprayed into the motherboard slots and the boards reseated a few times.

CAUTION: The RF-5020 was originally intended for use with one of the Harris matching 20 watt, 125 watt, 400 watt, or 1kW amplifiers, which have extensive overcurrent, static, and RF protection built in for the receiver-exciter. The 5020 was not originally intended for connection directly to an antenna. The unit has an open front end and has proven sensitive to high static and ambient RF conditions. Damage to the A2 module can result if the unit is operated in high static electricity environments (antenna in blowing snow, thunderstorms), or high RF conditions (nearby transmitters). Avoid these types of conditions and disconnect the antenna when the unit is not in use.

Accessories:

Plug-in options:

RF-5110MD High-Speed Data Modem
RF-5120AFC Automatic Link Establishment
RF-5121ALE Automatic Link Establishment
RF-5130LPC Digitized Voice (LPC) [RF-5130 uses LPC-10 Codec]
RF-5140FH Frequency Hopping
RF-5150DE Embedded Data Encryption

RF-302

Application:

Manual HF antenna tuner for the RF-301 transceiver. The RF-301 and RF-302, with associated accessories, comprise the GRC-165 HF set (Army nomenclature). The Navy called the tuner the CU-1457, with the entire set known as the AN/URC-58. The Army sets are O.D. in color; the Navy sets are battleship gray.

Tech Specs:

Frequency range: 1.6 to 16 MHz

Power handling capacity: 100W PEP (Really! See text)

Size: 5" x 7.5" x 15.5" including knobs and connectors

Weight: 8 pounds

Active devices: 1N277 rectifiers (2 total) in VSWR bridge

Power requirements: none

Antenna connection:
Input—SO-239
Output—single wire ceramic antenna post and ground screw

Whip antenna: Originally designed to tune 9, 16, or 35 foot whips

Other antennas: 75 and 150 foot long wire antenna kits were also available

Tech Manuals: NAVSHIPS 0967-034-8000 "RF-302 Antenna Coupler",

Construction Style: rugged chassis and panel construction with three-sided, gasketed outer sheet metal case. All components reparable/replaceable.

RF-302, top view, cover removed

Description:

The RF-302 is a fairly straightforward manual tuner used with the Harris RF-301 HF transceiver, military nomenclature GRC-165 or AN/URC-58. It will match a variety of balanced or unbalanced antennas at up to 100W PEP from 1.6-16 MHz. The original range of the RF-301 transceiver was 2 to 15mHz. The case is dust- and splash-proof, an attribute most definitely not shared by the usual tuners employed by hams.

The previous illustration shows the 10 and 28uH roller inductors, switched capacitor array (upper left), VSWR bridge module (lower left), and inductor drive mechanism (right).

Detail of roller inductor drive mechanism

This is a nicely-built tuner!

Operation:

Feed the tuner with 50-ohm coax and connect the antenna to the ceramic insulator. For whips working off a vehicular ground and fed with a single wire, keep the feed wire as short as possible -- maximum 2.5 feet. The author managed to keep it down to 20" on his pickup installation. Run a ground wire from the ground lug to a good, solid ground on the vehicle chassis.

Operation section adapted from the manual, author's comments in the []:

The RF-302 is initially tuned to a transmitting frequency by using a low power RF input and adjusting the controls for a minimum reflected power indication on the built-in directional wattmeter. No external meter is required for tuning. After the RF-302 has been tuned to a frequency and the control counter settings recorded, it may be quickly retuned to that frequency without an RF input by positioning the controls to the pre-logged settings. The paragraphs below detail the procedures for tuning, with or without prelogged information.

Note: The RF-302 should be tuned with transmitter power set at about one third maximum output. When the RF-302 is not properly tuned, it does not provide a proper load for the transmitter. Plate current in the transmitter final stage will be excessive and damage to the transmitter may result.

PRINCIPLES OF OPERATION.

In order to understand the method of tuning the RF-302, it is helpful to know how it operates. Two variable coils are tuned such that the antenna impedance at the operating frequency is matched to the 52-ohm transmitter output impedance for efficient power transfer to the antenna. The coils are adjusted with the FINE TUNE and LOAD controls. The inductances are at maximum with the dials set to 000. The COARSE TUNE control switches fixed capacitors in series with or across the antenna to ground in such a manner that the antenna can be properly tuned by the coils. The COARSE TUNE control can be thought of as altering the reactance of the antenna to make it easier to match.

The meter on the RF-302 is used to monitor forward power (power to antenna) and reflected power (power reflected from antenna because it is inefficiently matched). A toggle switch on the RF-302 panel is used to select FORWARD or REFLECTED power meter operation. By adjusting the RF-302 for a minimum reflected power indication, the antenna is tuned for peak efficiency.

A zero REFLECTED power indication means that exactly 52 ohms resistive load is presented to the transmitter output, i.e., a VSWR of 1/1.

TUNING PROCEDURES.

When tuning the RF-302 with an RF input, a single tone signal (such as a CW signal, or an unmodulated AM carrier) should be used. Before attempting to tune the RF-302, tune the transmitter to the desired frequency by following the procedures in the transmitter instruction manual.

Note: Although it is possible to perfectly tune most antennas and bring the REFLECTED power indication to zero, it should be noted that such precise tuning isn't essential. It is usually considered adequate and no appreciable loss of effectiveness will be evident if the REFLECTED power is kept less than 5% of the FORWARD power.

When no previous tuning information is known, the procedure "TUNING WITHOUT KNOWN CONTROL SETTINGS" should be used to tune the RF-302. Once the RF-302 is tuned, record the control settings for future use.

TUNING WITH KNOWN CONTROL SETTINGS. When the control settings for the frequency and the antenna setup in use have been previously recorded, proceed as follows:

a. Set the RF-302 controls to the positions recorded in the charts for the desired frequency.

b. Set up the transmitter for 50W output at the desired frequency.

c. Key the transmitter.

TABLE 3.1 - TYPICAL CONTROL SETTINGS.

	35 FT. Whip* SB-V35			16 FT Whip* SB-V16			9 FT Whip* SB-V9		
F	CT	FT	L	CT	FT	L	CT	FT	L
2.0	6	338	103	7	360	290	7	346	264
2.5	5	339	177	6	364	229	7	368	445
3.0	1	339	067	5	363	169	5	305	057
3.5	1	348	278	5	375	336	5	330	247
4.0	1	348	420	5	382	449	1	293	067
4.5	1	347	525	1	374	336	1	321	214
5.1	1	325	621	1	374	447	1	338	339
5.5	1	325	663	1	376	503	1	351	402
6.0	1	335	717	1	386	557	1	365	463
6.5	1	338	775	1	394	603	1	366	517
7.0	2	296	800	1	395	643	1	374	559
7.5	2	292	798	1	395	677	1	374	596
8.0	2	133	857	1	396	707	1	382	624
8.5	4	234	560	1	396	735	1	389	649
9.0	6	254	753	1	396	763	1	394	669
9.5	6	345	795	1	394	792	1	398	688
10.1	6	371	819	1	392	828	1	402	709
11	5	308	767	1	377	907	1	406	735
12	3	102	691	4	382	682	1	410	761
13	3	348	736	4	369	725	1	413	785
14	3	368	769	4	221	760	1	413	806
14.999	3	382	800	6	400	862	1	406	825

F - FREQUENCY (MCS)
CT - COARSE TUNE
FT - FINE TUNE
L - LOAD

d.

RF-302 Rough Tuning Chart

Set the RF-302 meter switch at REFLECTED POWER.

e. Carefully check LOAD and FINE TUNE controls to see if a slight readjustment will improve the tuning (reduce the reflected power).

f. After tuning, increase power and operate.

3.3.2 TUNING WITHOUT KNOWN CONTROL SETTINGS. When the control settings for the desired transmitting frequency are not known, a search will have to be conducted to find settings which will produce the lowest VSWR. The VSWR is lowest when the ratio of forward power to reflected power is greatest.

Typical settings of the RF-302 controls for different types of whip antennas are listed in table 3.1. These settings will be useful in indicating approximate settings for starting the search.

[*Note:* Use the chart to get preliminary settings, along with tuning for maximum receive noise, which is always a valid method].

When table 3.1 cannot be used, set COARSE TUNE at 1, FINE TUNE at 360, and LOAD at 900, then proceed as follows:

Note: The lowest numbered COARSE TUNE switch position should be used that will allow correct tuning. It is sometimes possible to tune in more than one position. The lowest numbered position will provide the most efficient operation.

a. Set meter switch at REFLECTED POWER.

b. Key and tune transmitter for 50W output at desired frequency.

c. Rotate LOAD control over complete range or until a null indication is obtained on the power meter.

d. If no null indication is obtained, advance COARSE TUNE switch one position and repeat step c, continuing until a null indication is obtained.

Note: A sharp rise in reflected power may occur just before a null, due to the fact that the power from the transmitter builds up quite rapidly as the proper tuning point is approached.

e. When a null indication is obtained, alternately adjust LOAD and FINE TUNE controls for minimum reflected power.

f. Record the control settings for future use.

g. Increase to full power and operate.

[*Note:* a good deal of tuning can usually be avoided by adjusting the tuner for maximum receive noise before applying any power.]

Additional Comments:

See the MSR-8000 section for a picture of the tuner in use in the author's mobile installation. A 20" high-tension wire goes through holes in the cab and bed to an AB-450 antenna bracket mounted just inside the bed wall. The bracket has a ceramic insulator which accepts the sectionalized 15" PRC-47 whip for stationary ops below 14mHz.

The so-called "tuners" that are supposed to handle 1.5kW (offered by manufacturers who will not be mentioned but are widely known) are junk. Try and put a gallon and a half through one of them even a hair off resonance and they'll go off like flashbulbs. Sure, most people are careful to pre-tune to receiver noise, etc. (including yours truly). But there is no margin -- bump the tune or load control and BLOOIE! The Harris RF-302 is a real tuner that will handle an honest 100 watts under most any conditions.

What To Do If It's D.O.A.:

If the tuner seems to work okay but there is no indication on the FORWARD/REVERSE power meter, the 1N277s in the VSWR sense module may have popped. The author's tuner tended to shuck the chains off the drive gears until he readjusted the tension on them. No problems since then. If one or both falls off, remember that when the counter reads "000," the roller inductances should be at maximum.

Erratic operation may be caused by dirty switch contacts or dirty roller coils/pickups. These are silver plated in this tuner and may accumulate tarnish. A little carefully applied Tarn-X often fixes this.

Caution: Remember to thoroughly rinse all traces of Tarn-X off the contacts after cleaning them! Otherwise, unwanted corrosion will result in other parts of the tuner.

R-389

Application:

General purpose LF/VLF receiver for fixed service applications. Part of the family of receivers that also includes the R-390, R-390A, and R-391. An application mentioned in the T.M. combines the R-389 with Frequency Shift Converter CV-115/URR and a teletypewriter set for RTTY use.

Tech Specs:

Frequency range: 15 to 1500 kHz in two ranges 15-500 (using 5 automatically selected ranges) and 500-1500kHz (using 2 automatically selected ranges)

Receiving modes: CW, MCW, AM voice, RTTY, SSB

Type of tuning: manual or motor-assisted tuning

Size: 19" x 10.5 x 14.5"

Weight: 82 pounds

Audio output:
600 ohm unbalanced line -- 500mW
600 ohm balanced line -- 10mW
600 ohm phones -- 5mW

Type of circuit: double conversion superhet. First I.F. at 10mHz, second at 455kHz

I.F. selectivity: .1, 1, 2, 4, and 8kHz

Active devices: Vacuum tubes (36!), of types 6AK6, 6BH6, 6BJ6, 6C4, 12AT7, 12AU7, 26Z5W, 6BA6W, 6BE6W, 6082, and 3TF7 ballast.

Power requirements:
115/230VAC +/- 10%, 48 - 62Hz
250W total, 150W with oven heater off

Audio input/output devices: Any 600ohm accessories

Antenna connection:
Unbalanced—random wire
Balanced—125 ohms

Tech Manuals:
TM 11-855, "RADIO RECEIVER R-389/URR", description, theory, troubleshooting, repair, alignment, schematics

Construction Style: Heavy 19" rackmount panel-and-chassis arrangement consisting of sub-

chassis along similar lines to the R-390/R-390A. Subchassis are repairable but, like the R-390A, the RF subchassis will be just as difficult to work on.

Description:

This receiver is a fairly rare item. The author's example shown is the only one he's ever seen in 25 years of lurking at hamfests. That may just mean he happened to miss other units, but it's an indication there aren't many floating around. This one was made by Collins in 1951 and has serial #725. The T.M. says the power supply, I.F., and A.F. subchassis will interchange between the R-390, R-391, and R-389. It would be interesting to see if an I.F. subchassis out of an R-390A with the mechanical filters would also interchange.

At first glance, the panel bears some similarities to its close relative, the R-390. Further study will reveal

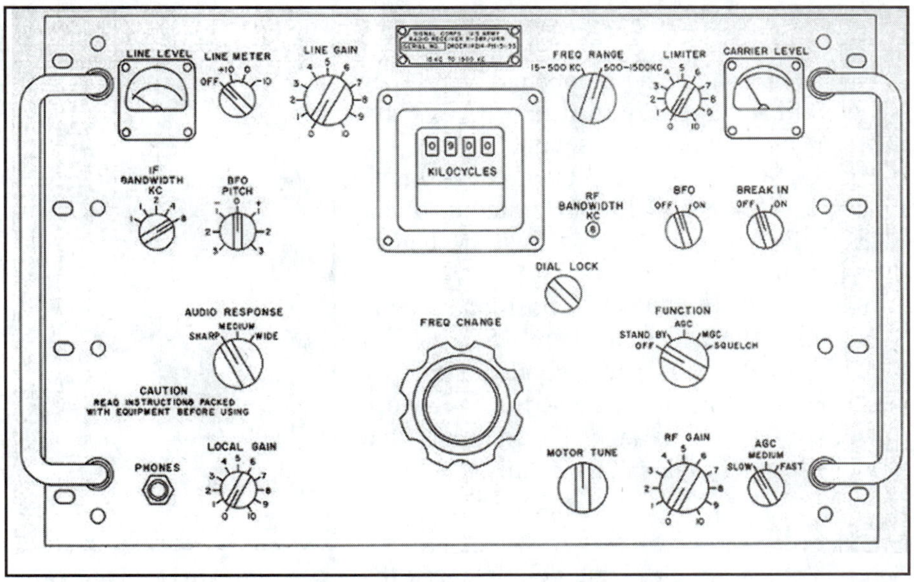

R-389 front panel

many differences, both electrical and mechanical. If you think the R-390 is an electromechanical beast, wait until you see this thing. The R-389 is absolutely astonishing in its complexity. There are two drive motors in the unit. The first is activated when the MOTOR TUNE control is rocked left or right momentarily, and spins the FREQ CHANGE knob rapidly. This was added because of the very large tuning range of the set, one hundred times the lowest received frequency. That may be no big deal with today's synthesized broadband receivers. But to do it back then with gears and slug-tuned coils was nothing short of miraculous.

In fact, "lock to lock," from 500kHz to 1500kHz is about 200 turns of the FREQ CHANGE knob. Without the motor drive, the term "carpal-tunnel syndrome" would have been coined in the 1950s!

The second drive motor automatically sets up the tuned circuits for five different ranges when operating between 15 and 500kHz, and two different ranges when operating between 500kHz and 1500kHz. It also drives the RF BANDWIDTH KC indicator dial, which displays the maximum width of the passband to which the I.F. may be set at the operating frequency of the receiver. The seven different ranges and the corresponding RF bandwidth that appears in the window are as follows:

R-389 Bandwidth Ranges:

RANGE (kHz)	RF BANDWIDTH (kHz)
15 - 27	1
27 - 55	2
55 - 117	4
117 - 242	8
242 - 500	8
500 - 865	8

The actual switchover points may vary a few kHz either way, depending on how the mechanical alignment was set up.

The RANGE switch selects one of two main ranges 15-500 or 500-1500kHz, and also activates a mask revealing one of two mechanical digital counters.

Where you'd find the CAL position on an R-390 FUNCTION switch, there's SQUELCH, which enables a reasonably good carrier operated squelch. Not much good for SSB work, obviously. The R-389 lacks the calibrate function of the other radios in the family.

Operation:

Setup is pretty simple. First, double check that switch S801 on the power supply is set to the correct line voltage. The switch can be reached from the bottom of the chassis -- it's the toggle next to the main power transformer.

Note: Be sure the fuse in the holder marked "AC 3A" is correct for the line voltage you are using: 117VAC—3A 125V fuse; 230VAC—3A 250V fuse.

The original power cord is the CX-1358/U and has a standard AC plug on one end and an oddball 4-pin female connector on the other. If you don't have the connector, you'll have to clip lead or solder. The AC connections at POWER jack J-102 on the rear panel are pins A, D.

Possible audio output choices:

a) 600 ohm headphones in PHONES jack.

b) 600 ohm headphones between LOCAL AUDIO terminal 7 and PHNS terminal 8 on back panel.

c) 600 ohm speaker between LOCAL AUDIO terminals 6 and 7 on back panel.

Don't hook up an antenna yet. Set the FUNCTION switch to AGC and let the set warm up for 30 minutes or so before checking the following. If the CARRIER LEVEL meter is not zeroed, set the RF GAIN to max and zero the CARRIER LEVEL meter by adjusting R537 on the I.F. subchassis (top, left hand side as you look at the front panel):

The front of the radio is to the left in the illustration. The CARRIER LEVEL meter zero adjustment is very touchy -- rock it back and forth a few times before setting it. Sometimes the carbon film in the pots gets worn right at the "sweet spot" and you have to replace

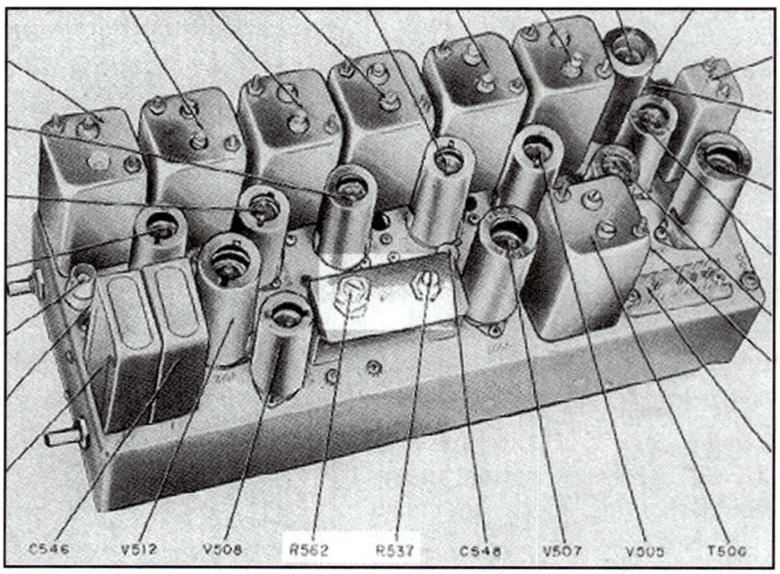

I.F. subchassis, showing carrier level meter zero adjustment R537, and overall I.F. gain adjustment R562.

it. R537 is 15 ohms. If the I.F. gain seems low, you can try adjusting R562 in small increments, remembering where it was set. If local AM broadcast stations are kicking the meter up 60-80dB, it's a good bet that dialing in any more I.F. gain will be counterproductive.

Plug in a set of headphones (PL-55 type plug) and connect an antenna of some sort.

LINE LEVEL meter: Indicates level of balanced-line output.

LINE METER switch: In OFF position, switch disconnects LINE LEVEL meter from balanced-line output. In +10 position, add 10 decibels (db) to LINE LEVEL meter vu reading; in 0 position, read LINE LEVEL meter directly in vu; in -10 position, subtract 10 db from LINE LEVEL vu reading.

LINE GAIN control: Controls level of audio signal applied to balanced-line output terminals.

FREQ RANGE switch: Selects either 15- to 500-kc range, or 500- to 1,500-ke range, and moves dial mask for displaying counter for that range.

LIMITER control: In the 0 position, limiter does not operate. As knob is turned clockwise, peak signal impulses are cut off to reduce noise interference. Increased clockwise rotation gives more reduction of signal peaks.

CARRIER LEVEL meter: Indicates level of incoming rf signal. Indication of 0 db when RF GAIN control is fully on corresponds to an input signal of 2 to 5 microvolts.

IF. BANDWIDTH switch: Selects width of the passband in kc for 455-kc if. amplifier stages.

BFO PITCH control: Changes tone of cw signals.

RF BANDWIDTH indicator: Shows maximum width of pass band to which I.F. BANDWIDTH control may be set at operating frequency of receiver.

BFO switch: In ON position, places beat-frequency oscillator in operation.

BREAK IN switch: In ON position, permits break-in operation when installation provides for it.

AUDIO RESPONSE switch: Varies response of audio amplifier. In SHARP position, 800-cps signal is loudest. In MEDIUM position, 3,500-cps signal is highest frequency heard. In WIDE position, all frequencies passed by I.F. amplifiers are heard.

DIAL LOCK: Rotates clockwise to lock FREQ CHANGE control.

FUNCTION switch: When rotated to any position other than OFF, connects receiver to power and selects desired receiver function. The positions and functions are as follows:
STAND BY: Receiver disabled but filaments remain lighted and vfo remains on; receiver ready for instant use.
AGC: Gain is controlled automatically for normal reception.
MGC: AGC disabled; gain is controlled manually by RF GAIN or an external gain control.
SQUELCH: Squelch circuit is connected for silencing receiver when signal falls below threshold determined by setting of RF GAIN control.

LOCAL GAIN control: Controls level of audio-frequency signal heard in phones or loudspeaker.

FREQ CHANGE control: Tunes receiver to the desired frequency within the selected range and operates the frequency dial.

MOTOR TUNE switch: Controls tuning motor that operates the FREQ CHANGE control. Its use aids in making large frequency changes.

RF GAIN control: Controls gain of rf and if. amplifiers. When squelch circuit is operative, controls squelch threshold.

AGC time constant switch: Determines speed of change in gain of receiver for a certain change of signal strength. In FAST position, the time constant is .01 second; in MEDIUM position, .49 second; and in SLOW position, 4 seconds.

OVEN switch: Screwdriver adjustment on back panel. In ON position, turns on vfo
Heater for increased frequency stability.

OPERATION INSTRUCTIONS (from the TM); see author's comments in the []:

Section 1. OPERATION UNDER USUAL CONDITIONS

Warning: The voltages employed are sufficiently high to endanger human life. Every precaution should be taken by personnel to minimize the danger of shock. See that one of the GND terminals on the back panel (9 or 10) is grounded.

21. Starting Procedure

Caution: The power supply of the receiver must be set to the correct ac input voltage. Be sure that all the external connections to the receiver are satisfactory for the desired type of operation.

a. If the receiver is operated under low-temperature conditions, or in a location where there are large changes in temperature, set the screwdriver adjusted OVEN switch on the back panel to ON. When the set is operated in a temperature-regulated building or when maximum frequency stability is not required, set the OVEN switch to OFF.

[The oven increases the power requirement of the entire receiver by more than 50%. In a controlled indoor environment, you will probably not need it].

b. Turn the FUNCTION switch to AGC. Allow the receiver to warm up for several minutes before operating. If the FREQ RANGE and FREQ CHANGE controls are positioned to the same channel on which operation was last performed, the operational procedure may be continued without delay. If further tuning is required, carry out the procedure given in paragraph 22.

22. Tuning Procedure
To tune Radio Receiver R-389/URR to the desired frequency, proceed as follows:

a. Turn the FREQ RANGE control to the range that includes the frequency to be received.

b. Rotate the FREQ CHANGE control to the exact frequency to be received. The MOTOR TUNE control may be used to aid in tuning rapidly to the desired frequency.

Caution: Do not turn the FREQ RANGE switch from the low range (15-500 KC) to the high range (500-1500 KC) if the FREQ CHANGE control has been turned to its highest point. Turn the FREQ CHANGE control to about 15 kc below its highest point before changing ranges.

23. Voice Reception [These directions are for AM mode, of course]

a. Set the BFO to OFF, the LINE GAIN to 0, the RF GAIN to 10; the LOCAL GAIN to 5, the IF. BANDWIDTH to 1, the AUDIO RESPONSE to MEDIUM, the AGC to MEDIUM, and the LIMITER to 0 (off).

b. Adjust the FREQ CHANGE control to obtain the highest reading on the CARRIER LEVEL meter while the desired signal is being received. If it is difficult to tune in the station, turn the IF BANDWIDTH switch to 2 or 4 and tune for best reception. Return the IF. BANDWIDTH switch to 1 and adjust the FREQ CHANGE control for the highest reading on the CARRIER LEVEL meter.

c. If the receiver is subject to vibration or bumping, tighten the dial lock.

d. Adjust the LOCAL GAIN control to obtain desired volume level in the headset and/or loud-speaker.

e. If noise is excessive, rotate the LIMITER control clockwise, as required.

[The "noise limiter" in this series of radios is useless. It introduces enormous amounts of distortion and simply isn't very effective at all].

f. If the signal is fading rapidly, set the AGC switch to FAST.

g. If interference is encountered, set the IF BANDWIDTH switch to obtain the clearest reception.

h. To quiet the receiver between transmissions set the FUNCTION switch to SQUELCH with no signal tuned in. Rotate the RF GAIN control counterclockwise from full-on position, as necessary, to reduce high-level noise. Do not reduce the gain enough to lose the desired signal.

Note: Do not use the SQUELCH position of the FUNCTION switch if desired signals are weak or subject to fading.

[The squelch doesn't work for SSB signals].

i. When the balanced-line output circuit is used to feed telephone lines or other equipment, set the LINE METER switch to the required range and adjust the LINE GAIN control to obtain the desired indication on the I.INE LEVEL meter.

j. If the break-in relay is connected to transmitter control circuits and the receiver is to be disabled during periods of transmission, set the BREAK IN switch to ON.

24. Tone-modulated Radiotelegraph Reception

For reception of tone-modulated radiotelegraph signals, operate the controls in the same manner as outlined in paragraph 23 for voice-modulated signals. Set the IF. BANDWIDTH control at 2 or 1 if it is necessary to reduce interference.

a. Some operators prefer to use the beat-frequency oscillator (bfo) when receiving tone-modulated signals. This practice is particularly useful in the following cases:

(1) When the received signals are too weak to produce good reception, carefully tune the FREQ CHANGE control for the highest reading on the CARRIER LEVEL meter. Turn the BFO switch to ON and adjust the BFO PITCH control for a comfortable tone. Set the AUDIO RESPONSE switch at MEDIUM or SHARP.

(2) When severe interference makes reading the signals difficult, carefully tune the FREQ CHANGE control for the highest reading on the CARRIER LEVEL

meter. Turn the IF. BANDWITH control to 1 or .1 and check the tuning of the FREQ CHANGE control. Turn the BFO switch to ON and adjust the BFO PITCH control for a comfortable tone. Set the AUDIO RESPONSE switch at MEDIUM or SHARP.

b. When adjusting the AUDIO RESPONSE switch, note the fact that only frequencies close to 800 cycles per second are heard when it is set in the SHARP position. The setting of the BFO PITCH control will result in a tone of approximately 800 cycles per second. If a tone higher or lower in frequency is desired, set the AUDIO RESPONSE switch at MEDIUM and adjust the BFO PITCH control to a more pleasing pitch.

c. Do not use the SQUELCH position of the FUNCTION switch when receiving telegraph signals. Use either the AGC or the MGC position.

26. Unmodulated Radiotelegraph Reception

Operate the receiver controls in the same manner as for voice reception, with the following exceptions:

a. Set the BFO switch to ON.

b. Adjust the BFO PITCH control to obtain comfortable pitch.

c. If signal interference is encountered, set the IF. BANDWIDTH switch to the next lower position. To obtain the greater degree of selectivity, set the IF. BANDWIDTH switch to 1 or .1 and the AUDIO RESPONSE switch to SHARP. Readjust the BFO PITCH control to obtain the loudest signal.

d. For manual gain control only, set the FUNCTION switch to MGC and control the sensitivity with the RF GAIN control.

e. When receiving machine code transmission, turn the FUNCTION switch to AGC and turn the AGC switch to SLOW.

f. To reduce effects of fading, turn the FUNCTION SWITCH to AGC and turn the AGC switch to SLOW. To obtain full sensitivity, rotate the RF GAIN control to position 10.

g. Do not use the SQUELCH position of the FUNCTION switch when receiving continuous-wave signals. Use either the AGC or the MGC position.

*[**NOTE:** ADDITIONAL INSTRUCTIONS FOR SSB RECEPTION FOLLOW*]

Like the R-390, the R-389 is perfectly capable of receiving SSB signals. Simply turn on the BFO and set it one hash mark or so in the negative direction from zero for UPPER sideband and a mark or so in the positive direction from zero for LOWER sideband. This is assuming an IF BANDWIDTH KC control setting of 2 kHz. For narrower I.F. bandwidths, use less displacement from zero; for wider bandwidths, use more.

Note: Remember you shouldn't use I.F. bandwidth settings any wider than those shown in the RF BANDWIDTH KC window.

You'll have to ride the RF gain control to keep from overdriving the detection circuits. Since the R-389 has no calibrator, you won't be able to do a calibrate operation with the BFO offset. This means the indicated frequency will be a little off during SSB reception.

27. Stopping Procedure

a. When the receiver is not to be used for a short interval, but is to be maintained in a state of readiness, turn the FUNCTION switch to STANDBY. Use the STANDBY position for only short periods of time. When it is used for about 15 minutes or longer, vacuum tube life may be shortened.

[***CAUTION:*** Since more pointed warnings along these lines were issued concerning R-390s, the author would suggest using the STANDBY position for one purpose only: for a minute or so at start-up to let the filaments warm up before applying B+. It's just too easy to forget and leave the radio on STANDBY for long periods, trashing the tubes.]

b. When the receiver is not to be used for a long interval, turn the FUNCTION switch to OFF.

Additional Comments:

There's all sorts of stuff to listen for. WWVB on 60kHz, CW beacons, data stations (including RTTY), big CW stations down around 15kHz. Those on the coasts will hear lots of beacons. The author remembers "LG" on 233kHz from his time in Long Beach.

The author wound 200 turns of fine enameled wire off old TV tube coils on a wooden frame and has picked up lots of stuff. Connect the loop with regular shielded audio cable. The loop will be directional, and you'll need to find a compromise between maximizing the signal and nulling the local noise. Experiment!

Tip: Shorting pin 5 on the rear panel to ground when the BREAK IN switch on the front panel is set to ON will disable the receiver. Use this with a companion transmitter.

What To Do If It's D.O.A.:

This is one complicated piece of gear! Serious trou-

Author's VLF loop

bleshooting will need the TM. Here's some things you can try:

1) Check the fuses

2) Check the tubes (this should be done anyway with new/unfamiliar equipment). Bear in mind that tubes may check "good" on a tester and not function correctly in the circuit. When in doubt…substitute.

3) Check the 3TF7 ballast tube regulator for continuity. If it opens up, the filaments on the BFO and VFO tubes will quit and the receiver will basically go dead.

4) Check power supply voltages. There should be 180VDC B+ at J601 on the side of the power supply subchassis.

5) Check the filament circuits. There are several series filament strings where one filament failing will cause several tubes to quit working. It's probably best to check them individually, especially where it appears several tube filaments have conked out.

6) Make sure DIODE LOAD terminals 15 and 16 on the back panel are jumped together, and RF GAIN terminals 1 and 2 are similarly jumpered together.

PAL-350 (RFA-1 and PSP-350)

Technical Materiel Corporation RFA-1 Linear Amplifier

Application:

HF linear amplifier and power supply made by Technical Materiel Corporation in the early 1960s. The designation PAL-350 refers to the RFA-1 and PSP-350 taken together.

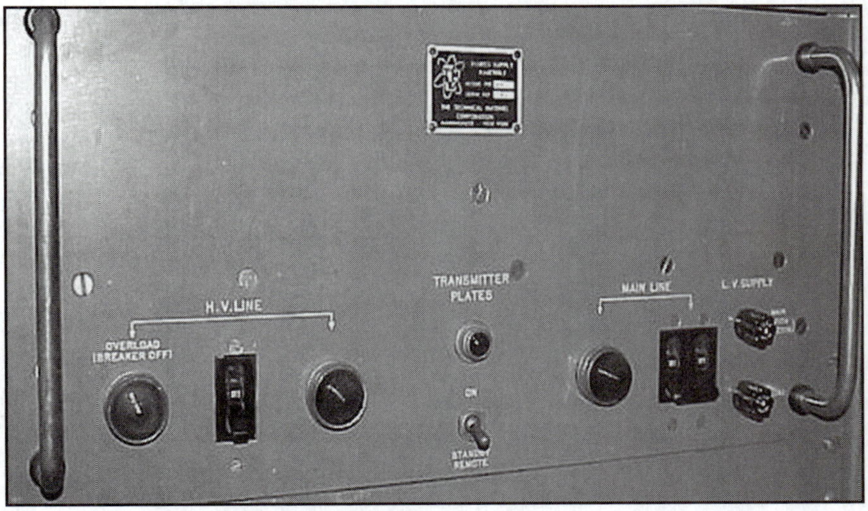

Matching PSP-350 power supply

Tech Specs:

=RFA-1 Linear=

Frequency range: 2 to 32 MHz continuous, bandswitched

Output power: 350 watts PEP, 200 watts CW or FSK/RTTY

Output impedance: matches 30 to 1000ohms at 0-degree angle. Matches 50 to 700 ohms at +/- 45deg angle. (All unbalanced to ground)

Input impedance: 70 ohms coaxial

Input drive: 100mW for rated output

Signal to distortion ratio: better than 40dB down relative to PEP output

Harmonic suppression: Second harmonic at least 50 dB and all others at least 50dB down from PEP output.

ALDC: Automatic load and drive control is provided to limit distortion during high drive peaks or load changes

Size: 5.25" x 19" x 17"

Weight: 30 pounds

Active devices: vacuum tubes, 6CL6, 6146, 4CX250B (2)

Power requirements: voltages usually supplied by the matching PSP-350 power supply.
Plates: 2000V
Screens: 360VDC regulated
P.A. and driver bias: -150VDC regulated
Driver stage plates: 150V

Cooling: filtered, forced-air blower system

Antenna connection: SO-239 (some amplifiers have been seen with Type C)

Tech Manuals: "TECHNICAL MANUAL for LINEAR POWER AMPLIFIER MODEL PAL-350" (TMC document)

Construction Style: rugged, overbuilt chassis and panel. All components are accessible and repairable.

=PSP-350 Power Supply=

Input power: 115/230V, 50/60Hz, single phase

Output voltages: 2000V, 360VDC regulated, -150VDC regulated, 150V

Size: 8.75" x 19" x 13"

Weight: 60 pounds

Active devices: vacuum tubes, 5R4GY, 0A2(2), 0B2(2), 866(2)

Tech Manuals: Covered in PAL-35 manual listed above

Construction Style: rugged, overbuilt chassis and panel. All components are accessible and repairable.

Description:

The RFA-1 amplifier is conservatively rated and extremely well-made.

All components are of the highest quality, and the construction is impeccable. If you think the Commander amplifiers are the ne plus ultra, check out TMC gear.

TMC PAL-350, top view, covers removed

Careful measurements made by a friend on his GPT-1K (the 1kW TMC amp with the PL-172/8295 in the final) showed spur and harmonic figures 10dB better at rated output than those given by TMC. The author believes there's no reason that wouldn't be true of the RFA-1—the amplifier designs are very similar. Your mileage may vary, of course. Even the best output stage can be defeated by gross mistuning!

The RFA-1 amplifier has a 500-watt analog, the RFE-1, which looks identical. In fact the author and his friends haven't been able to find any circuit details that are different, other than the substitution of 4CX350As, and a beefier power supply (PSP-500) capable of running the RFE-1 to 500W PEP output. Taken together, the RFE-1 and PSP-500 comprise the PAL-500. Both setups are differentiated from amateur linears using similar tubes in that they can be driven to full output with 100mw of drive.

Note: Unless noted otherwise, the PAL-350 and PAL-500 are functionally equivalent. Either amp will operate from either power supply provided you pay attention to the caution below.

Caution/Tip: When using a PSP-350 to power a RFE-1, you must set the HV breaker trip to 360mA. It's okay of course to power an RFA-1 with a PSP-500 if you similarly set the PSP-500 HV overcurrent trip to 360mA. The point here is that the PSP-350 has a lower capacity than the PSP-500, and won't deliver enough poop to run the RFE-1 version of the amp to the full 500W.

Operation:

The RFA-1 is generally powered by the matching PSP-350 supply, which provides all the necessary voltages for the linear from 115/230VAC. The interconnect cable between the supply and linear is a fairly rare bird, and the connectors are not easily found. You

DPX-13 HV1 34S at left. DPX-13 HV1 34P at right

may have to change out the connectors and fabricate a new cable. The connectors on the cable are as follows:

RFA-1 end	DPX-13 HV1 34S
PSP-350 end	DPX-13 HV1 34P

This cable is wired pins to corresponding pins, except for Pin 6, which is not wired on either side. Be sure to use high-tension 10KV wire for the A1 to A1 connection, which is the 2000V B+. Here are the detailed pin-outs for those who are minus one unit or the other:

Pin 1	part of metering circuit (P.A. plate voltage)
Pin 2	115VAC (filament transformer and blower)
Pin 3	360VDC (P.A. screens)
Pin 4	part of metering circuit (P.A. plate current, see Note below)
Pin 5	-150V regulated (bias for P.A. stages, adjustable at back of amplifier)
Pin 6	not used
Pin 7	115VAC (filament transformer and blower)
Pin 8	-150V regulated (bias for driver stages)
Pin 9	+150V unregulated (driver plates)
Pin 10	control interlock
Pin 11	control interlock
Pin 12	chassis ground
Pin A1	2000V B+

Note: Pins 10 and 11 are connected in series with cover interlock switches in both the P.A. and the supply. If the loop is broken the system won't function. The loop can be defeated by pulling UP on the interlock switch(s) in the unit when the covers are removed.

Important Note Concerning Pins 1 and 4: Simply connecting 2000V to Pin A1 on the amplifier will get B+ to the P.A. okay, but will not actuate the P.A. plate current meter. The plate current metering circuitry is split between the RFA and PSP. You need both units, interconnected as above, to get a functional plate current meter.

Pin 4 on the RFA-1 connector is connected directly to one side of the plate current meter, which appears in the parts list as a true 500mA meter. The other side of the meter is on chassis ground. Pin 4 on the PSP-350 connector goes through a 56 ohm resistor to chassis ground. The "hot" side of the resistor is connected to the center tap of the HV transformer.

It's actually a little more complicated than this, but the point is that the meter does NOT appear capable of insertion directly in series with the positive side of the 2000V line! If you power your RFA-1 with your own supply, you will need to devise a similar scheme to read current in the negative return for the B+.

The interconnections described thus far provide DC operating voltages, metering circuit returns, and 115VAC only. The control circuits are accessed on terminal strip E101 at the back of the power supply.

Terminal Strip	E101
Terminal 1	Remote actuation
Terminal 2	Chassis ground
Terminal 3	115VAC antenna relay
Terminal 4	115VAC antenna relay
Terminal 5	Remote interlock
Terminal 6	Remote interlock

Note: Ground Terminal 1 to key the system

CAUTION! TMC used 115VAC T/R relays in their systems. When the system is keyed, there is 115VAC on terminals 3 and 4 of the PSP-350 or PSP-500!

Note: Jump terminals 5 and 6 if the PSP-350 is used without equipment containing interlocks.

A last observation for those building your own supplies. TMC has included very effective over-current protection that trips the HV breaker on the PSP-350 when the total plate current goes over 360mA (assuming it has been adjusted correctly for that value). You will need to design in effective overcurrent protection to keep from lunching your 4CX250Bs!

The PSP-350 was originally supplied with 866 mercury vapor HV rectifiers. While 3B28s substitute directly, the real solution is to chuck the hollow state devices altogether and substitute 1N2637 solid state plug-in replacement diode stacks. Last time the author checked, RF Parts still stocked them. They can also often be found surplus.

The interconnections between the PAL-350, PSP-350, and associated SBE-series exciter (see separate SBE section) or your own exciter are straightforward.

Here are the installation and operation sections adapted from the TM; author's comments in the []:

INSTALLATION REQUIREMENTS

a. As shown in the illustration, interconnect J102 and J102 using power cable assembly W101 (supplied) [supplied if you were verrrrrrry lucky...]

b. Connect the AC power cable (supplied) to J101.

c. Connect a coaxial cable such as RF-8/U or RF-11/U depending upon antenna or lead-in, to OUTPUT connector (see figure 2-1).

d. Interconnect PAL-350's associated exciter's output to PAL-350's RFA-1 INPUT connector.

[*Note:* The LOW INPUT connector J202 on the PAL-

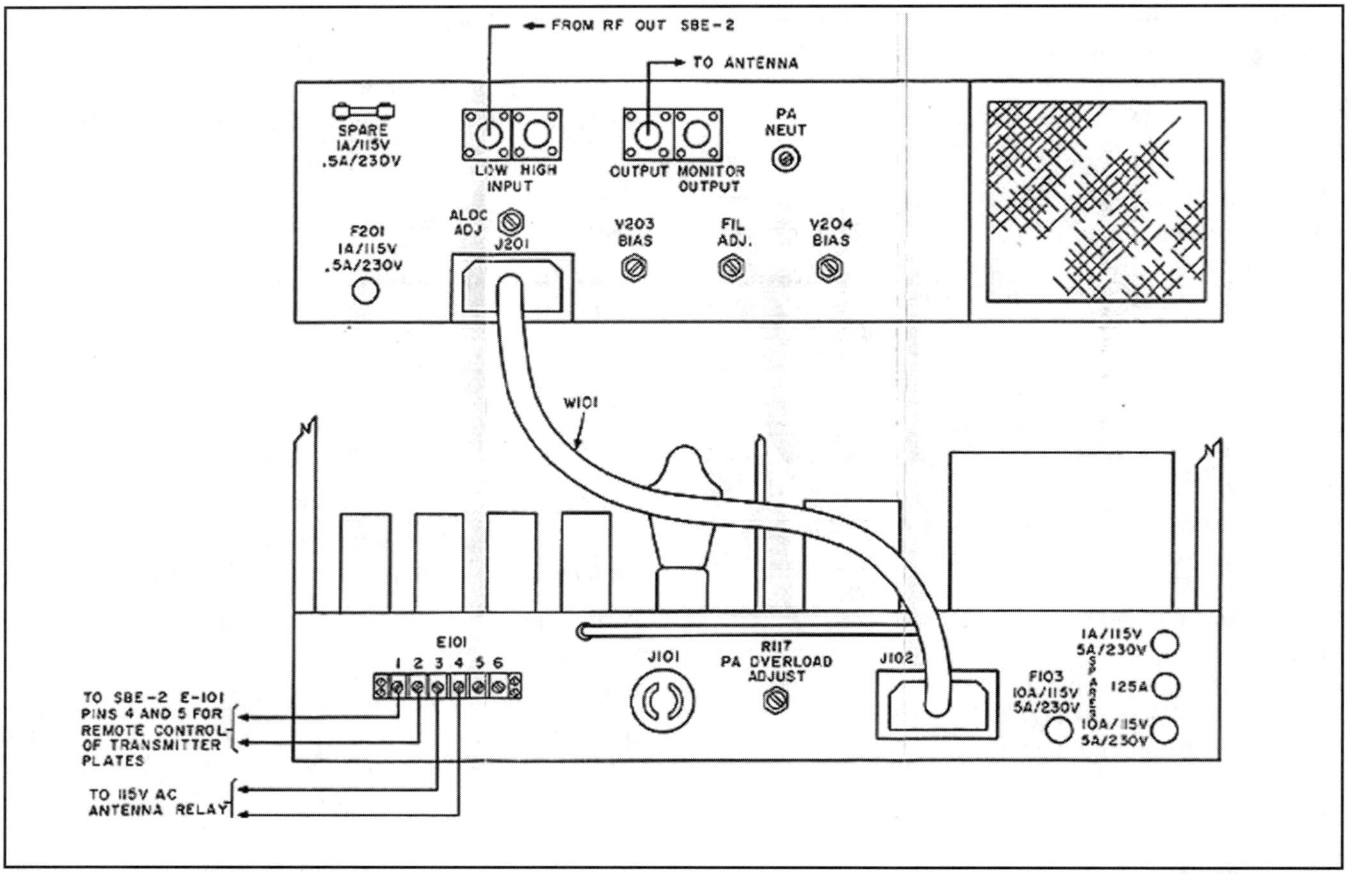

PAL-350 system interconnections

350 accepts 100mW drive for compatibility with a wide variety of exciters such as the SBE-2 or SBE-3, RT-618/URC-35, RF-1310, RF-5020 etc. TMC also provides a HIGH INPUT connector J203 which accepts one watt or less for those with slightly hotter drive sources. In any case, you may still need to devise some sort of attenuator].

DESCRIPTION OF CONTROLS

a. LINEAR POWER AMPLIFIER, RFA-1.

(1) DRIVER/BAND switch: Switches in and out various coils to change the resonate frequency of the driver plate tank.

(2) DRIVER/TUNING control: Varies the driver tuning capacitors to resonate the circuit to the desired frequency.

(3) PA/BAND switch: Sets the PA tank to the proper band by selecting the appropriate tap on the PA tank coil.

(4) PA/TUNING control: Varies the main PA tuning capacitor to resonate the circuit to the desired frequency.

(5) PA/LOADING control: Adjusts the amount of coupling to the load.

(6) MULTI METER switch: Permits selective indication of the PA filament voltage, the PA screen grid current, and the PA DC plate voltage; also, the RF voltage at the input and output of the PA, and the RF voltage at the PAL-350's output.

(7) PA PLATE meter: This meter indicates PA plate current.

(8) MULTI METER indications:

Section	Item
PA	HV - PA DC plate voltage
	ISG - PA screen grid current
	FIL - PA filament voltage
RF	IN - RF PA plate voltage
	DR - RF drive to PA stage
	OUT - RF output voltage

b. POWER SUPPLY, PSP-350.

(1) MAIN LINE ON/OFF switch and circuit breaker: In the ON position, this switch applies 115-volt or 230-volt AC power to the transmitter and provides line overload protection.

(2) HV LINE ON/OFF switch and circuit breaker: When the TRANSMITTER PLATES switch is in the ON position, the HV LINE switch (ON position) applies voltage to the final plates. It also provides overload protection when tripped by excessive PA or screen grid current.

(3) TRANSMITTER PLATES ON/STANDBY-REMOTE switch: In the ON position this switch applies plate and screen voltages to the driver, ALDC and final plates (HV LINE switch in ON position). When the HV LINE switch is in the OFF position, this switch applies plate and screen voltages to the ALDC and driver stages only. The STANDBY REMOTE position permits remote control of these functions.

PRELIMINARY ADJUSTMENTS.

a. Before operation of the equipment in any new installation, it is necessary to allow a twenty-minute warm-up period to vaporize any liquid mercury which might have gathered on the rectifier tube elements. [You won't have to worry about this if you've replaced the 866s with 3B28s or solid state plug-ins. You will, however, still have to wait for the time delay relay in the HV circuits to latch in]. Proceed as follows:

(1) Make cable connections as shown in image above.

(2) Turn HV LINE breaker to OFF.

(3) Turn TRANSMITTER PLATES switch to STANDBY.

(4) Turn MULTI METER switch to PA FIL.

(5) Turn MAIN LINE switch to ON. Allow to remain ON for twenty minutes.

(6) MULTI METER should indicate 30 (equivalent to 6 volts).

[*Note:* The author has never trusted the MULTI METER filament reading. Filament voltage on the 4CX250B finals should be 6.0 volts. Even minor departures from this value can shorten the tube life. A true AC RMS reading wattmeter is best for measuring the tube filaments, but you can get by with a standard VOM. The author measured them for years by pulling the bottom cover, pulling the final tube airbox cover, and measuring the voltage directly on a tube socket. Finally he got frustrated with this and mounted a test point on the back of the amp].

If reading is not correct, use the FIL ADJ. control located on the rear of the RFA-1 chassis for adjustment.

[*NOTE:* Steps 7, b.2, and b.3 below don't have to be done unless you are setting the PA bias, generally a one-time procedure when finals are swapped out, etc.]

(7) Rotate bias potentiometers R213 and R221 and the ALDC potentiometer R228 to the fully counterclockwise position. All these controls are located at the rear of the RFA-1 unit.

b. To operate, proceed as follows:

(1) Carry out steps 6 through 20 in Table 3-1, "Tune Up on Exciter's Carrier." [You don't have to do this if you are just setting the PA bias!]

(2) Switch MULTI METER to read high voltage. Reading at this point will be 3000 volts approximately.

(3) Slowly adjust potentiometer R213 until V203 is drawing 90 mA plate current, using the plate current meter as the indicator. Adjust R221 until total plate current, V203 and V204, is 160 mA. The high voltage will now be 2000 volts approximately. Lock the adjustment potentiometers.

[It's interesting that the settings for the two pots are unequal. The total is 160mA, which would suggest setting the first pot to 80mA, then the second to 160mA. But the author has confirmed that the bias voltages on the two tubes end up almost exactly the same when TMC's procedure is used. It's possible that as the bias is changed from maximum (cutoff) downward on the second tube, the loading on the bias supply is changed so as to equalize the two voltages when the total plate current is 160ma].

(4) Carry out steps 21 through 25 in Table 3-1, "Tune Up on Exciter's Carrier."

TUNE UP ON EXCITER'S CARRIER

Step	Control	Operation	Purpose
1	HV LINE	OFF	No high voltage on RFA
2	TRANSMITTER PLATES	STANDBY	
3	MULTI METER	PA FIL	Preparatory to steps 4, 5
4	MAIN LINE	ON	1/2-hour warmup
5	MULTI METER	Check 30	Indicates 6 volts (filaments)
6	HV LINE	OFF	No high voltage on RFA
7	TRANSMITTER PLATES	STANDBY	
8	PA/TUNING	0	Reference tune up point
9	PA/LOADING	0	Reference loading point

10	MAIN LINE	ON	Applies low voltage to RFA
11	MULTI METER	PA FIL	Multimeter should read 30
12	DRIVER BAND	Proper band	
13	PA BAND	Proper band	
14	MULTI METER	REF DR	Preparatory to step 18
15	Turn on associated exciter which should be tuned and set for proper mode of operation; adjust exciter for low output level.		
16	DRIVER TUNING	Preset per table below.	
17	TRANSMITTER PLATES	ON	
18	DRIVER TUNING	Readjust Step 16 until peak reading is obtained on MULTI METER. Keep peaks below 40 by operation of associated exciter's output control.	
19	Do not readjust DRIVER/TUNING unless another frequency is wanted; Reduce associated exciter output to 0. Check to make sure load or antenna is connected to J205.		
20	HV LINE	ON	Place high voltages on RFA
21	Increase associated exciter's output slowly, observing PA PLATE meter. Meter should indicate idling current of 200ma.		
22	Adjust PA/TUNING	Dip PA PLATE METER with MULTI METER in RF DR position	
23	Adjust PA LOADING as follows: Observe the RF DR on the MULTI METER as the PA LOADING is increased. As the meter reading increases, readjust the PA TUNING for a dip in plate current as before. Continue this loading procedure until the values indicated in the charts immediately below are observed. Use Chart #1 for single tone input and Chart #2 for two tone input.		

Chart #1—SINGLE TONE INPUT used for tune-up
Observed on MULTI METER:

Position	Reading	Actual value
ISG	15	-5ma
RF DR	18	18V
RF PL	10	500V
RF OUT	7	70V

Observed on PA PLATE METER: 240ma.

Chart #2 -- TWO TONE INPUT used for tune-up
Observed on MULTI METER:

Position	Reading	Actual value
ISG	12 to 10	-8 to -10ma
RF DR	25 to 30	25 to 30V
RF PL	18	800V
RF OUT	14	140V

Observed on PA PLATE METER: 300ma. Combinations that may be applied: speech and/or tones, etc. RF output measurements can indicate a wide range of voltages since the actual loading of an antenna may involve a great variety of impedances and reactances.

24 Readjust PA/TUNING per Step 22.

25 Use the OUTPUT CONTROL of the associated exciter to keep the RF PL reading on the MULTI METER from ever exceeding 800 volts. This must be done regardless of the various input combinations that may be applied.

DRIVER TUNING PRESETS BASED ON DESIRED OPERATING FREQUENCY
DRIVER TUNING control setting

Driver Band	1	2	3	4	5	6	7	8	9
2-4 MHz >	2	2.25	2.5	2.5	2.75	3	3.5	3.75	4
4-8 MHz >	4	4.5	5	5.55	6	6.5	7	7.5	8
8-16 MHz >	8	9	10	11	12	13	14	15	16
16-32 MHz >	16	18	20	22	24	26	28	30	32

[This chart is just to get you close on the DRIVER TUNING setting before you start tuning up. In practice, the author just hunts until he gets a peak response!]

For the most part, the tune-up procedure is accurate and effective. Use it along with your previous experience tuning amplifiers.

While the amplifier does have internally derived automatic load and drive control (ALDC), the adjustment is located on the back panel! In the alignment procedure section of the manual, TMC states that for SSB operation, "the control should be advanced as far clockwise as possible without introducing obvious distortion." Also: "While the unit is still fully tuned at some frequency, turn ALDC adjustment potentiometer slowly clockwise until output will just begin to drop off. Increase the drive slowly from the exciter; output must increase only slightly."

The problem with this procedure is that the ALDC adjustment tends to vary (sometimes significantly) with frequency. The author found that running around in back of the blasted rack to mess with the ALDC every time he switched bands was not an option. Not to mention the fact he couldn't see the plate current meter while doing the adjustment. The author has devised a procedure for relocating the ALDC control to the front panel, between the tune and load controls.

During tune up, the author leaves the ALDC control fully CCW (no ALDC) and tunes the amp to full output (350W) in the normal fashion. Then he advances the ALDC until the power out dips to about 300W or so. Generally, this is about the right RF compression level to keep from popping overcurrent breakers. Your mileage may vary. If you use a speech processor that keeps the drive level constant, you may get by with less ALDC, or even none at all.

Setting the PLATE OVERLOAD control. This is R116 on the back of the PSP-350. It should be set to trip at 360 mA plate current for all operating conditions. For adjustment purposes, this current can be obtained by overdriving on single tone input.

Additional Comments:

These amplifiers have the advantage of accepting a wide variety of external anode tube variants. In short, you can stuff in any tubes of the correct form factor with 6V filaments and rock 'n' roll:

4CX250B: originally specified
4X150: first of the 150 series, no ceramic
4X150A: milspec 4X150, some have ceramic in the base, some don't
4X150A/FAA: higher spec version made for the FAA
4X250: non-ceramic older version (?), not usually seen
4CX250R: ruggedized version of the 4CX250B series
4CX250BC: premium version of the 4CX250B (listed by Eimac as requiring closer filament regulation, but have worked fine in the author's transmitter)
4CX250BT: a 4CX250B with a tapped hole in the center pin for inverted operation
4CX350A: 350 watt plate dissipation

Two PAL-350s have been in nightly service in the author's radio room for over a decade. They are conservatively-rated, overbuilt, and a distinct pleasure to use.

What To Do If It's D.O.A.:

Over two decades or so of use (the last with nearly nightly operation), the author has had a few failures, which he'll share now.

1) The glass oil-filled plate bypass and inline plate capacitors tend to leak over time. It's a good idea to check them a couple times a year by running a finger under them. If you detect any oil, the cap(s) should be replaced. The author has replaced one cap that definitely quit working, and a second that was functional but obviously leaky.

WARNING: It should go without saying that you must shut down, completely disconnect the HV, and use a shorting bar from the tube anodes to ground before doing any poking around inside the linear!

2) Two of the three big 4uF 3000V HV filter caps went bad in the author's PSP-350. One was shorted and the other was open. This is bound to happen in forty-plus year-old equipment! He replaced all three with modern (and much smaller) equivalents that weren't full of Pyranol.

3) One of the 7 henry HV chokes failed with a partial short about the same time as the HV cap failures -- the two might have been related. The author contacted Peter Dahl, who made him a drop-in replacement for about $150 shipped.

The three replacement non-PCB caps are shown—they have the same specs as the originals at about half the size. The black Peter Dahl choke replacement is shown next to two of the original parts. Visible at top center is one of the 1N2637 solid-state rectifier replacements.

4) If you get a bum final tube with the plate shorted to the screen, the 10 ohm resistor in series with the screen supply will blow up ka-BOOM! with a sound like a pistol shot. The author figures the HV breaker can't react fast enough. This is an easy fix—just replace the resistor and pitch the tube. It's a heckuva lot better than losing an inductor in the screen supply circuits!

Filament test point modification. The little red "FIL. ADJ." hash mark on the PAL multimeter for setting the final tube filament voltage is not very accurate. When Eimac says the filament voltage is supposed to be 6.0V they don’t mean 5.9V or 6.1V. In addition, Eimac

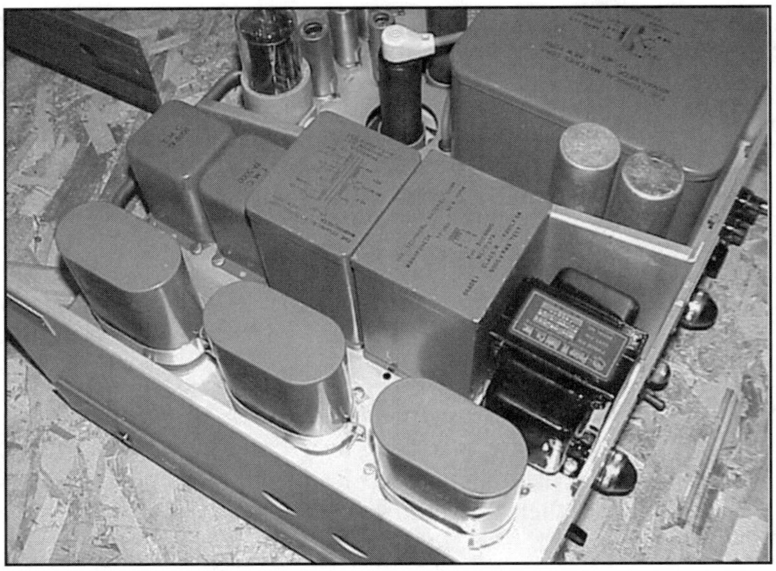

Author's PSP-350, showing fixes

doesn't want you running the tubes without cooling air —even just on filament power. The author got tired of pulling the airbox cover and hanging a voltmeter off the filament line just long enough to set the filaments, so he mounted a binding post on the rear panel.

Location of filament test binding post

The post is connected to the two terminals that are connected together on the filament adjust pot. That's it! Measure the AC voltage between the post and chassis. Setting the filaments after tube changes is a snap now.

Upon hooking up his second PAL-350 system, which consists of an RFA-1 and PSP-500 supply, the author noticed something very strange. With the bias pots set fully CCW— maximum bias, tubes cut off— the plate current meter read 35-40mA upscale. This reading persisted even with the two finals removed from the amplifier!

Sure enough, there was another source of current draw in the metering circuit. TMC did a really screwball thing with the PSP-500— they designed the HV circuit so the plate current meter on the linear reads plate current PLUS bleeder current. This makes no sense, as the operator is not interested in the bleeder current, but only the plate current. Reading the sum just clouds the issue.

In the PSP-500 the shunt resistor R107 and the end of the VR tube string both go to chassis ground instead of to the center tap of the HV power transformer, as in the PSP-350. As a result, this completes a path back through the plate current meter to R106, and reads the bleeder current, even when no plate current is being drawn. This is baloney, and had to be fixed.

The author re-wired his PSP-500 to operate like the PSP-350 in the areas that determine how plate current is sensed. This modification is fairly simple and is presented here. All components are clearly labeled on the PSP chassis.

At gaseous regulator V103
1) Remove ground connection from Pin 7
2) Connect black wire directly to pin 7

At R102
1) Cut black lead at ground end of R102 THAT GOES INTO HARNESS. This is the lead that goes to Pin 7 of V103. LEAVE the lead from the ground lug to R102.

2) Connect the cut wire to the two terminals of R116, the PA OVERLOAD ADJUST pot, that are connected together, and that also connect to brown wires. (The brown wires throughout the chassis are connected to the center tap of the HV transformer).

At C109
1) Cut black ground lead at C109 that goes from the cap to R107.

2) Pull it free of the harness and connect it to the same spot on R116 as above.

Shunt resistor R107 and the end of the VR tube string are now both returned to the center tap of the HV transformer secondary, instead of directly to ground. When used with the PSP-500, both of the author's RFA-1s now read plate current only (with the exception of maybe 5ma of stray induced meter voltage).

RA-6790/GM

Application:

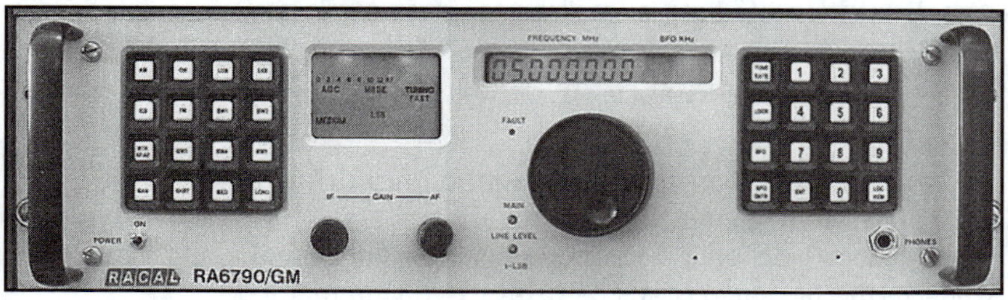

General purpose synthesized rack mount HF receiver. The author has replaced the stock membrane-type keypads with pushbutton-types on his personal unit. More on this in the text. The military nomenclature is R-2174.

Tech Specs:

Complete listing from the Racal service manual:

Frequency Range: 0.5 kHz to 29.999999 MHz

Frequency Resolution: 1 Hz increment

Frequency Tuning: By keyboard entry or continuous tuning with selectable rates, FAST (1000 Hz), SLOW (30 Hz), and FINE (1 Hz) increments; BFO continuous in 10 Hz increments.

Frequency Indication: 8 digit electronic readout of tuned frequency to 1 Hz; 3 digit and sign readout of BFO relative to IF center ±8.0 kHz.

Frequency Stability: ± 5 parts in 108 per 10degC temperature increment over the range of 0degC to 50degC using internal 5 MHz reference oscillator. Provision for an external 1, 5 or 10 MHz reference input/output. 0 dBm nominal into 50 Ohms.

Detection Modes: CW/A1 Continuous Wave; CW/A2 Modulated Continuous Wave; USB/LSB (upper/lower sideband) A3A, A3H, A3J, A2A, A2H, A2J; AM/A3 Amplitude Modulation; A4 (Facsimile) ISB/A3B Independent Side Band (optional); FM/F3 Telephony.

Gain Control Modes:
1) AGC
Control Range: An increase of 110 dB above AGC threshold will result in a change of output of less than 3 dB.

Threshold Range (preset): -113 dBm to -100 dBm

Time Constants: Attack: 20 msecs; Decay Short < 30 msecs; Medium 200 + 100 msecs; Long 3.75 seconds ± 1.25 msecs

2) Manual/Automatic Gain Control: Provision is made on the front panel to select, and by use of the RF Gain Control, to manually control the AGC threshold anywhere within the range of 110 dB above the preset AGC threshold level.

Input Impedance: 50-ohm nominal, 2:1 VSWR Type N Connector

Synthesized BFO: 455 ±8 kHz in 10 Hz steps

Noise Figure: < 15 dB

Dynamic Range: RF > 180 dB/Hz

Sensitivity
(500 kHz to 30 MHz):
1. SSB: -113 dBm (0.5 uV) for 10 dB (S+N)/N Ratio.
2. AM: -99 dBm (2.5 uV) for 10 dB (S+N)/N Ratio in a 6 kHz bandwidth.

Overall Selectivity (Standard Mechanical Filter Complement): A wide variety of mechanical and crystal filters is available for optional requirements such as general purpose, low ripple, low shape factor, controlled delay, or linear phase. The standard filter complement is provided by six mechanical filters that are supplied with the receiver. This includes two sideband filters and four symmetrical filters. The seventh filter slot is linked in order to provide a fifth symmetrical bandwidth defined by the selectivity of the 20 kHz roofing filters. The -3 dB and -60 dB bandwidths are defined as follows:

LSB/USB: -3 dB >450 Hz to >3000 Hz; -60 dB >-600 Hz to <4300 Hz

The remaining five bandwidths are symmetrical:

	-3 dB	-60 dB
BW1	>300 Hz	<2 kHz
BW2	>1 kHz	<4.5 kHz
BW3	>3.2 kHz	<8 kHz
BW4	>6 kHz	<14 kHz
BW5	>20 kHz	<80 kHz

Intermodulation (Out of Band): For signals 100 kHz or more from receiver tuned frequency the third order intercept point is greater than +30 dBm. Second order intercept point is greater than +60 dBm. Below 1.5 MHz these limits may be exceeded.

Intermodulation (In Band): Better than -50 dB for two -10 dBm input signals within the IF passband when measured at the IF or line AF output.

Cross Modulation: The level of a 30% modulated signal, 50 kHz off-tune necessary to cross modulate an on-tune carrier to a depth of 3% shall be greater than +21 dBm (2.5 volts).

Blocking
1. On Tune: Less than 10% distortion for +13 dBm (1 volt) 30% Modulated AM input signals.
2. Off Tune: 1 dB on a 30% modulated on-tune signal when in the presence of a +23 dBm (3 volt) unmodulated carrier 50 kHz off-tune.

Reciprocal Mixing: The apparent noise appearing at the receiver input when in a 3 kHz bandwidth, caused by a 0 dBm signal 100 kHz off tune is less than -100 dBm.

Image and Spurious Rejection: Greater than 80 dB, for signals at least ± 50 kHz from tuned frequency.

Internal Spurious Responses: < -124 dBm

Outputs
1. IF: Frequency 455 kHz, Impedance 50 Ohms. Level -10 dBm nom. Connector BNC.

2. Following outputs available at rear panel audio connector (25 pin Type D).
AF: 100 Hz to 16 kHz for -3 dB.
a. 1W nominal into 8 Ohm load. Distortion < 3% at 500 mW.
b. Monitor: Metered AF line output. 1 mW, 600 ohms balanced <2% distortion. All receiver modes selectable at front panel.
c. Line 1. AF line output. 1 mW, 600 ohms balanced <2% distortion. Operable only with ISB option. All modes select at front panel except LSB.
d. Line 2. AF line output. 1 mW, 600 ohms balanced <2% distortion. Operable only with ISB option, OSB mode.
AGC: Diversity Connection with ground which provides dc. voltage 10 volts to 4 volts to signal levels between threshold and +110 dB. Similar connection for ISB channel when fitted.
Fault: Indication of fault condition is available at the rear panel.

3. Phones: 30 mW into 600 Ohm load. Distortion <3% at 10 mw. Connector: Front Panel Phone Jack

4. REFERENCE OUT: Selectable TCXO Reference frequency of either 1, 5, or 10 MHz (selected by links on A8 assembly). Connector: Rear panel BNC

Environmental
1. Operating Temperature: 0degC to 50degC
2. Operating Humidity: 10% to 85% non-condensing.
3. Altitude: Operation to 15,000 ft.

Power Consumption: Less than 40 Watts (nominal)

Power Requirements: 115/230 Vac ± 10%, 48 Hz to 420 Hz, single phase.

Dimensions: 5.25 in. (13.33 cm) x 19 in. (48.3 cm) x 18.5 in., 19 inch rack mount
Weight (approximate): 32 lbs. (14.5 kg)

Active devices: Solid state, generally available ICs and semiconductors.

Tech Manual: "TECHNICAL MANUAL FOR RA6790/GM HF RECEIVER RCI 84249 (R-2174, R-2174A)"; specs, description, operation, troubleshooting, repair, schematics.

Construction Style: Single-sided, single layer reparable PC boards with easily reachable leaded components and ICs.

Description:

The Racal RA-6790 (R-2174) is an advanced general purpose solid state HF receiver that has been available on the surplus market for several years. In many ways it is a very excellent radio; in others (in the author's opinion) it's utterly miserable.

It's excellent because of the extreme stability and precision it brings to HF operations, and the wide variety of filter options it supports. Also because of the complete control it gives over the BFO, which is tunable in 10Hz steps, making it possible to quantitatively tune in data signals where offsets must be known precisely.

That said, the following points accumulate on the miserable side:

1) Membrane keypads. These things are the worst idea ever to be used on electronic equipment. Failures on the 6790 appear to be due to the poor support the pads get with Racal's mounting method. The author eventually punted after repeated failures and purchasing one new pad (which failed in less than a year), and successfully replaced both with pushbutton-type pads.

2) Under-spec tantalum capacitors. Racal saved a few pennies by specifying tantalum caps with working voltages just barely exceeding, or right at, the impressed voltage. As a result, there have been instances of catastrophic failure in which the caps literally explode with a report akin to a .22cal round going off. These caps are cheap, but the destruction often induced elsewhere by their failure is not. When a tantalum cap failed in the power supply of the author's unit, the resulting power spike on the -15V

line destroyed the matched FETs in the First Mixer Module.

3) Poor heat sinking on the First Mixer transistors. Certain transistors in the 1st Mixer module run outrageously hot, and are constant failure points. Later production receivers have the transistors heat-sinked to the mixer lid.

4) I.F. filter blow-by. Most 6790s encountered are supplied with crystal filters, the standard ones being 3.24kHz symmetrical, 400Hz/1.4kHz/6.0kHz asymmetrical, and sometimes 16kHz asymmetrical. While these filters exhibit excellent group delay characteristics and sound nice, their skirts aren't as steep as mechanical types and, more importantly, they exhibit some pretty annoying blowby. For instance, if you are listening to a signal of average strength or higher in USB mode and tune upward, you'll hear the signal fall out of the passband at 3.5kHz or so above the virtual carrier frequency, as it should. Then it will reappear weakly another 2kHz further upband, as if the filter was 6+ kHz wide with attenuation at the upper end.

The 6790 is the author's main HF receiver, and has been since he finally got all the bugs out. But it has definitely been a love-hate relationship.

Operation:

Operation of the 6790 necessarily centers around the filters since they determine the overall utility of the receiver. The standard 3.24kHz symmetrical SSB filter is too wide for amateur SSB ops, especially in crowded conditions. The author highly recommends installing a narrower filter for SSB. He decided on a 2.1kHz filter from Sherwood Engineering in Denver. Check out Bob Sherwood's web page at:

<http://www.sherweng.com>

These filters are not cheap, but after you install one, you'll think you have a whole new receiver. The author's current filter setup is:

USB/LSB: original 3.24kHz symmetrical filter
BW1: 400Hz asymmetrical
BW2: 1.4kHz asymmetrical
BW3: 2.1kHz asymmetrical
BW4: 3kHz asymmetrical
BW5: 6kHz asymmetrical

The 2.1 kHz filter is used most often, by offsetting the filter in CW mode -- either manually or automatically using the author's passband tuning software.

RACAL 6790 SETUP. These are basic instructions, meant to get you going with a minimum of hassle, and assuming a functional receiver to start with. Verify the frequency standard switch on rear panel is set to "INT" to enable the internal standard. Connect a speaker to Pins 6 and 18 of J3 AF OUT connector on the back panel, or phones to the front panel PHONES jack.

J3 AF OUT connector pin-outs:
Line 1 Output (Used only with ISB Option). Provides USB output during ISB operation; AM/FM/CW/SSB output during non-ISB operation.
1 Output
2 Center Tap
14 Output

Line 2 Output (Used only with ISB Option). Provides LSB output during ISB operation.
3 Output
16 Center Tap
15 Output

Monitor Line Output. Provides AM/FM/CW/SSB output during non-ISB operation; provides switch controlled selection of USB or LSB during ISB operation.
4 Output
5 Center Tap
17 Output

18 Loudspeaker Output. Same as Monitor Line Output.
6 Signal Ground

7,8,9 Ground
21 Main IF Diversity AGC Monitor/Input
22 ISB Lower Sideband Diversity AGC Monitor/Input
23 Fault Indicator (Low indicates Fault)

Selected pages in the operation section from the Racal Technical manual are reproduced here; see author's comments in [].

3.3.2 RA6790/GM Receiver/System Interface Check

The purpose of this check is to verify proper operation of the receiving system once the Receiver is installed in its operating location. Prior to performing this check, ensure that the Receiver Options Installation Check(s) outlined in Paragraph 3.3.1 have been completed. [The assumption is that the receiver is operative]. To perform the Receiver/System Interface Check:

a. Physically install the Receiver in its operating location.
b. Connect the system RF signal source (antenna; output from multicoupler, etc.) to the RF IN jack J1 on the receiver rear panel.
c. Connect the system remote control device (if used) to connector A6A1W1J1 on the receiver rear panel. [More on remote control via the serial interface later…]

d. Connect the system audio line to the AF OUT connector J3 on the receiver rear panel. [Pins 16 and 8 are the speaker output].
e. Connect the system external reference frequency source (if used) to REF IN/OUT connector J7 on the receiver rear panel and set the INT/EXT REF switch S2 to EXT.

NOTE: If the Receiver's internal reference is to be used, connector J7 is not used and S2 is set to INT. [Use the INTERNAL standard to start with, to cut down the operational variables].

f. If the Receiver IF is to be sampled for system use, connect the sampling cable to IF connector J2 on the receiver panel.

NOTE: If the IF is not to be sampled, connector J2 is not used.

g. Connect a suitable ground wire to the GROUND terminal on the receiver rear panel.
h. Connect a headset or plug-in speaker to the PHONES jack on the receiver front panel.
i. Connect the Receiver to a suitable power source by connecting the power line cord to the A10J1 assembly on the receiver rear panel.

CAUTION: Ensure that the voltage value printed on the pc wafer in the A10J1 assembly matches the available line voltage as outlined in Section II, Paragraph 2.3.4.1. Failure to match the Receiver to the available line voltage may result in blown fuses and/or receiver errors.

j. Energize the Receiver by switching the POWER ON switch on the receiver front panel to the ON position.
k. Observe the Mode LCD. The word REMOTE should not be present. If the word REMOTE appears in the display, momentarily depress and then release the LOCAL/REMOTE pushbutton switch on the righthand keypad. The word REMOTE should disappear. If it does not, refer to Section V of this manual. [In that case, there is a fault].
1. Initialize the Receiver. (Refer to Paragraph 3.3.1 for initialization procedures.)

[***NOTE:*** They are referring to running the BITE test: Depress and hold LOCK on the right keypad. While depressing LOCK, depress the AM button on the left keypad. Release both buttons. Observe the FREQUENCY MHZ LCD. Digits in the display should be rapidly changing and should continue to change for about a minute. If, during this period, the display stops changing and displays a two-digit number, the BITE has found a receiver fault].

The Receiver will initialize itself in approximately one minute. If after initialization, only a two digit number is displayed in the frequency LCD, and the word REMOTE remains in the Mode LCD, refer to Section V of this manual. [BITE has found a receiver fault].

m. Energize (as required) the remaining equipment in the system.
n. Tune the Receiver to a known signal such as a local AM radio station (refer to Paragraph 3.3.3.1 for tuning procedures) and optimize reception so that the signal is clear and intelligible in the headset or speaker. If no signal is heard, try a different signal. If the signal still cannot be heard, check the antenna, multicoupler (if used), antenna cables, RF IN connector, etc. If still no signal is heard, refer to Section V of this manual. [Probable receiver fault].
o. Verify that the AF output from the Receiver is present and is being processed by the system AF analysis equipment.
p. Verify that the IF output (if used) is present and is being processed by the system IF sampling equipment.

NOTE: If either the IF (if used) or AF signals are not present at the input to the analysis equipment, check the appropriate cables and/or connectors. (Especially verify the wiring of mating connector for J3 AF OUT on the receiver rear panel).

q. Using the remote controller, manipulate receiver functions. (Refer to Paragraph 3.3-3.2 for remote control operation.) If the Receiver does not respond to commands from the remote controller, check cables and connections, especially the mating connector for A6A1W1J1 on the receiver rear panel. [More on remote control ops using the serial interface later…]

Once satisfactory results have been obtained in steps j through q, the Receiver System Interface Check is complete and the system is operational.

3.3.3 LOCAL/REMOTE Control Selection

There are two possible control conditions for operation of the RA6790/GM HF Receiver. In the first or LOCAL control condition, the receiver functions are controlled by selecting modes, levels, bandwidths, etc., using controls physically mounted on the Receiver itself. In the second or REMOTE control condition, the functions are selected by using a coded message sent by a remote operating terminal. It should be noted that when the Receiver is placed in the REMOTE condition, the controls on the Receiver itself are non-functional. Determination of which condition the Receiver is in (ie. LOCAL or REMOTE) can be made simply by checking the lower-right corner of the receiver's mode LCD. If the Receiver is in REMOTE, the word REMOTE will be displayed. If the Receiver is in LOCAL, no word will be displayed.

Switching between the two conditions is achieved by pressing and releasing the LOCAL/REMOTE pushbutton switch on the righthand keypad.

3.3.3.1 LOCAL Control Operation

The following procedures are to be used to operate the RA6790/GM HF Receiver in the LOCAL control condition:

a. Ensure that all desired options have been installed and that the Receiver Options Installation Check has been performed in accordance with Paragraph 3.3.1.
b. Turn POWER ON switch to ON position. Indications that the Receiver is energized include:

(1) Edgelight illuminated.

(2) Some data shown in Liquid Crystal Displays (LCDs).

(3) Sounds in audio device (headphones or speaker). If Receiver does not energize when POWER switch is toggled, refer to Section V of this manual.

c. Establish LOCAL control. Momentarily depress the LOCAL/REMOTE pushbutton switch on the righthand keypad while observing the lower-right comer of the mode LCD. Depress the switch again. Note that the display will show the word REMOTE on alternate pressings of the switch, while REMOTE disappears on alternate pressings.

When the Receiver is under LOCAL control, the word REMOTE is not shown in the display. If the word REMOTE cannot be alternately displayed and erased, refer to Section V of this manual.

d. Initialize the Receiver. The RA6790/GM Receiver has a built-in capability to self determine the values of the IF bandwidth filters installed, and to assign these filters to particular bandwidth pushbuttons (BW1 through BW5 on the lefthand keypad). These bandwidth assignments normally result in the narrowest bandwidth filter being assigned to the BW1 pushbutton with increasingly wider bandwidths being assigned to BW2, BW3, etc. in ascending order.

To initialize the Receiver:

(1) Press and hold the LOCK pushbutton switch on the righthand keypad.

(2) While holding down the LOCK pushbutton, press the AM pushbutton switch on the lefthand keypad.

(3) Release both the LOCK and AM pushbuttons simultaneously. The normal indications that the Receiver is in the initialization process are:

(a) The word REMOTE will appear in the Mode LCD.

(b) The frequency LCD will begin displaying a series of rapidly changing digits in both the FREQUENCY and BFO portions of the display.

(c) Rapidly changing sounds and tones in the audio device (speaker or headset).

(d) Changing modes and levels in the mode LCD.

Indication that the initialization process is complete is obtained by observing the word REMOTE in the mode LCD. When REMOTE is erased and the Receiver returns to the displayed values that were present prior to the initialization process, the process is complete. (NOTE: The Initialization process takes approximately one minute to complete.) If during the process the display stops changing and a two-digit number appears and remains in the FREQUENCY LCD, BITE has detected a problem in the Receiver. Refer to Section V of this manual.

e. Select Desired Frequency (00.500000 MHz to 29.999999 MHz). There are two methods for selecting any desired frequency. The first of these is to enter the frequency digits directly using the ENTER pushbutton switch and the numeral pushbutton switches on the righthand keypad. The second is to select a tune rate and manually tune to the desired frequency using the manual tuning knob. The procedures for each of these methods are as follows:

(1) Direct Entry:

(a) Momentarily depress and release the ENTER pushbutton switch on the righthand keypad.

(2) Manual Tuning:

Ensure that the word LOCK is not present in the mode LCD. If it is present, momentarily depress then release the TUNE RATE pushbutton switch on the righthand keypad. The word LOCK should disappear. If LOCK cannot be removed from the display, refer to Section V of this manual.

f. Select Tune Rate. This is accomplished by alternate depressions of the TUNE RATE pushbutton switch. Observe the right side of the Mode LCD. One of three displays should be present under the word TUNING: SLOW, FAST, or no display (refer to Paragraph 3.2.9). These three displays correspond to three rates of change of frequency caused by rotation of the Manual Tuning Knob as follows:

No display: Fine tuning, 1Hz steps
SLOW: Slow tuning, faster than fine tuning. (Usually 30Hz steps)
FAST: Fast tuning. (1kHz steps)

Rotate Manual Tuning Knob to desired frequency changing tune rates as desired.

EXAMPLE:
Receiver is set at 20.179000 MHz, and the desired frequency is 10.853000 MHz.

(1) Select FAST tune rate.

(2) Tune to approximately 10.8 MHz, using manual tuning knob.

(3) Select SLOW tune rate.

(4) Tune to approximately 10.85 MHz using Manual Tuning Knob.

(5) Select fine tuning rate.

(6) Tune to 10.853000 MHz using Manual Tuning Knob.

g. Select Detection Mode. (AM, CW, USB, LSB, ISB U/L, FM). Detection modes are selected by momentarily depressing and then releasing the appropriate pushbutton switch on the lefthand keypad. A discussion of operation in each mode follows.
h. AM.

(1) Depress and release the AM pushbutton switch. Indications that the Receiver is in the AM mode are: the word AM under MODE in the Mode LCD; a one or two digit number next to BW KHZ in the upper right comer of the Mode LCD.
If unable to achieve these indications, refer to Section V of this manual.

(2) Select desired IF bandwidth (BW1 through BW5). IF bandwidths range from narrowest (BW1) to widest (BW5), with values of the IF bandwidth filters actually installed in the Receiver. To select a bandwidth, depress and release one of the bandwidth selection pushbutton switches (BW1 through BW5). The bandwidth of the filter selected will appear next to the BW KHZ words in the mode LCD. If no number appears for a particular bandwidth then either no filter has been installed for that position or there is a problem with the Receiver.

NOTE: If a full complement of filters (7) has been installed and a no-number bandwidth appears, re-initialize the Receiver. If the condition persists, refer to Section V of this manual. If less than a full complement of filters (less than 7) has been installed, the no-number condition on one or more bandwidths is normal and should be disregarded except to ensure that, for operating, a bandwidth with a number displayed must be selected.

(3) Select Gain Control (AGC, MAN, SHORT, MED, LONG). There are three choices for IF Gain Control:

(a) Automatic (AGC)
(b) Manual
(c) Automatic with Manual Threshold

A discussion of these choices follows.

i. Automatic (AGC). AGC automatically establishes and maintains a desirable IF signal level. Through the use of operator selectable response times, it also establishes the time it takes for the AGC circuitry to respond to IF signal level changes. To select AGC:

(1) Observe the mode LCD under the word AGC. If the word MAN is displayed, momentarily press and release the MAN pushbutton switch on the lefthand keypad. the word MAN should disappear. In addition, another word (LONG, MED, or SHORT) should be displayed under AGC on the display.

(2) Select the response time by depressing and releasing the desired pushbutton switch (LONG, MED, or SHORT) on the lefthand keypad. The Receiver is now under automatic gain control.

j. Manual. Manual gain control allows the operator to set a desired IF signal level, and disables the automatic response and adjustments to changes in IF signal level. To select manual gain control:

(1) If the word MAN is not present under the word AGC on the Mode LCD, momentarily depress and release the MAN pushbutton switch on the lefthand keypad. The word MAN will appear on the display.

(2) If under the word MAN another word (LONG, MED, or SHORT) is present, momentarily depress and release the pushbutton switch on the lefthand keypad which corresponds to the displayed word. Once the word MAN is the only displayed word under AGC on the mode LCD, the Receiver is under manual gain control.

(3) Observe the level meter display in the upper left corner of the Mode LCD. If the word RF is displayed proceed to (4). If the word AF is displayed, momentarily depress and release the METER RF/AF pushbuttons switch on the lefthand keypad to change the display to RF.

(4) Once RF is displayed in the mode LCD and the Receiver is under manual gain control, the IF GAIN potentiometer may be adjusted to the desired signal level by rotating the knob while observing the RF level indication on the meter in the mode LCD.

k. AGC with Manual Threshold. This gain control capability allows the operator to manually set the desired IF signal level which the AGC automatically maintains. Response times are operator selectable through the LONG, MED or SHORT pushbutton switches on the lefthand keypad. To select AGC with manual threshold:

(1) Select manual IF gain control as directed in j.

(2) Set desired signal level as desired according to j (4).

(3) Momentarily depress and release the desired response time pushbutton switch (LONG, MED or SHORT) on the lefthand keypad.

The Receiver is now under AGC with manual threshold control. Indications that the Receiver is in this condition are:

(a) The word MAN is displayed under the word AGC in the mode LCD.
(b) The word LONG, MED or SHORT is displayed under the word MAN in the mode LCD.

1. CW. Select the CW mode by depressing and releasing the CW pushbutton switch. Indications that the Receiver is in the CW mode are:

(a) The word CW under MODE in the mode LCD.
(b) The appearance of digits under BFO KHZ on the FREQUENCY LCD.

Select BFO (Beat Frequency Oscillator) Frequency (-8.0, 0, +8.0 kHz). The BFO may be operated in either of two conditions:

(a) BFO - adjustable
(b) BFO CENTER - non-adjustable

NOTE: BFO functions can be controlled by the operator only in the CW mode, and are disconnected from their controls in all other modes.

m. The BFO condition allows the operator to tune the BFO manually to achieve the desired audio frequency for received CW signals. The BFO CENTER condition allows the operator to zero-beat (fine tune the Receiver to the exact RF) the incoming CW signal. To select BFO Frequency (adjustable):

(1) Observe the BFO kHz portion of the FREQUENCY LCD. If a three digit number and "+" or "-" sign is displayed proceed to step 2. If a single zero preceeded by a "+" sign is displayed, momentarily depress and release the BFO CENTER pushbutton switch on the righthand keypad. This should cause the display to show a three digit number preceeded by either a "+" or "-" sign.

(2) Observe the mode LCD and note the display under the word TUNING. If the word BFO is displayed, proceed to step 3. If there is no word displayed under tuning, or if SLOW, FAST, or LOCK is displayed, momentarily depress and then release the BFO pushbutton switch on the righthand keypad. The word BFO should now be displayed under the word TUNING on the mode LCD.

(3) Rotate the manual tuning knob while observing the BFO KHZ portion of the FREQUENCY LCD. The numbers displayed should change as the knob is rotated.

(4) Select the desired BFO Frequency (+8.0 kHz to -8.0 kHz) either by watching the BFO KHZ display while rotating the knob until the desired BFO Frequency is attained, or by listening to the audio signal while rotating the knob until the desired audio frequency is heard.

n. To select BFO CENTER (for zero-beating the incoming signal):

(1) Observe the BFO KHZ portion of the frequency LCD. If a single zero preceeded by a "+" sign is displayed, proceed to step (2). If a three digit number preceeded by a "+" sign is displayed, momentarily press and then release the BFO CENTER pushbutton switch on the righthand keypad. A single zero preceeded by a "+" sign should now be displayed.

(2) Observe the mode LCD and look under the word TUNING. If there is no word displayed under TUNING or if SLOW or FAST are displayed, proceed to step (3). If either the word LOCK or the word BFO is displayed, momentarily depress and then release the TUNE RATE pushbutton switch on the righthand keypad. The word LOCK or BFO should disappear.

(3) The incoming signal may now be fine-tuned by slowly rotating the manual tuning knob until the audio tone from the signal is nulled.

(4) Momentarily depress and then release the BFO CENTER pushbutton switch and observe that a three digit number preceeded by a "+" or "-" sign appears in the BFO KHZ portion of the FREQUENCY LCD.

(5) Momentarily depress and then release the BFO pushbutton switch on the righthand keypad. The word BFO should appear in the mode LCD.

(6) Slowly rotate the manual tuning knob until the desired BFO Frequency is obtained.

(7) Momentarily depress and then release the LOCK pushbutton switch on the righthand keypad. This disables both the manual tuning and BFO adjustment functions preventing loss of the signal if the manual tuning knob is inadvertently rotated.

(8) Select IF bandwidth as outlined in h. (2).

(9) Select gain control as outlined in h. (3).

o. USB or LSB

(1) Select either upper sideband or lower sideband mode by momentarily depressing and then releasing

the appropriate (USB or LSB) pushbutton switch on the lefthand keypad. Either USB or LSB should appear under the word MODE on the mode LCD.

(2) All other functions in USB or LSB are identical to the AM mode (except for IF bandwidth, which is fixed in either the USB or LSB mode).

p. ISB U/L (available only if ISB option A5 assembly is installed).

(1) Select the independent sideband mode by momentarily depressing and then releasing the ISB U/L pushbutton switch on the lefthand keypad. The word I-USB or I-LSB should appear under the word MODE on the mode LCD. Subsequent depressing and releasing of the ISB U/L pushbutton will alternately display I-USB or I-LSB.

NOTE: If the ISB option is not installed, depressing the ISB U/L pushbutton switch will have no effect on receiver operation.

(2) Other functions are identical to those of the AM mode.

NOTE: Each sideband is independently controlled in this mode. Therefore separate AGC settings are possible for both I-USB and I-LSB.

q. FM.

(1) Select the FM mode by momentarily depressing and then releasing the FM pushbutton switch on the lefthand keypad. The word FM will appear under the word MODE in the mode LCD.

(2) Other functions are identical to those of the AM mode.

r. Forced IF Bandpass Filter Bandwidth Setup

In case BITE results are unsatisfactory, the forced IF bandpass filter bandwidth setup can be initialized by simultaneously depressing the LOCK and ISB U/L pushbutton switches, and observing the AUX indicator displayed in the MODE LCD. This process accesses RAM within the microprocessor circuit card assembly (A6A2) for loading of the IF bandpass filter bandwidth complement. It should be noted, however, that this setup cannot be accomplished during the BITE routine. In addition, the setup can be terminated (if desired) before all seven filter slots FL1-FL7 have been changed, and when all seven filter slots have been loaded, the AUX indicator will go off (disappear) automatically, thus, disabling any further changes and resetting the receiver to normal operation.

(1) Symmetrical Filter Entry

NOTE: Entry of symmetrical filters is accomplished using the 0-9 numeral pushbutton switches, in 100 Hz steps, delimited with the ENTER pushbutton switch. For example, to enter a .4 kHz (400 Hz) filter, the "4" and "ENTER" pushbutton switches would be depressed; to enter 20 kHz, the "2", "0", "0" 1 "ENTER" pushbutton switches would be depressed (200 x 100 Hz = 20 kHz).

(a) Depress and release the appropriate numeral pushbutton switches (0-9) on the right-hand keypad, then depress and release ENTER pushbutton switch.

NOTE: In the case of blank filter slots, load by depressing the ENTER pushbutton switch only.

(2) USB and LSB Filter Entry

(a) Depress and release USB and/or LSB pushbutton switch(s) on the left-hand keypad, and observe AUX indicator displayed go off in the MODE LCD.

NOTE: In the case of blank filter slots, load by depressing the ENTER pushbutton switch only.

(3) Terminating Forced IF Bandpass Filter Bandwidth Setup:

(a) Depress and release ISB U/L pushbutton switch on the left-hand keypad, and observe AUX indicator displayed go off in the MODE LCD.

The preceeding should be enough to get you started in good shape. Anywhere the narrative says "refer to Section V of this manual," they're telling you to go to the Maintenance section. Serious troubleshooting is beyond the scope of this book.

Additional Comments

The tech manual is really necessary to get the most out of your 6790. To see about obtaining a complete manual copy, try emailing Max-Gain Systems:

<http://www.mgs4u.com>

The author has made several modifications to his 6790. These were necessary to return the receiver to usability, or to ensure its continuing usability.

KEYPAD REPLACEMENT. The author got disgusted with constantly fixing/replacing the keypads on his radio. The following outlines how he replaced the MODE (left hand) pad. Later, he replaced the FREQUENCY pad. The panel holes are 3" by 3", which is a standard keypad size. Grayhill 16-button pads were found at the local surplus place. These phone-type pads will have to be electrically modified to give sixteen separate switches and one common ground. The author modified his pad by cutting traces to isolate the sixteen buttons, then jumpering together the other side of all the buttons so there is one common return. The author's keypad had a plastic stub on each corner, probably meant to be shoved through mounting holes, then melted. The author just drilled holes in the lexan behind the 6790 panel, then put a glob of epoxy on each stub after mounting the pad.

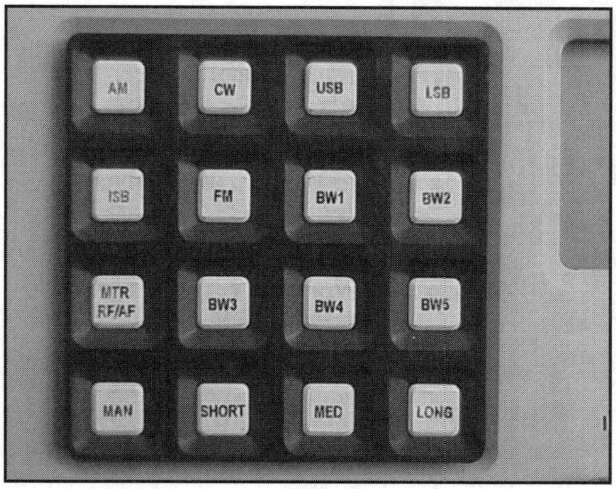

Mode Pad mounted on panel

Paint over the telephone markings on the keys with white-out. Then make new key labels by typing text into Word, printing it out, then cutting out and affixing to buttons with clear acrylic lacquer.

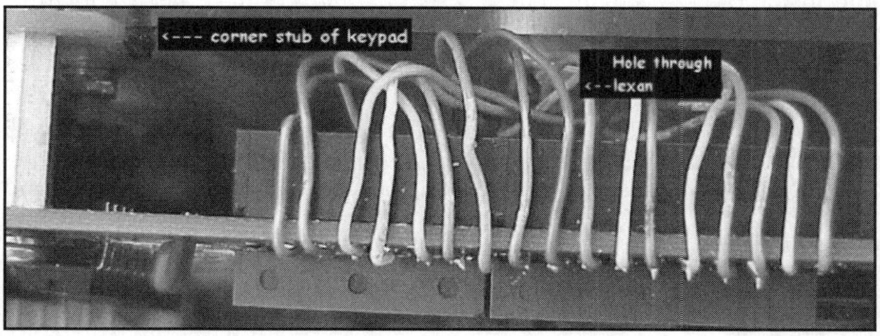

Method of wiring in new Mode keypad

Drill a big hole in the lexan for the wires, run them through, and solder them to the PCB behind the original connector. The new pad is just mounted right to the lexan, via the four plastic corner stubs. The pin-outs for A9J3, the mode pad connector, as given on the schematic, are completely wrong. Here is the correct way to wire the mode pad:

Function Button vs. A9J3 pin:

LONG	17
BW5	16
LSB	15
BW2	14
MED	13
BW4	12
USB	11
BW1	10
SHORT	9
BW3	8
FM	7
CW	6
MAN	5
METER RF/AF	4
ISB U/L	3
AM	2
ground	1

Replacement of the FREQUENCY keypad is done in a similar fashion.

MIXER FAN ADDITION. Q2, Q3, and Q4 in the First Mixer all run obscenely hot. On top of that, Racal has sealed up the module tight for RF considerations, leaving the heat nowhere to go. This is a great prescription for a future meltdown.

The typical computer processor fan is only about 1/2" tall, and will fit nicely into the area shown. The air gets through at the arrow, and through the space between the tops of the filters and the cover to get to the transistors in question at left. The "5V Unregulated" on the Racal power supply is really about 11VDC nominal. With the fan connected and drawing 80mA, the voltage only dropped a few tenths of a volt.

Hint: To locate the fan, lay it inside in the correct position, then use a carpenter's square against the main chassis edges to lay down index marks even with the fan holes on surrounding module covers. Then replace the First

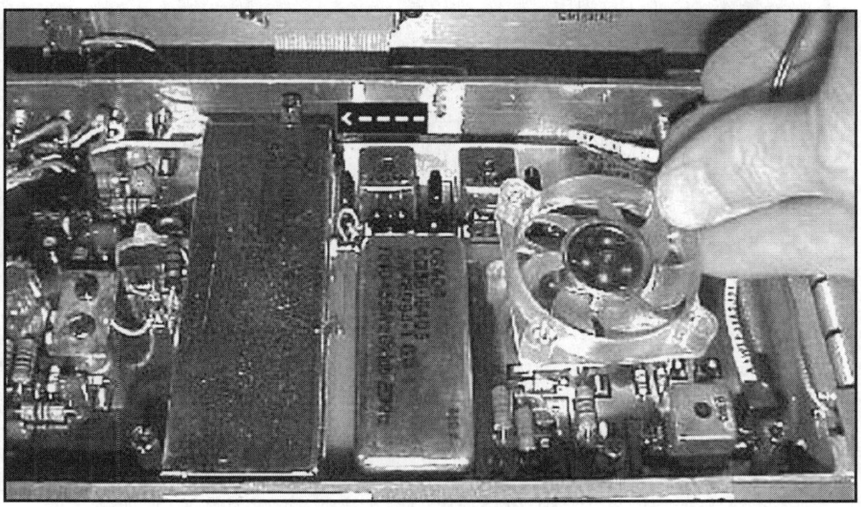

Enough room in the mixer module for a CPU fan

Fan mounted on Mixer Module lid

Mixer cover and make locating marks using the square and the marks previously made. Once the four fan holes are drilled, the main hole can be traced using the inside of the fan housing.

Note the use of common aluminum screen for RF screening; this seems to work fine. The lid material is very hard, probably stainless steel. The best way to make the big holes is drill holes around the inside of the circumference, cut across the holes with a Dremel, then finish up the hole with a power reamer and a hand file.

Location of hole drilled through chassis for fan 12VDC power

Run power from the "Ground" and "5V UNREG" terminals inside the power supply directly through to mixer compartment. Drill a hole at the point shown at the arrow straight through the power supply chassis and the casting below, which brings the wire out inside the mixer compartment at a convenient place to hook to the fan. You'll get a very un-6790-like sound when you turn your receiver on now – the faint whirring of a fan. This mod has been in service for two years with no hint of degraded mixer performance.

FIRST MIXER FET FIX. The SD215DE FETs inside the mixer can, are electrically fragile. The author's failure was due specifically to a failure of the -15V output in the receiver power supply, causing the voltage to spike and take out the devices. The spike was due to a bad tantalum bypass capacitor on the -15V three terminal regulator.

Note: The author replaced all the bypass caps on the three terminal regulators in the power supply with 50VDC rated components. Highly recommended.

Make sure the mixer can in the A2 module is what actually failed. Q1, Q2, and Q3 have been known to go bad—as mentioned above, they run ridiculously hot. FET failure symptoms include a very high noise level during no signal conditions, and 3-4uv sensitivity or worse. With no antenna hooked up, the author's receiver sounded like an unsquelched FM set. Once you've determined the mixer can is bad, pull the can from the board and desolder the lid. Use a hefty soldering iron and push an X-Acto knife under the lip, while prying up and moving the iron next to the knife.

Unsolder the top wires first on both sides, pull the board out slightly, then desolder the rest of the wires. Once the board is out, pull the FETs. You may be able to find the discrete FETs at Radio Shack's catalog

FET board inside mixer can, before any work

sales. Note that if you use separate FETs you won't be guaranteed they are matched. The trimmers inside the can are for matching them but there is no data in the tech manual on how to do that.

A better solution is replacing the discrete FETs with an SD5000N quad-FET chip. That's because the four FETs share the same substrate and will be matched. The author obtained this chip from Radio Shack several years ago. It was a cheaper solution at $5 plus shipping, but definitely tougher to install mechanically.

FET board, with added leads

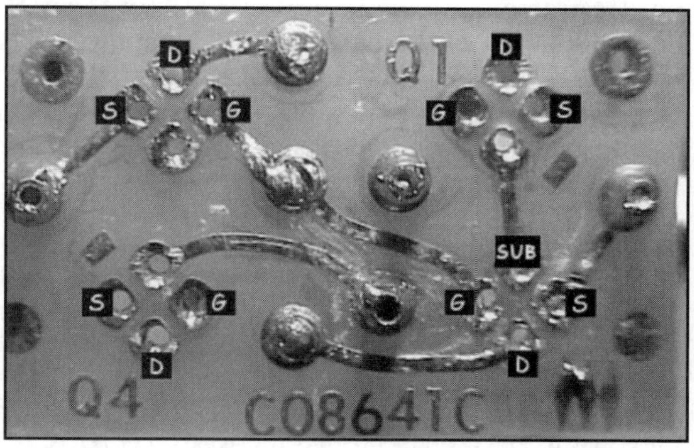

FET board, following removal of FETs, with lead holes labeled

Leads soldered to SD-5000 chip

Note this is the Q1 side, which has enough room to mount the chip "dead bug style" after the thin portion of the chip leads are cut off. The (D)RAIN, (G)ATE, and (S)ource for the four FETs are labeled, as well as the common (SUB)STRATE connection. The four substrate leads on the original FETs were connected together, but the hole labeled "SUB" in the illustration is the most convenient for wiring to the chip. You'll need to install the chip in the inverted position after adding fine wires to the board.

The author solders to IC pins by first tinning them very rapidly, and separately tinning the wires to be attached to the pins. He then rapidly touches the iron to the pin and wire using no extra solder.

It takes some finagling, long-nose pliers, and some pretty fine work to shoehorn this in.

Do not mess with the four trimmers (three of which are visible). Their original settings will probably not correspond exactly to the perfectly matched FET quad. The author has not noticed any degradation in mixer performance in any case.

FET board, re-installed into mixer can

Note: The mixer can must be covered to avoid picking up synthesizer hash. Instead of resoldering the cover, you may want to pursue this simple solution in case you have to get in there again! Simply lay a thin piece of cardboard in the can for an insulator, then wrap two layers of tin foil over the top and tape it down on the sides with duct tape. This will shield the mixer components. If the FET(s) ever fry again, the chip can be reached with no de-soldering, and thus without removing the mixer can from the board.

SERIAL INTERFACE SETUP. If your 6790 came with an A6A1 board, it can be remotely controlled via RS-232 link. The A6A1 board attaches directly to the microprocessor board, both of which are located on the inside left-hand chassis wall.

First, you'll need a connector. Some interface (A6A1) boards have a cheesy telecom connector for J1. The military version of the radio accepts a push-and-turn connector. The nomenclature is MS3476L16-26S. See the Appendix C for possible sources.

Cable construction data for PC to Racal 6790 RS-232 interface:

6790	FUNCTION	CONNECTION AT PC
A	system ground	not used
B	data out a	to pin 3, receive data on RS-232 at PC
C	data out ground	to pin 7, ground, on RS-232 at PC
D	data out b	not used
E	ground	not used
F	data in a	to pin 2, transmit data on RS-232 at PC
G	data in ground	to pin 7, ground on RS-232 at PC
H	data in b	jumpered to J
J	ground	jumpered to H
K	RX# d1-1	grounded to b (little b)
L	d1-2	grounded to b
M	d1-4	grounded to b
N	d1-8	grounded to b
P	d2-1	grounded to b
R	d2-2	grounded to b
S	d2-4	grounded to b
T	d2-8	grounded to b
U	parity select	left open (no parity)
V	even/odd parity	left open (don't care)
W	baud rate b4	leave open (high)
X	baud rate b3	leave open (high)
Y	baud rate b2	leave open (high)
Z	baud rate b1	grounded to b
a	ground	not used
b	ground	used for all pin grounding above
c	system ground	not used

Notes:

1) GROUNDED is a LOW or a ZERO as given in the table above and in the manual.

2) This setup sets the receiver number to 00 (set via pins K through T).

3) IMPORTANT! If you use BASIC, be sure to jump RTS to CTS, and DSR to DTR at the computer. This is because the RACAL engineers neglected to implement those signals in their interface. I/O implemented in BASIC will time out if it doesn't see CTS from the remote device. Jumping RTS to CTS ensures CTS will always be asserted.

4) Be sure to check your A6A1 (Serial Board) link configurations to see they're set up for RS-232C/188C/423 operation: install LK1, remove LK2, install LK3, install LK4, install LK5, remove LK6. Once you've finished the cable, set up your terminal software as follows: 9600 baud, 7-bit characters, 2 stop bits, odd parity.

For an initial test, type

$00S2 <return> (dollar, zero, zero, capital S, two <return>)

This should switch the 6790 into REMOTE mode. The word "REMOTE" should appear in the LCD display just as if you had pushed the REMOTE/LOCAL button.

If this doesn't happen, it's troubleshooting time – and there are a LOT of things that can go wrong! Suggestions:

1) Did you type "dollar sign, zero, zero, big S, two" then <return>? The input is case sensitive. Small "s" will NOT work. Any mistakes must be cleared by hitting <return>, then retype.

2) Double-check the interface parameters at the PC side.

3) Double-check your cable wiring. The cable data above is set up to hook data in to data out, etc. No "null modem" is needed.

4) Link 1 on the Processor Board (A6A2) must be REMOVED when the Interface Card (A6A2) is installed. Check this – the previous owner might have goofed.

5) Make sure Notes 3 and 4 above have been accomplished.

Note: there are two different interfaces being tested here – the one at the 6790 consisting of the A6A1 and your cable, and also the one at the PC. Both have to be set up correctly for success. I've found in the past that sometimes it helps to first independently confirm the PC interface is working by seeing if it will talk to another computer using the correct parameters. If you've done all of the above, and it still doesn't work, the serial board might be bad. Substitution is probably the only check in that case.

Assuming "REMOTE" appears okay, continue on by entering

$00F11.175 <return>

which sets the frequency to 11.175000mHz.

Then you can input

$00S1 <return>

which will switch back to LOCAL mode.

At this point, read the section in the manual on Remote Control – then you can go bananas entering all sorts of neat commands to change mode, frequency, BFO offsets, etc. The author has implemented a rudimentary passband tuning program for the 6790 that is fairly effective against strong adjacent stations. It works by switching to CW mode and offsetting the receive frequency up or down by 1.6kHz (half the 2.1 kHz filter width), and offsetting the BFO in the opposite direction (depending on which sideband is in use). The listener can then move the passband up or down by moving the receive frequency and BFO in opposite directions in 10Hz steps until the offending signal moves out of the passband.

The program was written in BASIC, with the usual BASIC I/O – nothing fancy.

What To Do If It's D.O.A.:

First, run the BITE test. You may simply have a misplaced filter. However, faults in this receiver are seldom that elementary. No one with a 6790 can get by without the service manual—sooner or later you will "git bit."

This is one of those radios that you really need to check out at the time of purchase. Two day swap meets like Dayton or Orlando are best because they often give you a chance to really check out the receiver—at least on major functions. That is, if the seller will consent to a check-out period. If he doesn't, that should tell you something too. The other alternative is to buy a unit from a dealer like Toronto Surplus, etc. The cost may be higher, but you have to weigh that against having to fix major faults. This radio is fragile enough without having to mess with big problems to start with!

Another fault often encountered is a dead or weak memory backup battery. The battery is a 2.4V (nominal) nicad battery located on the A6A2 microprocessor board (top side of the chassis, just behind the front panel on the right hand side—the A6A2 is mounted to the side of the chassis). The A6A2 and A6A1 serial interface will be connected together if your radio has the serial card installed. If the battery can't hold a charge, the 6790 will not retain any previously input settings. Marginal batteries are worse because they can cause all sorts of erratic operation.

The author has also found that if the serial interface is installed and the interface jumpers are correctly set up to enable the interface and the connector with all the various jumpers is not installed in the back panel serial connector A6A1W1J1, his particular 6790 will go into "slow down" mode. The BITE test will take ten minutes to run; the display will respond several seconds after the knob is turned, etc. Once the connector is reinstalled, the receiver returns to normal. This is true whether or not the serial interface is actually in use.

TIP: Get the tech manual!

PRC-6

Application:

Low-power hand-held FM portable radio set used for voice communications over short distances.

Tech Specs:

Frequency range: 47 to 55.4mHz on one crystal-controlled channel (200 kHz channel spacing). Units are often encountered set up on 51.0 MHz.

Crystal type: CR-23/U, 4.300mHz below the desired operating frequency

Type of modulation and power out: FM, 250mW

Transmit bandwidth: wideband (as much as 50-plus kHz if you yell loud enough)

Size: 14.75" x 4.75" x 4.25"

Weight: 3.5 pounds, less battery

Active devices: 13 sub-miniature vacuum tubes, of type 5678, 2G21, 5672, 5676, 3B4

Power requirements:
Receiver filaments—1.5 VDC 440ma
Receiver B+—45 VDC 13mA
Transmitter filaments—1.5 VDC 1A
Transmitter low B+—45 V 14mA
Transmitter high B+—90 VDC 28mA

Audio input/output devices: integral mike and headphone; also H-33 handset

Whip antenna: 24" flexible tape-type

Other antennas: 50-ohm BNC jack

Tech Manuals:
TM 11-296 "RADIO SET AN/PRC-6, OPERATIONAL AND ORGANIZATIONAL MAINTENANCE", operating and user maintenance only, no theory or schematics

TM 11-4069 "RADIO SET AN/PRC-6, FIELD MAINTENANCE", theory, troubleshooting, repair, alignment, schematics

Construction Style: Relatively lightweight die-cast outer case closes clamshell-style. Miniature leaded components are wired in a very compact chassis, point-to-point and between sub-mini tube sockets (see text). Radio is certainly repairable, but is a challenge to work on.

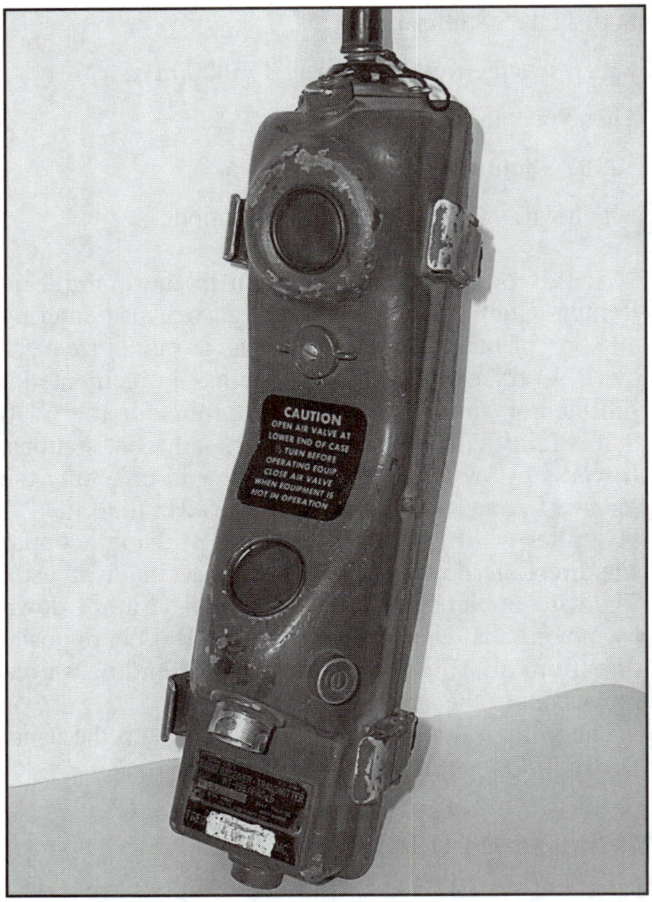

Description:

The PRC-6 replaced the BC-611 in the early 1950s and was another radio contributing to the move to FM for hand-held and manpack short-range radios. Transistors were in their infancy following their invention at Bell Labs in 1949, so the PRC-6 ended up being possibly the final word in vacuum tube circuit miniaturization. The entire 13-tube transceiver is contained on a chassis construction 5.5" x 2.75" x 2.25", about .02 cubic feet.

This is about as tightly as you can jam regular leaded components into a sky-wired/point-to-point chassis. This compactness is made possible by another engineering marvel, the sub-miniature vacuum tube.

This is a complete pentode device squished into a glass tube about the diameter of a pencil!

Operation:

Batteries are the main challenge with portable gear from this era, and the PRC-6 is no exception. Most of the space inside the case is reserved for the original

Battery BA-270/U, which supplied multiple voltages and is no longer available. There are some sites on the web that advertise DC-DC converters allowing you to power the set from a single DC voltage. There are also sites with schematics for do-it-yourselfers. Try a web search on "PRC-6 DC-DC converter" or similar. If you don't plan on using the radio heavily, stuffing it with ten or eleven 9V alkaline batteries tapped in the

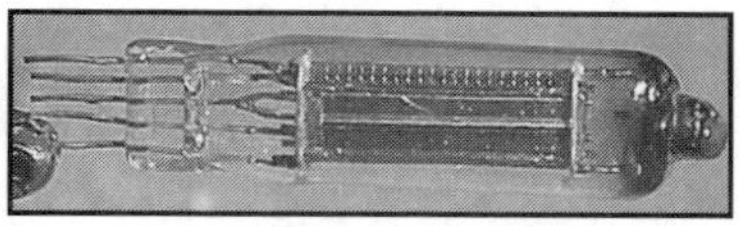

5672 sub-mini vacuum tube, about actual size

↑ *PRC-6 chassis, removed from radio, top and bottom views* ↓

PRC-6 chassis, installed. chassis cover removed.

center of the string, and two alkaline D-cells in parallel for the filaments will work fine.

Once you have batteries installed, here are the condensed operating instructions located on the inside front cover of the 1955 TM. Author's comments in []:

TO RECEIVE

1. Screw antenna into connector on top of case.
2. Open AIR VALVE (turn to left).
3. Turn switch marked EXT.-OFF-INT. to INT. and wait a minute for tubes to warm up.
[Typically, the set is ready almost immediately].
4. Hold handy-talkie with earphone to your ear. Adjust knob marked VOLUME to get hissing noise in the earphone. Handie-talkie is now in "receive" condi-

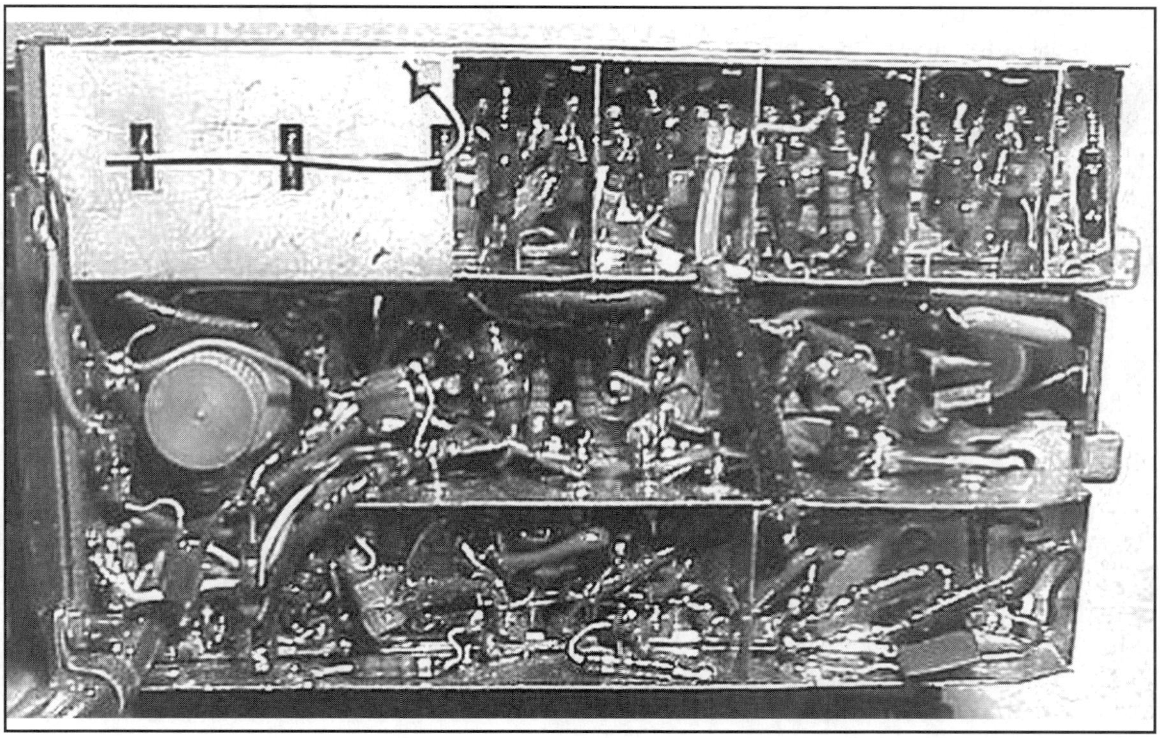

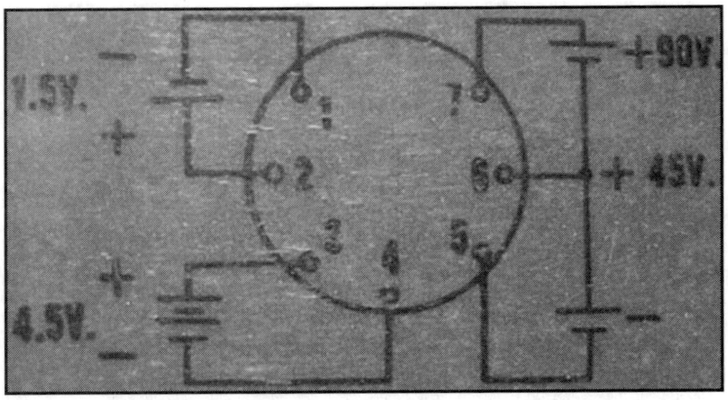

Power pin-out, PRC-6 seven-pin power connector

tion; noise decreases when voice signals are received. [When the BC-611 AM transceiver was designed in about the early 1940s, there was no term for a totally hand-held radio. "Walkie-talkie" had not yet been coined. In the movie "Kelly's Heroes," Clint Eastwood asks for a BC-611 with, "Give me that hand radio." The author has always been impressed with that bit of authenticity.]

To SEND
5. Before sending, listen to be sure no one else is "on the air."
6. Press PUSH-TO-TALK button on handie-talkie and hold it in. Hissing noise will stop.
7. Keep your lips close to the microphone; talk in normal tone. While talking, you will hear your own voice (sidetone) in earphone.
8. When you are through talking, release PUSH-TO-TALK button and listen for answering signals.

TO USE HANDSET
9. Plug handset cord into connector on handie-talkie.
10. Follow operating instructions 1 through 8 EXCEPT:

a. In step 3 turn switch marked EXT.-OFF-INT. to EXT.
b. Use handset push-to-talk switch, handset earphone, and handset microphone.

TO TURN SET OFF
11. Turn handie-talkie switch marked EXT.-OFF-INT. to OFF.
12. Close AIR VALVE (turn to right)

Tip/Caution: Unless you have a good reason to change the frequency, it's a good idea to leave the radio set up on 51mHz. That's the generally accepted FM military calling frequency; sort of like the way 3885 is used for the calling frequency when using AM gear. You'll have a much better chance of talking to other PRC-6s on 51mHz.

The alignment procedure in the Operator's TM uses a gadget called an Alignment Indicator ID-292/PRC-6, which is plugged into the test jack after removing the Pin 1 to Pin 7 jumper. The counter settings for 51.0mHz should all end up around 55 or so. The author's radio, which seems to work fine, has the following settings:

Counter 1, Crystal Oscillator: 58
Counter 2, First RF: 53.5
Counter 3, Second RF: 57
Counter 4, Power Amp: 50.5
Counter 5, Transmitter Oscillator: 55
Counter 6, Transmitter Doubler: 53.5
Counter 7, Antenna loading: 56

If your radio came without a crystal, and the counter settings are all around 55, try putting in a crystal for 51mHz. The actual crystal frequency is 46.7mHz. You may be able to get your radio to work on other frequencies by presetting the counters, then tweaking them during operation. But the author can't guarantee anything without going through the actual channel alignment procedure. Best to leave it on 51mHz.

The AN/PRC-6 in (more conventional) use

Tip: At purchase time, check the frequency of the installed crystal against the chart located inside the case. For instance, if the supplied crystal is 45.5mHz, the operation frequency is 49.8mHz, the counter settings should all be around 61, and you are the proud owner of a PRC-6 capable of monitoring users of old cordless phones!

TM 11-4069 has a channel setup procedure using a TS-505 (or any vacuum tube voltmeter) for those who need to re-crystal their units. Try Fair Radio Sales for this TM.

Additional Comments:

The PRC-6 was the author's very first mobile radio. He used a 55" inch piece of 3/4" aluminum pipe mounted on the bumper, fed with coax and connected to the 50-ohm connector on the radio.

What To Do If It's D.O.A.:

Verify that the counter settings are within reason given the crystal frequency. They should be within a few counts either direction of the specified settings.

One or more tube filaments may be open. They're really tough to see! Resistance checks are the best way to find popped filaments. The author has an assortment of sub-mini tubes he's collected over the years, and generally attacks problems in PRC-10s, PRC-6s, etc., by tube substitution, after trying to match operational symptoms to a tube function on the schematic.

V1: 1st RF amp
V2: 2nd RF amp
V3: Mixer (Receiver)
V4: 1st IF amp (Receiver)
V5: 2nd IF amp (Receiver)
V6: 3rd IF amp (Receiver)
V7: Limiter (Receiver)
V8: Audio (Receiver)
V9: Crystal oscillator (Receiver)
V10: Modulator (Transmitter)
V11: Oscillator (Transmitter)
V12: Doubler (Transmitter)
V13: Power amp (Transmitter)

Accessories:

Not often seen, the Crystal Kit CK-6/U is the complete 43 channel crystal kit.

The ID-292 appears to be nothing more than a specialized voltmeter.

While the AT-249 is the direction-finding antenna originally made for use with the PRC-6, the AT-1082 unit (for the PRC-25/77) is a more modern and capable unit. It's also more widely available. (See AT-1082 section).

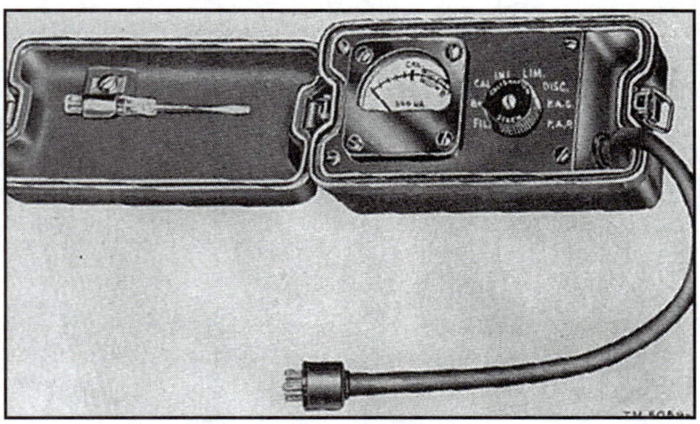

Channel Alignment Indicator ID-292/PRC-6

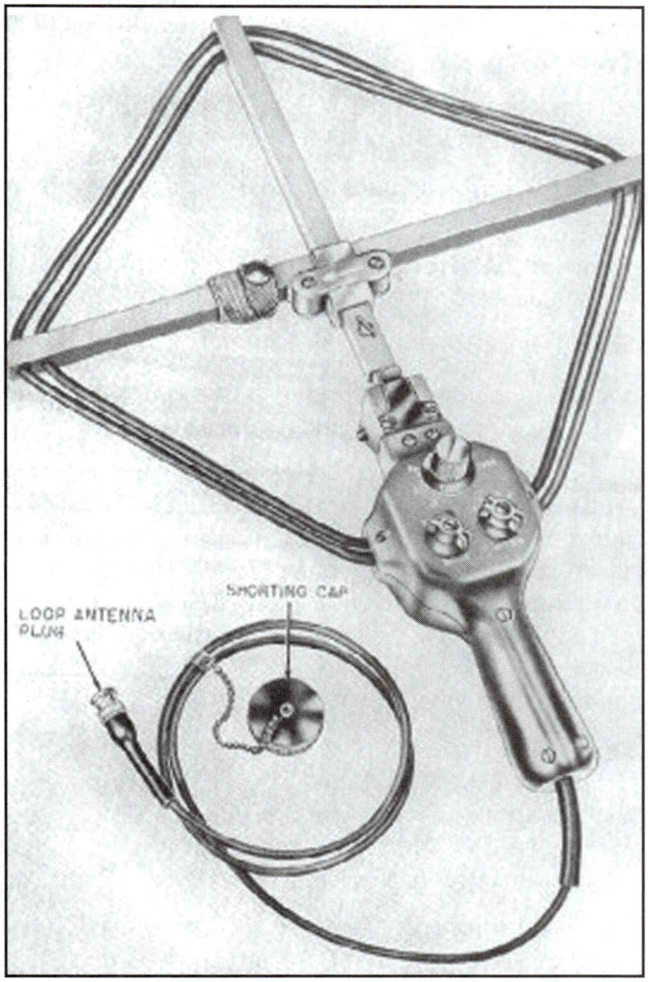

Antenna AT-249/GRD

PRC-8, PRC-9, PRC-10, PRC-8A, PRC-9A, PRC-10A, PRC-28

Application:

Radios in this family were portable, low-power FM sets designed in the mid-fifties and used through the early stages of Vietnam until they were replaced by the PRC-25. They were pack-mounted or installed in vehicles to provide voice comm over relatively short distances. They also functioned as direction finding (DF) units when used with a suitable DF antenna. They were quite advanced for their day, using sub-miniature vacuum tubes which were, in some cases, sealed in replaceable plug-in modules. A PRC-9A is shown above. The PRC-28 is a single-channel crystal-controlled version, and is seldom encountered.

Tech Specs:

Frequency range: PRC-8, 20-27.9 MHz; PRC-9, 27-38.9 MHz; PRC-10, 38-54.0 MHz

Type of modulation and power out: FM voice. PRC-8, 1.2W; PRC-9, 1.0W; PRC-10, .9W

Transmit bandwidth: 20 kHz-plus "maximum" deviation (see text)

Size: 3.25" x 10" x 17" with battery box

Weight: 11 pounds, RT unit, battery box, and tape whip only. About 20 pounds with battery and accessories.

Active devices: Vacuum tubes. PRC-8/9/10s use sub-mini tubes 1AD4, 5678 (five of the seven 5678s are in sealed plug-in modules), 5676, 5672, and one 5A6 9-pin miniature tube. PRC-8A/-9A/-10As use sub-minis 1AD4, 5678 (four of the 5678s are in sealed plug-in modules), 6286, 5672, and one 5A6 9-pin miniature tube.

Power requirements:
Originally used with Battery BA-279/U in the pack configuration, or with Amplifier/Power Supply AM-598/U and 24V battery or vehicular system. Actual voltages are 1.5, -6, 67.5, and 135VDC.

Audio input/output devices: H-33 handset, LS-166 speaker, H-63 Headset/Boom mike and GSA-6 Chest Set, or pretty much any audio accessory with a U-77 plug.

Antenna connection: BNC 50-ohm jack, whip antenna block with separate screw-in points for short tape whip or long whip.

Whip antennas: AT-272A short ribbon whip, consisting of flexible gooseneck and flat metal 29.5" tape whip, total length 38"

Long whip, consisting of Antenna Spring AB-129/PR and Antenna AT-271, total length 10' 1".

Other antennas: Long Wire AT-984A/G

Antenna tuner: Whip antennas are automatically handled by variable tuned circuits within the sets.

Tech Manuals:
TM 11-5820-292-10, "OPERATOR'S MANUAL, RADIO SETS AN/PRC-8, -8A, -9, -9A, -10, -10A, and -28". Operator's manual only, no theory or schematics.

TM 11-4065, "RADIO SETS AN/PRC-8, -9, AND -10, FIELD MAINTENANCE". Theory, schematics, troubleshooting, repair, and alignment. (see text)

Construction Style: Mixture of point-to point sky-wire, PC board, and sealed module construction. Repairable, except perhaps the sealed I.F. cans (unlettered and "A" models) and the discriminator can ("A" models only).

Description:

Considering the fact they contain 16 tubes, the PRC-10 and its variants are very compact and ingenious designs for their time. Construction is a panel-and-chassis arrangement that slides into a gasketed outer case:

At upper left in the chassis top view, sealed modules are visible for the AFC, receiver oscillator, mixer, 2nd

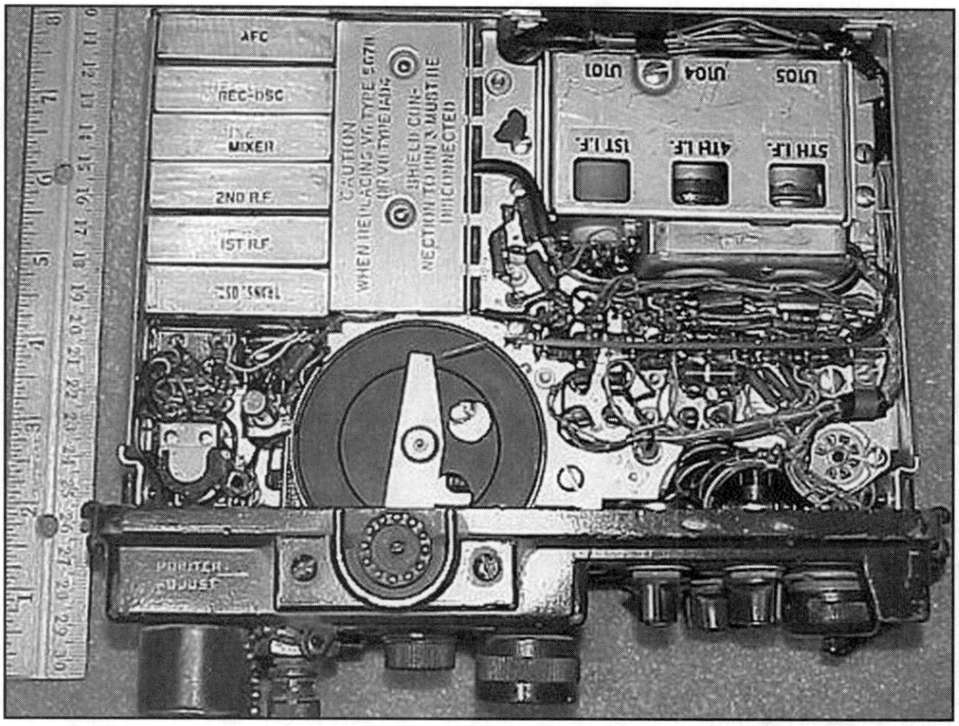

PRC-10 chassis top view

RF, 1st RF, and transmit oscillator functions. These are a little tricky to work on because they have to be completely removed from the chassis and opened. The screw-on plate just to the right of the modules covers the sub-mini tubes associated with each module. At upper right are the 1st, 4th, and 5th I.F. stage plug-in modules. All I.F. modules are identical.

At lower left is an open frame T/R relay. The "A" models used a sealed relay—probably a good idea in climates like Vietnam! The socket at lower right is for access to test points during troubleshooting and alignment. The sky-wired components at middle right show this set's vintage. Component packing is just about as dense as it can get for this type of construction.

In the bottom view, the 2nd and 3rd I.F. and discriminator plug-ins are visible at the upper left. The relatively large open frame relay at lower left is the squelch relay, which is not present in "A" models. Apparently, the "A" models redesigned the squelch to operate completely electronically. The transmitter "final", a 5A6, is located at upper right. It's actually an electron-coupled modulated oscillator hooked to the antenna jack.

While the unlettered models had an AFC capability, the "A" models went one step further and actually introduced an active hunt process to get the set on frequency when the PTT is pushed. There's often a distinctive sound in the handset the operator's manual calls "motorboating" as the "A" models lock up. One of the I.F. modules has been replaced with a pulse generator module in the "A" models.

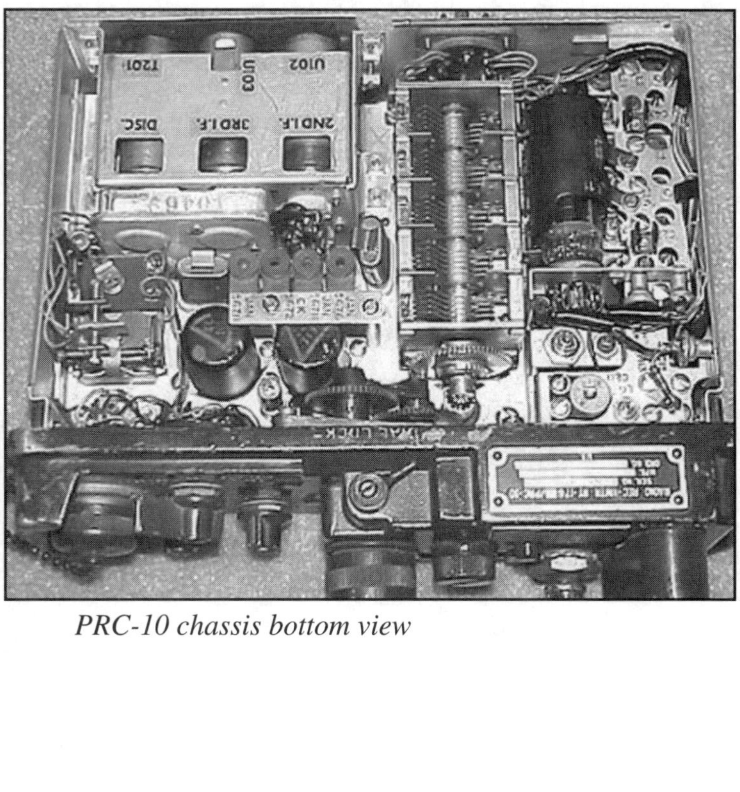

PRC-10 chassis bottom view

Operation:

The original Battery BA-279/U for this set is hardly ever found in usable condition. They haven't been made for a long time and the ones you find are usually dead. If you have access to an AM-598 power supply, you can check out the radio that way. Another option is to check the Internet for dealers selling DC-DC converter supplies for these radios that fit into the battery box along with a 12VDC battery.

The other option is to build up a battery pack with the necessary voltages. Pin-outs at the octal connector on the back of the radio case are as follows:

A:	+1.5VDC
B:	+1.5VDC
C:	no connection (chassis ground)
D:	+135VDC
E:	+67.5VDC
F:	-135VDC and -67.5VDC
H:	-6VDC
J:	-1.5VDC and +6VDC

CAUTION: The pins can be hard to read on the connector hanging off the rear of the case. If you have any question about which pins are which, look at the connector on the inside of the case and match pins with an ohmmeter. The tubes do not take kindly to getting 135V or 67V on their filaments! (There's a name for this—it's called "BINKING" the tube, because *bink* is the noise the tubes make as the filaments vaporize).

Once you have battery voltages hooked up, operation is a snap. These radios offered exactly what FM tactical packsets should—effective comm with an absolute minimum amount of controls to worry about.

Attach either the short or long whip -- you can't mix them up on the whip mount—and a handset and you're ready to go. Turn the SQUELCH to OFF to start with and the POWER switch to ON. Adjust the VOLUME to a reasonable level and advance the SQUELCH until it just closes.

Tune to the calibration point nearest your desired frequency to zero the dial. On the PRC-8, -9 and -10 it'll be near an even MHz point. On the PRC-8A, -9A, and -10A it'll be the nearest red hash mark. Turn the POWER switch to CAL & DIAL LITE and tune for zero beat in the audio. Move the dial pointer over the calibrate point with the POINTER ADJUST knob, then tune to your desired frequency. You can lock the dial if desired by rotating the DIAL LOCK. Transmit by hitting the PTT. That's it!

Note: You must wait for the motorboating sound (if any) to stop after hitting the PTT on the "A" models, indicating that the frequency has locked in. The author has found that some examples may not exhibit this sound but will still be in lock.

The internal circuitry will automatically match either the short or long whip. A 50-ohm antenna can be connected to the AUX ANT connector. Remove any whip antenna if using the 50-ohm connector.

Additional Comments:

Remember these radios produce wideband FM. All of the FM radios of this era were very wide compared to modern standards, and were typically rated at 20 kHz maximum deviation. But in reality, due to the modulation scheme, the louder you talk the wider they get. If you really bellow into the H-33 you can get as much as 50-60 kHz of deviation!

Tip: More like a friendly warning—before you get on 29.6 with a PRC-9 or 52.525 with a PRC-10, make sure you've done the deviation mod to the H-33 outlined in the RT-70 section. Otherwise you'll deviate the innards (the I.F., that is) out of the receiving station.

What To Do If It's D.O.A.:

Tubes, tubes, tubes. The author has found these radios are very reliable once you get them working. Most problems can be traced to vacuum tubes; they've been sitting in the radios anywhere from 40 to 50 years...plenty of time to get gassy.

Hint: For a more modern tech manual, try an Internet search on "PRC-10 manual". The author had some success.

Accessories:

The radios can also be used as direction finders with suitable DF loop antennas. They were also used in mobile installations with vehicular power supplies. See the AM-598 section.

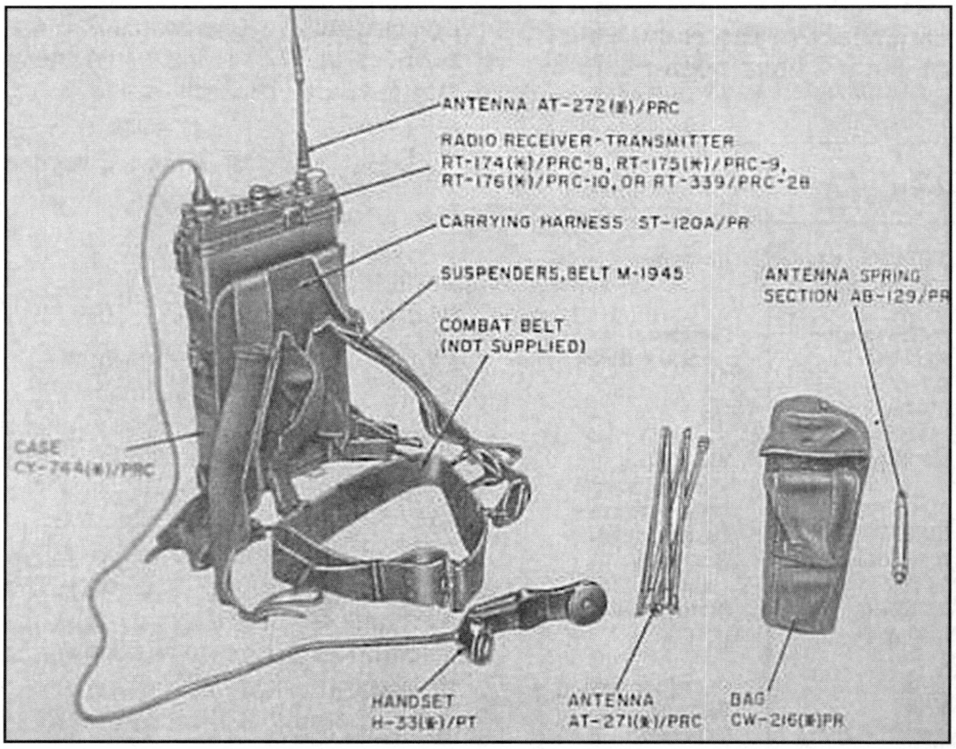

PRC-8/-9/-10 complete packset configuration

PRC-25, PRC-77

Application:

The PRC-25 (shown) is a short-range manpack FM transceiver that replaced the PRC-8/-9/-10 series. The PRC-77 is an all solid state upgrade identical in appearance and function.

Tech Specs:

Frequency range: 30.00 to 75.95mHz, synthesized in 50kHz steps (920 channels)

Type of modulation and power out: FM, 1.1 to 2 watts

Transmit bandwidth: ~9kHz

Squelch: 150Hz tone type (selectable ON-OFF)

Size: 4" x 11" x 11" not counting antenna

Weight: 20 pounds, as pictured, with 10 D-cell hi-cap nicad batteries installed

Active devices: solid state, early 60s vintage transistors. PRC-25 transmit power amplifier is a 2DF4 vacuum tube (see text). PRC-77 P.A. is transistorized.

Power requirements: originally BA-386/PRC-25 or BA-4386/U
PRC-25 - 15VDC at 1.4 amp transmit; 2.5VDC at 400mA (xmit filaments only)
PRC-77 - 15VDC at 1.4 amp transmit

Audio input/output devices: Any of the new family accessories such as the H-189 handset pictured, LS-454 (see LS-454 section), H-161, H-250, et al with U-229 plugs.

Antenna connection: 50 ohm BNC jack or whip antenna block

Whip antenna: AT-892/PRC, which consists of a steel tape whip and flexible gooseneck for a total length of 36".

AT-271A/PRC, which consists of a sectionalized tubular whip and AB-591 flexible base for a total length of 10' 1".

Other antennas: AT-912/VRC or AS-1729/VRC 10 foot vehicular whip. (For more on these, see RT-524 section)

Antenna tuner:
None on the 50 ohm BNC connector
Tuning is automatically accomplished internally when the PRC-25 is used with the tape whip or AT-271.

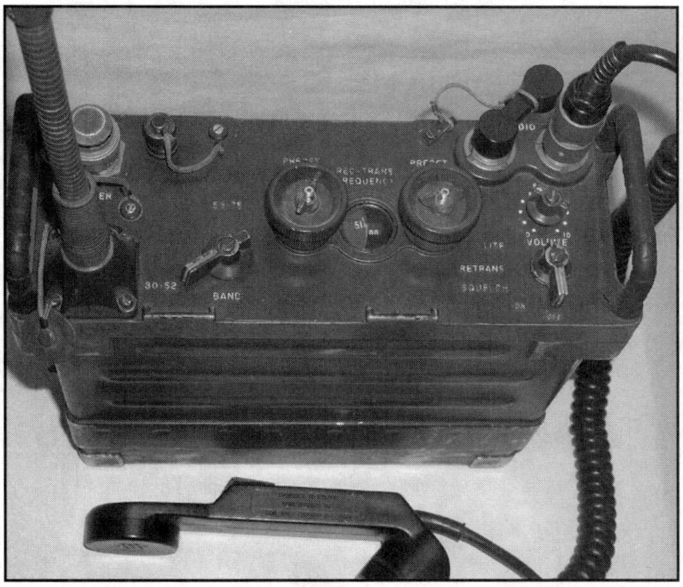

Tech Manuals:
PRC-25: TM 11-5820-398-35 "FIELD AND DEPOT MAINTENANCE MANUAL RADIO SET AN/PRC-25"; theory, troubleshooting, repair, alignment, schematics

PRC-77: TM 11-5820-667-35 "DIRECT SUPPORT, GENERAL SUPPORT, AND DEPOT MAINTENANCE MANUAL RADIO SET AN/PRC-77"; theory, troubleshooting, repair, alignment, schematics

Construction Style: Diecast heavy duty front panel and chassis assembly slides into a ribbed sheet metal outer case. Set is heavily modularized; in many cases the original extender cables will aid in troubleshooting and alignment. The resourceful surplus collector learns to do without them!

Description:

The PRC-25 is an excellent example of a transitional hybrid design between vacuum tube and solid state technology. The technology of the time included solid state devices for pretty much every function except the higher-level RF power-related applications. There were no solid state RF power devices at the time capable of doing what the designers needed in the P.A. stages, so they resorted to DC-DC converters and vacuum tubes. The PRC-25 transitioned to the PRC-77, which was very nearly the same radio, but with a solid state output stage taking advantage of the new devices available. The time difference was about four years, judging from the dates on the TMs.

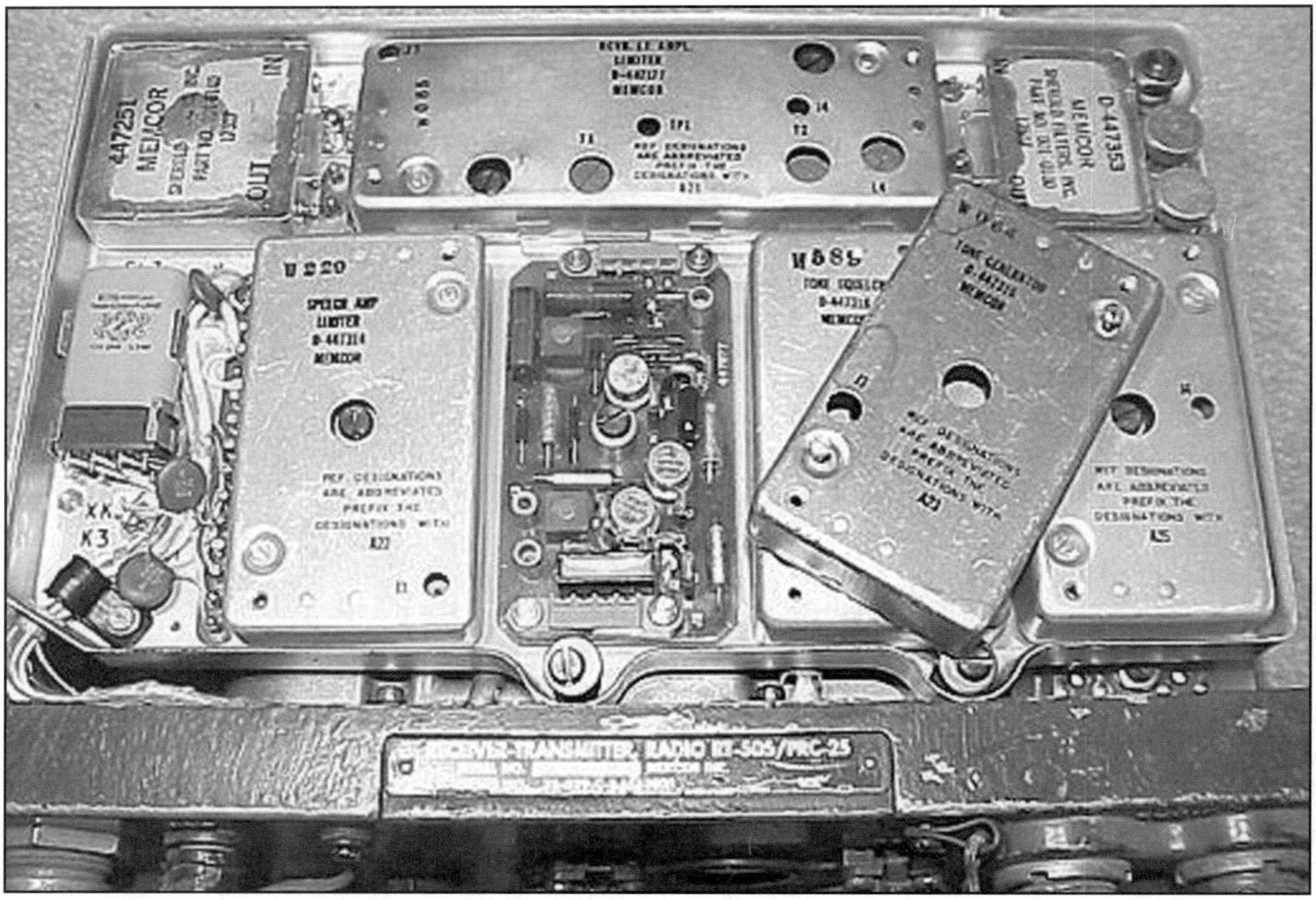

PRC-25 chassis, top, outer case and A23 module cover removed

PRC-25 chassis, top tray opened

The radio itself is yet another electromechanical tour de force, which is not all good because that makes it sort of tough to work on!

Components in the top modules are reachable for troubleshooting for the most part by removing the module covers. The top modules are located on a hinged tray that swings open after loosening the three hold-down screws.

The 2DF4 vacuum tube power amp stage is at top center. Note the two couplers connected to the enclosed variable caps marked "MEMCOR." They exit out of a gearbox connected to the two frequency entry controls.

PRC-25 chassis, bottom

Due to the indented placement of the bottom modules, troubleshooting is difficult without extender cards. Note the word impossible was not used -- measurements can always be taken by clipping probes to components leads, wrapping wires around module pins, and whatever. We as surplus nuts tend to have a LOT more patience than the average Army repair depot tech! The DC-DC converter module for the P.A. stage is at far right.

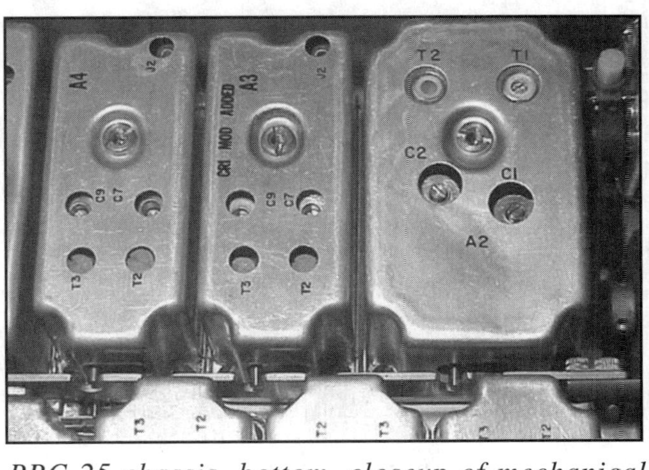

PRC-25 chassis, bottom, closeup of mechanical bandswitching mechanism

Moving the BAND control between 30-52 and 53-75 actuates a lever, changing band switches in six different modules.

Operation (PRC-25 and PRC-77):

Pin-outs at the power connector on the bottom of the R/T unit are as follows:

Pin A GROUND (-)
Pin B +12.5VDC
Pin C +2.5VDC (not needed with the PRC-77)

The author built a battery pack of ten hi-cap nicads, tapping two cells from the negative end to get the transmit filament voltage. Another possibility is an adapter for standard C-cells, generally available on the web. The adapter takes twelve C-cells, ten of which are in series for the +15 and two of which independently supply the transmit filament voltage. It will accept either primary or rechargeable cells.

Pin-outs for both the AUDIO connectors are:

Pin A GROUND
Pin B AUDIO OUT
Pin C XMTR ON (keyline)
Pin D MIKE INPUT
Pin E REMOTE RELAY CONTROL

NOTE: Pin E, the REMOTE RELAY CONTROL, is used when two PRC-25s are used for retransmission. The Function Switch on the receiving unit is set for squelched operation. When the squelch breaks, Pin E is grounded. The author has never tried this, but it appears when the correct cable is used, and the function switch on both radios is set for RETRANS, a signal coming in on one set will be retransmitted on the other and vice versa.

Once you're batteried-up, operation is a snap:

1) Connect a handset to either of the AUDIO connectors, and the tape whip, long whip, or 50 ohm antenna. Preset the VOLUME control about 1/4 way.

2) Turn the Function Switch to ON. You should hear a rushing noise in the handset.

3) Move the Function Switch to SQUELCH if squelched operation is desired.

Note: In the SQUELCH position, incoming signals must have 150Hz tone to be received.

4) Select the desired range 30-52mHz, 53-75mHz with the BAND lever.

CAUTION: Do not move the BAND lever while the radio is keyed. This caution was inserted in later versions of the TM. Nothing is mentioned about what will happen if you do this, but it can't be good.

5) Enter your desired frequency using the two large REC-TRANS FREQUENCY knobs.

6) Moving the Function Switch to the momentary LITE position will illuminate the dial.

7) To transmit, hit the PTT on the handset and jabber. You should hear sidetone.

That's all there is to it.

Additional Comments:

Symptoms commonly encountered with these radios when they drop out of alignment are:

1) Failure to lock up on frequency. The severity may range from failure to lock on just a few frequencies, to failure on many frequencies on either the high band or low band, or even total lock failure on both high and low bands.
2) The transceiver locks on frequency but a MHz high, MHz low, etc.

If you have a PRC-77 and experience any of these symptoms, the Direct Support Alignment Procedure from the TM may help, and is given here.

Note: While the 25 and 77 are very similar, the 25 alignment differs, and will require the specific PRC-25 TM. The following procedure addresses only a small fraction of possible PRC-77 faults, related to off-frequency operation. Before performing any steps, read the entire procedure through. The author's comments are in [].

[Direct Support Alignment Procedure—PRC-77 ONLY]
a. The direct support alignment procedures consist of aligning transmitter variable frequency oscillator module A38, synthesizer variable frequency oscillator module A39, and the sidestep oscillator (sso) in frequency control module A31. These procedures are contained in paragraphs 3-16, 3-17, and 3-18, respectively.
b. The following test equipment and tools are required:
(1) Digital Readout, Counter Electronic AN/USM-207. [Any 100mHz frequency counter can be substituted]
(2) Wattmeter AN/URM-120 (part of Coupler-Detector CU-754/URM_120). [A wattmeter is not strictly required]
(3) Dummy Load, Electrical DA-75/U (part of AN/T.MM-120). [Any 50-ohm dummy load will do]
(4) Multimeter ME-26B/U. [Any good high impedance VOM will work]
(5) Module extender for module A31.
(6) Alignment cover for module A31.
(7) Capacitor, 2 ,uuf +/- 0.5, for module A31 alignment (para 3-18).

[Items 5 and 6 will not generally be available to the surplus collector. The A31 alignment may not always be required; more on this below]

3-16. Alignment of Transmitter Variable Frequency Oscillator Module A38

Note: Before module A38 alignment, use Multimeter ME-26B/U and observe the apc voltage indication at test point J4 of module A31 as the transmitter frequency is varied (by turning the mc control). The apc voltage indication should be between 2.2 and 3.8 volts dc (maximum). If variation is greater, check A38 alignment as described below. [If you are experiencing any of the symptoms above the likelihood is very high that the A38 needs alignment]

a. Connect the dc power supply HP-721A or equivalent between A31J4 and ground. Set the voltage of the HP-721A to 3.0 volts.

Caution: Do NOT apply more than 8.0 volts at this test point.

[*NOTE: Instead* of applying this voltage, you can simply pull the A50 Transmitter Mixer Module. Make sure the radio is OFF before you do this! This effectively cuts the apc feedback voltage to the A38 and accomplishes the same thing as forcing in a constant apc voltage. In any case, the apc voltage must not be allowed to float during alignment because the feedback loop will constantly "chase its tail" and try to change your alignment inputs. Both the A38 and A39 modules are aligned as free running oscillators to get them close, after which the apc voltage should lock them in.]

b. Connect the AN/USM-207 between A49 pin 1 and ground.

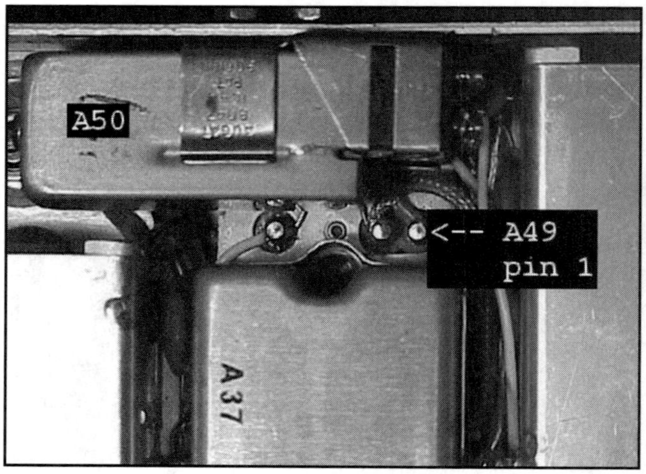

PRC-77 chassis beneath the fold-out module tray, showing A49 pin 1

To get to this alignment point, open the hinged module tray by loosening the three captive screws. The pin is toward the back of the chassis, between the A37 and A50 modules, as shown.

Connect the AN/URM-120 with the CU-754/URM-120 and DA-75/U to the RT-841/PRC-77 ANT connector J2.
c. Set the RT-841/PRC-77 BAND switch to 30-52 and the tuning controls to 30.00 mc. Connect Handset H-189/GR to one of the AUDIO connectors on the RT-841/PRC-77 and press the push-to-talk switch on the H_189/GR.
d. Adjust A38T1 until the AN/USM-207 indicates a frequency of 30.00 mc +/- 10 kc. [It's fairly easy to get this within a few kHz or so]
e. Set the RT-841/PRC-77 tuning controls to 52.95 me. (Depress the push-to-talk switch on the H-189/GR.)
f. Adjust A38C3 until the AN/USM-207 indicates a frequency of 52.95 mc +/- 10 kc.
g. Repeat the procedures given in c through f above several times to insure proper alignment of A38 on the low band. [As you repeat the steps, each successive tweak should get smaller until you can go back and forth between frequency settings without needing any more tweaks]
h. Set the RT-841/PRC-77 BAND switch to 53-75 and the tuning controls to 53.00 me. (Depress the push-to-talk switch on the H_189/GR.)
i. Adjust A38T2 until the AN/USM-207 indicates a frequency of 53.00 me +/- 10 kc.
j. Set the RT-841/PRC-77 tuning controls to 75.95 mc. (Depress the push-to-talk switch on the H-189/GR.)
k. Adjust A38C9 until the AN/USM-207 indicates a frequency of 75.95 mc +/- 10 kc.
l. Repeat the procedures given in h through k above several times to insure proper alignment of A38 on the high band.
[**m.** Turn the radio OFF and replace the A50 module if you removed it for the alignment]

3-17. Alignment of Synthesizer Variable Frequency Oscillator Module A39
a. Connect the AN/USM-207 between A44J2 and ground.

[***NOTE:*** *Instead* of grounding this point (or any of the other points specified in this para), you can simply pull the A12 Frequency Synthesizer System Second Mixer Module. Make sure the radio is OFF before you do this! This effectively cuts the apc feedback voltage to the A38 and accomplishes the same thing as grounding A44J2 or A12J2.]

b. Set the RT-841/PRC-77 BAND switch to 30-52, and set the tuning controls to 30.00 mc.
c. Connect test point A12J2 to ground and check the dc voltage at A39J3; 4 +/- 0.5 vdc should be present. [Or pull A12 as per the note above]
d. Adjust A39T1 until the AN/USM-207 indicates a frequency of 41.500 mc. [The adjustments in this paragraph are all very touchy! You may need to move the slugs/screws a little, then pull away your tool -- especially if it's metal.]
e. Remove ground from A12J2. [Unnecessary if you pull A12 per the note]
f. Set the RT-841/PRC-77 tuning controls to 52.95 mc.
g. Connect test point A12J2 to ground. [Unnecessary if you pull A12]
h. Adjust A39C14 until the AN/USM-207 indicates a frequency of 64.450 mc.
i. Remove ground from A12J2. [Unnecessary if you pull A12]
j. Repeat the procedures given in b through i above several times to insure proper alignment. [The settings should converge gradually after a few repetitions as with the A38]
[**k.** Turn the radio OFF and replace the A12 module if you removed it for the alignment]

[***NOTE:*** The A50/A12 trick was worked out by a Cincinatti Electronics tech who did actual PRC-77 alignment and troubleshooting back in the late 70s. The author brought one of his 77s back to life using this method.]

3-18. Alignment of Sidestep Oscillator in Frequency Control Module A31

[***NOTE:*** The following part of the alignment is a pain if you don't have the module extender or alignment cover. It may not even be necessary. Once you've aligned the A38 and A39, try the radio -- if you find representative channels are locking up okay and are within 1.5 to 2.0 kHz or so of indicated frequency, you're done. The actual acceptable frequency error for the PRC-77 was listed as +/- 3.5 kHz. This may seem very high, but remember that the deviation on this radio is 10 kHz +/- 3 kHz at 2000Hz! If you're seeing errors much beyond 3 kHz, you might want to try tweaking the A31.

a. Remove module A31.
b. Insert the module extender into J10.
c. Remove the cover from module A31 and place the alignment cover on the module.
d. Connect module A31 to the module extender.

[***NOTE:*** Most collectors won't have these gimmicks. They'll have to remove the A31, move the adjustment a small amount, replace the A31, and see which way the frequency error goes. Tedious, but effective.]

e. Set the RT-841/PRC-77 BAND switch at 30-52.
f. Connect the AN/USM-207 between A31J3 and ground through a small coupling (2uuf) capacitor.
g. Connect the AN/URM-120 (with CU_754/URM-120) to the RT-841/PRC-77 ANT connector J2.
h. Connect Multimeter ME-26B/U, set for rf voltage measurement, between A31J3 and ground.
i. Connect the H-189/GR to an RT-841/ PRC-77 AUDIO connector.
j. Press the H-189/GR push-to-talk switch and adjust A31T5 for an indication of 11.500 mc +/- 500 cps on the AN/USM-207.

[***CAUTION:*** Permeable material slugs like the one in T5 are *fragile*. Be sure you remove as much of the glyptol as possible before trying to turn it. Since the TM doesn't suggest using MEK or other solvents, don't try them. Solvents have been know to dissolve some slug types! It's best to carefully chip the glyptol away with a fine X-Acto knife.]

k. Adjust A31T6 for a maximum indication on the ME-26B/U. The minimum level of 3 volts RF is acceptable.

[***NOTE:*** The author recommends skipping step (k) to start with. The frequency adjust in step (j) is the more important. Do that first, then check the overall radio function.]

l. Remove module A31 and the module extender from J10.
m. Remove the module alignment cover from module A31.
n. Replace the cover on module A31.
o. Replace module A31.

***

In modern terms these radios are pigs, weight-wise, for what they do. The author really feels for the commo guys who had to carry one along with their weapon, ammo, food, and water. But it was the U.S. military's first real stab at a synthesized FM set and it came out pretty well, considering what the designers had to work with at the time! Incredibly, it stayed in production into the 90s, making it one of the longest runs (if not the longest) of any U.S. military radio -- a quarter century-plus.

What To Do If It's D.O.A.:

Once you get these radios going, they're fairly reliable. HOWEVER, the author has seen more than a few 25s and 77s that have wandered out of alignment and won't lock up, or transmit and/or receive on the wrong frequency. Absolutely the worst thing you can do is get in there and start indiscriminately screwdrivering to get the thing back on frequency. You WILL screw up the radio. This bears repeating.

CAUTION: Do not indiscriminately screwdriver/tweak internal adjustments attempting to get a PRC-25/-77 back on frequency or "peak it up." ***YOU WILL SCREW UP YOUR RADIO.***

Radios that are off frequency require a detail alignment, part of which is presented above. The various adjustments interact in a way that makes "alignment by instinct" impossible. The instructions given in the "Additional Comments" section for the PRC-77 addresses the most common failure the author and his group has encountered: failure to lock or off-frequency lock due to A38/A39/A31 moving out of alignment.

Accessories:

A useful accessory if you plan to use the PRC-25 in a vehicular application is the AM-2060/GRC Amplifier/Mounting Base, which provides shock mounting and audio amplification. The ST-138 Carrier provides the manpack configuration.

There are some readers who will remember (and probably not with great fondness) slogging this setup through innumerable rice paddies. That said, the author has spent many an enjoyable afternoon running /PM (pedestrian mobile) with his PRC-1099 on this particular ST-138. Note the extra padding on the straps. Yeah, yeah, yeah, the author's a wimp. But he's a comfortable wimp!

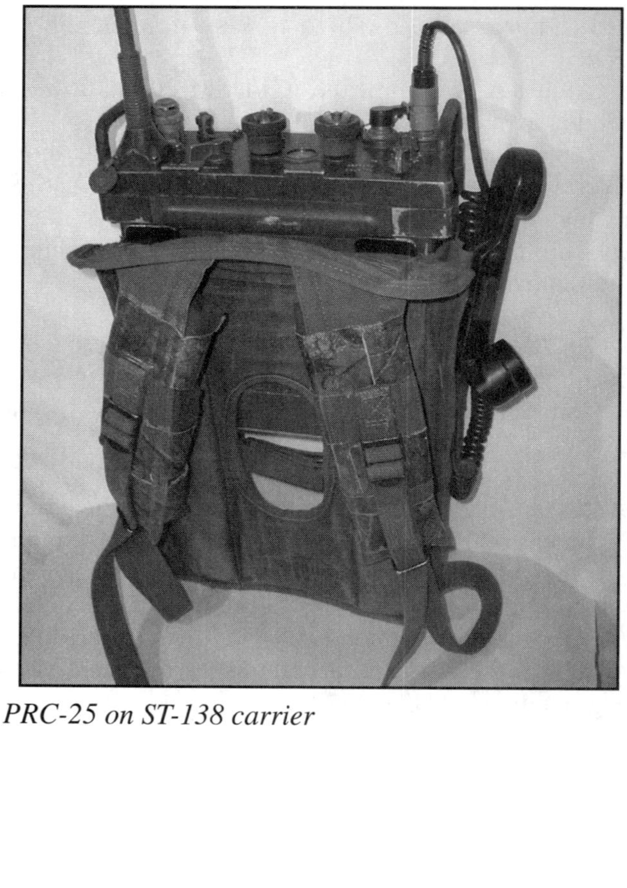

PRC-25 on ST-138 carrier

PRR-9, PRR-9(XE-9), PRT-4

Application:

Low power, short range FM comm on one or two crystal controlled channels. Late 60s Vietnam era vintage. The PRT-4 or -4A (shown at left) is the transmitter, and the PRR-9 (shown at right) or PRR-9(XE-9) is the separate receiver. The original idea was to give a full set to the squad leader, and a receiver only to each man in the rest of the squad. This radio was not well thought of by the troops, and was not widely used. Thousands were made, however and, for some strange reason, many hundreds were refurbished as late as the 1980s and painted in CARC green.

Tech Specs:

Frequency Range: 47-57 MHz on one crystal controlled receive frequency (PRR-9); and on two crystal controlled transmit frequencies (PRT-4).

Type of modulation and power out for PRT-4: FM, 450mW on Channel 1, 50mW on Channel 2. PRT-4A has the same power out, but also transmits 150 Hz tone to open the tone squelch on other radios such as the PRC-25, PRC-77, RT-524, etc.

Transmit bandwidth (PRT-4): 9 kHz or so.

Size:
PRT-4: 1.5" x 3" x 8" with retracted whip.
PRR-9: 1.25" x 3" x 7" not counting whip.
PRR-9(XE-9): 1.25" x 3" x 5.5" not counting whip.

Weight:
PRT-4: About 18 oz with battery
PRR-9: About 9 oz with battery

Active devices: The PRT-4 and PRR-9 "regular" models use 60s-vintage transistors like 2N987, 2N2614, and rectifiers like 1N270, etc.—no chips. But the PRT-4A uses a custom Delco chip for the 150 Hz tone generator, and the PRR-9(XE-9) uses three custom Delco chips (see text).

Power requirements:

PRT-4: 15VDC nominal, originally from Battery BA-399/U, but two standard 9V alkaline batteries will work (see text). Current drain: Channel 1, 150 mA max at 12V; Channel 2, 90 mA max at 12V
PRR-9: 6VDC nominal, originally from battery BA-505/U, but a single standard 9V alkaline battery will work (see text). Current drain: Unsquelched, 19.5 mA; Squelched, 12.5 mA

PRR-9(XE-9): 3VDC nominal (see text!). Current drain: Unsquelched: 22 mA; Squelched 8 mA

Audio input/output devices: Built-in mike (PRT-4). Both versions of the PRR-9 have an attached folded horn type speaker. They also have an earphone jack that accepts a common sub-mini mono audio plug. Supplied earphone: H-264/PRR-9.

Antenna connection: PRT-4: built-in collapsible 24" sectionalized whip. PRR-9: built-in non-collapsible 18" wire whip.

Tech Manuals:
TM 11-5820-549-12: "Operator and Organizational Maintenance Manual, Receiving Set, Radio AN/PRR-9 and Transmitting Set, Radio AN/PRT-4". Operation and high level maintenance, no schematics. Outlines how to change the crystal, etc.

TM-11-5820-549-35: "Direct Support, General Support, and Depot Maintenance Manual, Receiving set, Radio AN/PRR-9 and Transmitting Sets, Radio

AN/PRT-4 and AN/PRT-4A". Theory, troubleshooting, and schematics.

TM-11-5820-549-12-1: "Operator and Organizational Maintenance Manual, Receiving Set, Radio AN/PRR-9(XE-9)". Operation and high level maintenance for the integrated circuit version receiver, no schematics. Outlines how to change the crystal, etc.

TM 11-5820-549-50-1: "Depot Maintenance Manual, Receiving Set, Radio AN/PRR-9(XE-9). Theory, troubleshooting, and schematics for the integrated circuit version receiver.

Construction Style:
PRR-9 - conventional components mounted very compactly on one single-layer board with exposed solder traces. Repairable, but a bit of a challenge for the myopic!

PRR-9(XE-9) - same as the PRR-9, but even more of a challenge because the custom Delco ICs were made only for this radio and are long out of print.

PRT-4 - conventional components mounted very compactly on three single-layer boards, attached together into one assembly. Repairable, but a tough job.

PRT-4A - same as the PRT-4, with the addition of the custom Delco 150 Hz tone generator chip.

Description:

"Receiving Set, Radio AN/PRR-9, is a battery-operated, frequency-modulated (FM), crystal-controlled radio receiver which is designed for quick attachment and use on the standard helmet." This description straight out of the TM is humorous and ironic for a couple of different reasons. First, there aren't many surplus crazies who typically run around wearing Army helmets. Second, and more basically, these receivers won't fit on "standard helmets" any more. The U.S. changed to Kevlar helmets a while back, and the lip is too thick to allow attaching the PRR-9. So the whole point is moot anyway.

The PRR-9 has a miniscule audio output—5 mW—and uses an interesting folded horn type speaker to increase the audio efficiency. It's sort of a throwback to the old wind-up record player days.

The 18" whip comes in two versions -- one is held in place at the base by a spring and can be swiveled without loosening the thumbscrew; the other type won't move unless the thumbscrew is backed off.

Operation:

Quick setup and sanity checks for basic function will be covered first. For more detailed alignment info, see "Additional Comments" below.

PRR-9 folded horn speaker

The PRR-9 receiver (***not*** the XE-9 version) was designed to operate on 6 VDC. A battery pack of four AA or AAA alkalines works fine, but use of a single 9-volt is less awkward. See below for necessary adjustments if you wish to use a 9V battery.

Note: Most PRR-9s you encounter will be set up to receive on 51.0 Mhz. However, this is not always true, and you need to verify what's in there at time of purchase. The crystal for 51.0 MHz will be marked "40.30". The PRR-9 has a 10.7 MHz I.F., so the marked crystal frequency will always be 10.7 MHz below the actual receive carrier frequency.

Power connections for both receiver variants are as follows:
PRR-9
THICK post is positive (+), 6 to 9 VDC
THIN post is negative (-)

PRR-9(XE-9)
TIP is positive (+), 2.6 to 3.6 VDC
SLEEVE is negative (-)

WARNING: Do *not* hook up power to these units backwards—you will destroy them. Also, do *not* attempt to run the PRR-9(XE-9) on higher voltage than specified—you'll smoke it into the Stone Age. The author knows personally of two instances where other PRR-9(XE-9) owners attempted to use 6 VDC and fried their receivers.

Once you have power hooked up, rotate the knob on the end past OFF to turn on the unit. You may hear a quick squelch tail as the receiver squelches. Rotate the control fully CW until it clicks again; this disables the squelch and you should hear receiver noise. Turning the knob to OFF and back ON again will re-enable the squelch. That's all there is to operation. These things are easy to leave ON in squelched mode. Always check that you've turned the unit OFF before putting it away.

The PRT-4 and -4A transmitters were designed to operate from a 15 VDC battery, but two 9V alkalines in series work fine. The units have internal voltage regulators capable of handling the extra voltage no problemo and the author has never had any related trouble.

Two transmit channels may be selected with the CH-1 CH-2 switch. CH-1 operates at 500mW, and CH-2 operates at 50 mW.

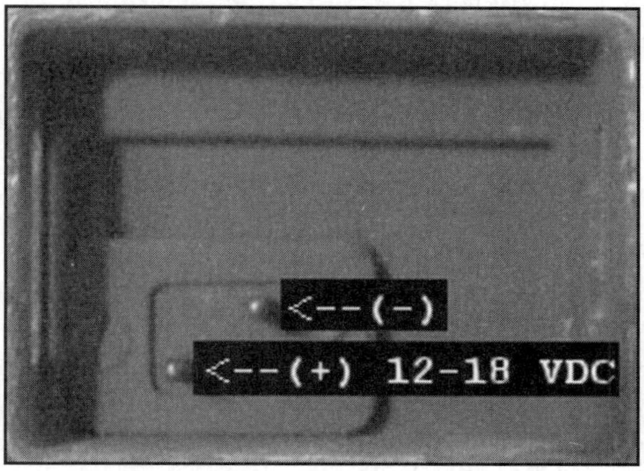

Power connections for PRT-4

Some PRT-4s have four pins, but only the two in the position shown are used.

Once power is hooked up, extend the whip to its full length. Set the TONE-VOICE switch override spring to center (detent position). Pick CH-1 or CH-2, push up on the VOICE side of the transmit switch and speak into the mike. You should hear your transmission on a receiver tuned to your frequency. Pushing up on the TONE side of the switch will transmit a calling tone. If you want to lock out TONE or VOICE operation, set the override spring accordingly.

Additional Comments:

There are some simple antenna tuning procedures you can do to maximize the efficiency of the PRT-4/PRR-9 combination.

Tip: The PRR-9 uses an inductively base-loaded 18" whip antenna. There's an adjustment for the inductor on the black antenna base on the end of the receiver. The author peaks his receiver antennas up by turning on a weak source of RF on the operating frequency, such as a sig-gen, synthesizer, or even just another FM set transmitting into a dummy load. Then he walks away until the sig becomes very weak in the unsquelched PRR-9, and adjusts the inductor screw for maximum quieting.

Tip: The PRT-4 uses a similar type of antenna, with an inductor adjust at the top right-hand side of the unit. The author peaks the transmitter by setting up a field strength meter antenna in parallel with the fully-extended PRT-4 whip, and peaking the transmitter antenna adjust for max.

To use a 9V battery in the PRR-9 (NOT the XE-9), the squelch must be readjusted, since the extra voltage tends to keep it open all the time. You should first perform the antenna optimization above. Then remove the receiver chassis from the case by loosening the two screws at the ON-OFF switch end, and the single speaker screw above the earphone jack. It may be necessary to gently pry between the end panel and case to work the chassis loose. There is only one variable resistor on the unit PCB—the squelch adjust.

Setup for adjusting squelch pot in the PRR-9

Shown is the typical adjustment setup used by the author. He uses the H-264 earphone, but any audio device with a sub-mini mono plug will work. Or you can always poke wires in the speaker pin holes.

Caution: Don't lay the chassis on anything metallic—there are exposed PC board traces!

Turn on the unit in squelched mode and adjust the squelch pot so the squelch just barely closes, then turn it a bit farther. Slide the chassis back into the case (without tightening the screws) and verify correct squelch operation. The squelch may operate differently with the chassis in the case. In fact, you may need to iterate the procedure a few times. The ideal setting is when you hear a short (1/2 sec or so) squelch tail when turning on the re-assembled receiver.

Before finally re-assembling the unit, take some time to marvel at the construction. This was state-of-the-art mil comm in 1967—a complete dual conversion, .5 uV sensitivity, 14 transistor FM receiver on a 5.25" x 1.375" PC board. Of course, it's far behind the surface mount marvels of today but it was quite something for its time.

The PRR-9(XE-9) is a very interesting bird. The contract number indicates 1969, which was around two years after the original PRR-9 contract. It was Delco's effort to more fully integrate the functions in the receiver by including more custom ICs, and adding a 150 Hz tone squelch. The ICs are interesting because they are almost certainly among the first to be used in relatively mass-produced articles. The three ICs and their functions are as follows:

Z1: RF Amplifier, First Mixer, Local Oscillator
Z2: 10.7 MHz I.F. Amplifier, Second Mixer, Second Oscillator, 455 kHz I.F. Amp
Z5: Limiter, Squelch, Audio Amplifier

Except for an outer case that's about 1.25" shorter, the (XE-9) version looks outwardly the same. The battery drain of the XE-9 version when squelched is an astonishingly low 8 mA at 3.2 VDC!

What To Do If It's D.O.A.:

These radios are pretty much in the class of "either they work, or they don't." In the author's experience, PRR-9s are fairly reliable. Once working, they almost never fail. If you get one that's bad, it may be more trouble than it's worth to work on it. At very least, you'll need the TM and schematics. If you get receive noise, but no receive, check whether there's a crystal installed at all. If the squelch misbehaves, try adjusting it per the procedure above.

The PRT-4 uses a phase-locked-loop to keep the VCO on frequency. Sometimes just sitting around will result in this part of the circuit getting out of adjustment. Newly reconditioned units have been known to prove defective. Symptoms will include off-frequency or unstable frequency on transmit. PRT-4s are seldom dead-on frequency -- remember these are relatively wideband (9kHz) and will often show up on a frequency counter as a few hundred Hz to a kHz off. This is normal. However, getting 52.6 when you should be on 51.0 is not. Sometimes these symptoms are present on one channel but not the other.

Sometimes the transmitters can be coaxed back into lock by carefully tweaking certain internal adjustments. Originally, alignment was done using the ID-1189/PR Indicator, Channel Alignment device. The TMs the author has all refer to this device, but give no specific alignment instructions. Apparently, the ID-1189 has the alignment instructions included.

C27: VCO adjust
TP1: Test Point One. Should be at or close to 0 VDC when the VCO is properly aligned.
R36: Frequency lock adjust for Channel 2 operation

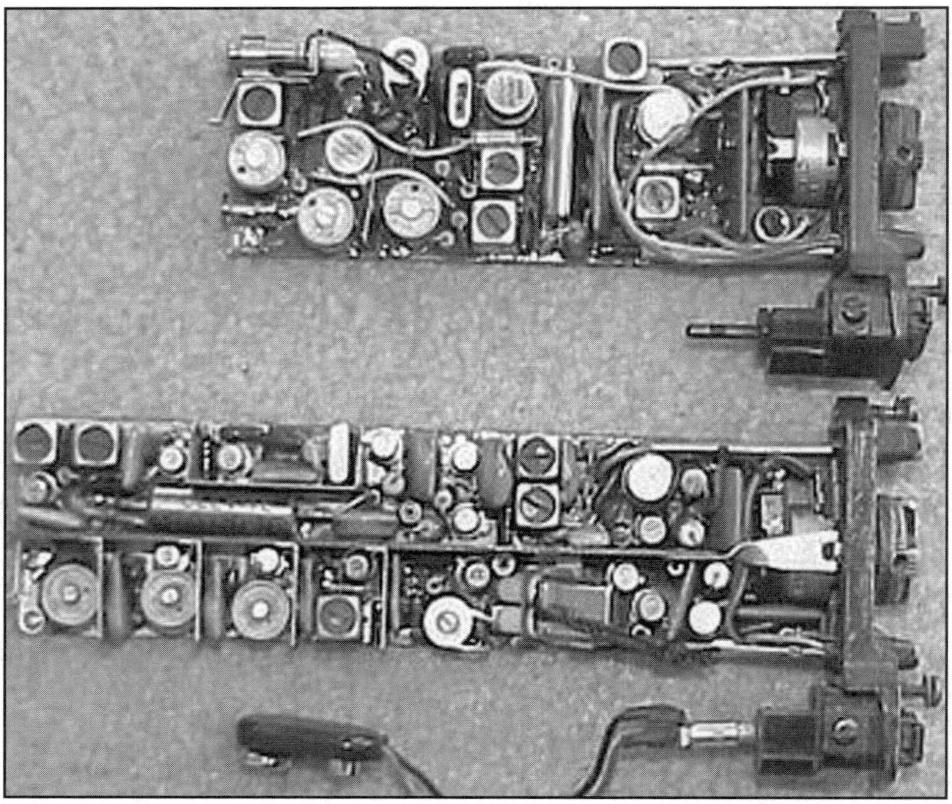

Comparison of the PRR-9(XE-9), top, and the PRR-9 chassis

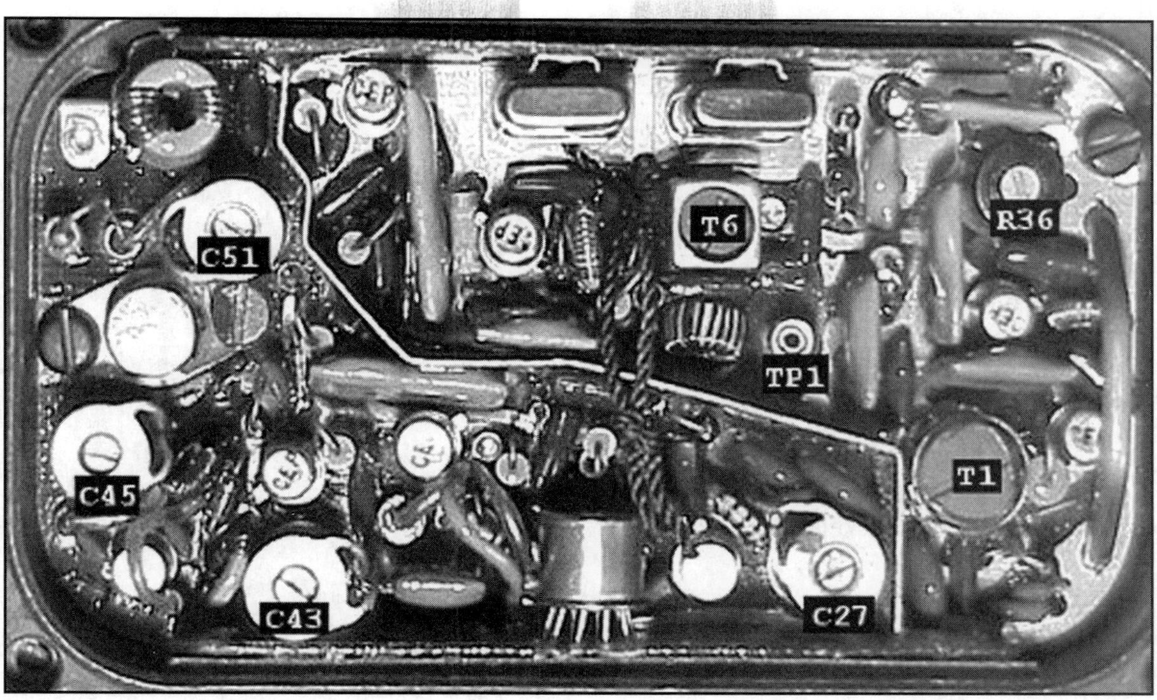

PRT-4A adjustment locations

T1: 10.7 MHz reference oscillator adjust

C43: Transmitter driver tune.
C45: Transmitter driver tank tune (Channel 1 only)
C51: Transmitter P.A. tank tune (Channel 1 only)
T6: 10.7 Mhz tuned circuit peaking (not frequency determining)

For off-frequency or erratic frequency on transmit, try carefully adjusting C27 in small increments while keying the transmitter in VOICE (carrier only). The author has found this is the adjustment that most often moves off. The voltage at TP1 should be pretty close to 0 VDC while transmitting when the VCO is aligned right.

If the 10.7 MHz local oscillator moves off, it can cause a problem. Unfortunately, T1 has a very fragile slug that's usually glued in like a sonuvagun with glyptol. Messing with T1 is a last resort. Once you get Channel 1 on, and if Channel 2 is malfunctioning, you can try some adjustment of R36.

Leave the rest of the adjustments alone, unless the power out is obviously way low after making the whip antenna adjustment above. Adjusting C43, C45, and C51 in small increments may bring the power up. T6 generally does not get out of alignment.

ALL of these suggestions are for when you don't have the channel alignment set, and your transmitter is really dead in the water. The best bet always is to see the thing work or get a guarantee from the seller. These radios are not easy to work on.

Tip: The author has encountered PRT-4s with bad mike elements, also. Your best line of defense against the D.O.A. effect with these radios is get a pair that are demonstrably okay. Good ones hardly ever fail.

Accessories:

There's really only one of any value -- the H-264, which is nothing but an in-the-ear earphone with a sub-mini audio plug.

PRC-127

Application:

This is a handheld FM VHF transceiver, basically a modified Bendix King commercial business band radio. The power was dropped from 5 to 2.5 watts and the scan switch was left off. These radios were used in the Desert Storm One, but it's interesting to note that G.I.s in Bosnia and elsewhere have been known to favor commercial Family Radio Service handhelds for squad-level comm instead of the semi-obsolete PRC-127...even to the point of paying for FRS radios out of their own pockets!

Tech Specs:

Frequency range: 14 preset channels from 136-160 MHz. The allowable channel spacing varies with the three versions:

PRC-127: 25 kHz
PRC-127A: 12.5 kHz (some documentation says 12.5khz, 15khz, 25khz or 30kHz)
PRC-127B: 5khz, 12.5khz, 15khz, 25khz or 30kHz

Note: The PRC-127B is the most likely version you will encounter. During programming of all versions, an attempt to enter a frequency on an invalid channel spacing will be disallowed.

Type of modulation and power out: FM, 2.5 watts

Design range: 3000 yards

Squelch control: Bendix's "CODE GUARD" system, which consists of either the usual CTCSS tones, or a three-digit value. Receive and transmit are separately programmable.

Size: 7.8" x 2.5" x 1.52"

Weight: 1.5 lbs, including battery

Active devices: late-80s-vintage semis and chips

Power requirements: 12VDC, from a 10-AA-cell battery holder, or rechargeable nicad or NiMH pack. Two packs were supplied originally.

Audio input/output devices: Speaker-mike

Antenna connection: top-mounted screw-in connector (3/8"-32 threads)

Whip antenna: AS-3960/PRC-127 10" screw-in rubber whip

Tech Manuals:
TM 11-5820-1048-10, "OPERATOR'S MANUAL, RADIO SET AN/PRC-127", basic operating instructions only, no programming instructions or schematics.

TM 11-5820-1048-24&P, "UNIT, DIRECT SUPPORT AND GENERAL SUPPORT MAINTENANCE MANUAL (INCLUDING REPAIR PARTS AND SPECIAL TOOLS LIST)", general functional description, programming info, high level testing.

Construction Style: Hard plastic outer case, fairly densely-packed circuitry on several stacked PC boards. Probably uneconomical to repair, given current prices on the surplus market for these radios.

Description:

The RT-1594/PRC-127 is a conventional business band handheld that's been slightly modified and drafted into military service. The programming buttons are normally covered by a plate made of the same dark gray, hard plastic material as the radio case. With the pad covered the radio presents just three controls, ON-OFF/VOLUME, SQUELCH, and the 14-position channel select switch. With the battery removed, the plate can removed for access to the programming keypad.

Operation:

You'll need to program one or more channels before using the radio. A programming sequence follows.

1) To reach the programming keypad and LCD dis-

play, remove the battery by pushing up on the catch on the right hand side, and turning the battery. Slide the cover plate down and off, then replace the battery.

2) Turn on the radio, and hold down FCN until PROG appears on the LCD along with a channel designation.

3) Enter the two-digit channel to be programmed. Channel 1 is "01", etc. Press FCN again. Press CLR to clear whatever is currently programmed.

4) Enter the receive frequency for the channel. If you goof, press CLR again and start over. Press ENT.

5) Enter the CODE GUARD value for the RX frequency. "0.0" will disable CODE GUARD and enable regular noise squelch. Entering a value XXX.X will cause that CTCSS tone to be required on an incoming signal to open the squelch. Pressing the # key and entering a value XXX will require reception of that CODE GUARD value from the other radio to open the squelch. When done, press ENT.

6) Enter the transmit frequency for the channel. If you goof, press CLR again and start over. Press ENT to enter the desired frequency.

7) Enter the CODE GUARD value for the TX frequency. "0.0" will disable CODE GUARD and no sub-audible tone or code will be transmitted. Entering a value XXX.X will cause that CTCSS tone to be transmitted. Pressing the # key and entering a value XXX will cause transmission of that CODE GUARD value. When done, press ENT.

8) Press CLR, then repeat steps 3 through 7 for the next channel. You can check the contents of a channel by going to the channel and repeatedly pressing FCN without changing anything. When all channels are programmed, cycle power on the radio to get out of programming mode.

Once the desired channels are programmed, operation is elementary.

1) Attach the whip antenna by screwing into top threaded connector.

2) Attach speaker-mike if desired.

3) Rotate OFF-VOL control to turn on radio

4) Adjust SQUELCH as desired.

5) Select desired channel with 14-position channel select switch.

6) Commence push-to-talk operations.

Caution: Make sure when you're through that the OFF-VOL control is fully in the OFF position. It's easy to bump this control and drain your batteries.

Additional Comments:

There are other firmware options that were present in the commercial version of the radio but not in the PRC-127.

What To Do If It's D.O.A.:

The author has not yet been able to find a depot-level tech manual for this radio. Without one, any sort of repairs beyond the most elementary or obvious will be really tough. The best bet here is to make sure you get a functional radio to start with. Reports from the author's surplus group indicate this radio, once working, is pretty reliable.

Apparently, it's possible to induce erratic operation or even "cause the radio to not communicate" by improper user-entered firmware settings. Many of these settings are not supported by the PRC-127, but can still be "turned on" during programming which may result in erratic operation. Here's how to check these settings and ensure they are set to the default values originally specified by to the Army:

1) Enter programming mode by holding down FCN until PROG and a channel designation appears. If "Ch 0" is not displayed, enter 00 and press FCN. There is obviously no channel zero -- specifying this channel reaches the firmware variables.

2) The first variable is a 7 digit (max) serial number. This can be set to any value and does not affect the operation of the radio. Press FCN.

3) This is the transmitter timeout length in seconds. "0" will disable the timeout. Press FCN.

4) This is the scan interval. SCAN is not implemented in the PRC-127 and the value will not affect operation. Press FCN.

5) Display will read "1-12345" and 1,2,3 should be steady and 4, 5 should be flashing. If this isn't the case, press the digit whose state you want to change. When done, press FCN.

6) Display will read "2-12345" and all digits should be steady (not flashing). If this isn't the case, press the digit whose state you want to change. When done, press FCN, and cycle power.

Steps 5 and 6 concern the firmware functions that, if mis-programmed on the PRC-127, can possibly cause erratic operation. For more information on these, search on the Web for the DS and GS TM listed earlier.

Another note from the DS and GS TM:

"NOTE: Certain programming errors can cause the radio set display and audio output to blank completely. One such error is the programming of a channel

with an invalid receive frequency. If the radio is turned on with the channel select control set to that channel, the display will blank and there will be no audio output. If there is a valid transmit frequency programmed for that channel, the radio will still transmit but give no indication that it is doing so. (Lack of a valid transmit frequency will cause the radio to emit a warning tone when the push-to-talk switch is keyed). If the display is blank when the radio is first turned on, first try switching to a different channel. If this produces a display, then you can enter the programming mode in the normal manner and correct any errors. If changing channels will not produce a display (no channels have a valid receive frequency), the following procedure is used to restore the radio to the normal operating mode."

a) Channel selector to 1.
b) Turn on radio.
c) Press and hold FCN key for 10sec.
d) R/T should beep for each of following keystrokes. If no beeps are present, R/T is defective. Press and release 1 key.
e) Press and release FCN key.
f) Press and release CLR key.
g) Press and release 1 key.
h) Press and release 5 key.
i) Press and release ENT key.
j) Cycle power. Display should now show "Ch 1". Audio will be available, and you can enter the programming mode in the normal manner.

The author has been in the software biz for a quarter century and, in his opinion, this situation is detestable. A company should never release a product it knows can be rendered unreliable by legal, allowed inputs from the user.

Accessories:

Shown are the nylon radio carry holster, automatic drop-in nicad battery charger, Nicad (or NiMH) battery pack, and 10-AA-cell alkaline battery pack. A speaker/mike (not shown) attaches to the side of the H/T.

AN/PRC-127 accessories

AM-2060

Application:

Mounting base/audio amplifier/antenna tuner control used with the PRC-25 or PRC-77 in a vehicular application. The above configuration of PRC-25 and AM-2060 is a VRC-53 set.

Tech Specs:

Power input: 24VDC +/- 3VDC (specified on the schematic)

Power output: 13VDC nominal regulated and 3VDC nominal regulated

Size: 6" x 15" x 13.5"

Weight: 20 pounds, less radio

Active devices: 60s-vintage transistors, no chips.

Tech Manuals: TM 11-5820-498-35 "DS, GS, AND DEPOT MAINTENANCE MANUAL RADIOS SETS AN/VRC-53, AN/VRC-64, AN/GRC-125, AN/GRC-160, AND AMPLIFIER/POWER SUPPLY OA-3633/GRC AND OA-3633A/GRC"; theory, troubleshooting, repair, schematics

Construction Style: Heavy diecast base with electronics enclosed in a water-tight compartment. Active electronics mounted on a single plug-in board.

Description:

The AM-2060, or OA-3633, is the base that allows mounting the PRC-25 or PRC-77 in a vehicular installation. The base slides into an MT-1029 shock mount, which is the part actually attached to the vehicle. The MT-1029 allows mounting either the AM-2060 or RT-524/RT-246 transceiver. (See RT-524 section).

Operation:

If you have vehicular mounting MT-1029/VRC, the power cord has red and black leads with spade lugs corresponding to +26VDC and -26VDC respectively. For those without a mounting base, here are the pin-outs at the AM-2060 rear POWER INPUT connector:

Pin J	+24VDC +/- VDC (through PWR ON-OFF front panel switch)
Pin A	GROUND
Pin B	"J3 circuits to MT-1029/VRC for connection to vehicular battery and associated intercom equipment (AM-1780/VRC, etc.)"
Pin D	Muting (grounded during transmit)
Pin H	Transmitter Off (keyline, ground for transmit)
Pin K	Audio out (220 mV RMS)
Pin N	Audio input (220mV RMS)
Pin S	Audio out (600 mW)
Pin T	Retransmit (pass through to Pin K of SET POWER)

NOTE: The author has no idea whatever what Pin B does. The schematic shows Pin B going nowhere once it enters the POWER INPUT connector, and includes the cryptic note given above.

1) Mount the radio in the base with the large thumbscrews on either side of the label. Note that the battery box vent must be positioned so it's out of the way.

2) Remove the jumper plug on the POWER connec-

tor on the PRC-25 and connect the CX-4655 interconnect cable between SET POWER on the AM-2060 and POWER on the radio.

Note: If you don't have the CX-4655, your situation is certainly painful, but not hopeless. The connectors on each end are identical, and are type BT06J 12-14P. Since the cable can be installed either way, the connections must be pin to pin. These pins are used: A, B, C, D, E, H, J, K, M, N

3) Turn the PWR ON switch to ON. Turn the SPKR switch to ON.

4) The radio may now be operated in the normal manner (see PRC-25 section), except advancing the PRC-25 VOLUME control will result in an audio level from the AM-2060 speaker considerably higher than from an LS-454 connected directly to the radio!

5) If you have connected either of the vehicular whips AS-1729 or AT-912, the tuned circuits in their bases will be automatically adjusted as you turn the ANT. FREQ. CONTROL on the AM-2060. See the RT-524 section for more on the whip antennas.

Additional Comments:

The VRC-53 is the vehicular set consisting of the PRC-25 and AM-2060.

Strangely enough, there's a different nomenclature for the PRC-25 and AM-2060 *if the PRC-25 contains a battery.* Then, it's called the AN/GRC-125. When the POWER connector plug is removed to connect the CX-4655 from the base, jumpers are removed that complete the power circuits to the battery. Power is then supplied from the base. The battery can remain installed in the battery box, but will be disconnected.

This is the "jerk and run" configuration, since the all that's necessary to switch to the manpack configuration is to pull the radio out of the AM-2060, replace the POWER plug, and strap it on the ST-138 carrier. The same concept was used on the PRC-104, and more recently with the PRC-138.

The corresponding nomenclature for the PRC-77/AM-2060 is VRC-64 and GRC-160.

For those unfortunates who end up with an AM-2060 without the necessary CX-4655 cable, one is pictured at right.

You never know when you might run across one in one of those grungy cardboard boxes you always see under the swap meet tables!

What To Do If It's D.O.A.:

The active electronics is located on a single pull-out board. Potentiometer R10, accessible with the compartment lid removed, appears to control both outputs.

Cable CX-4655

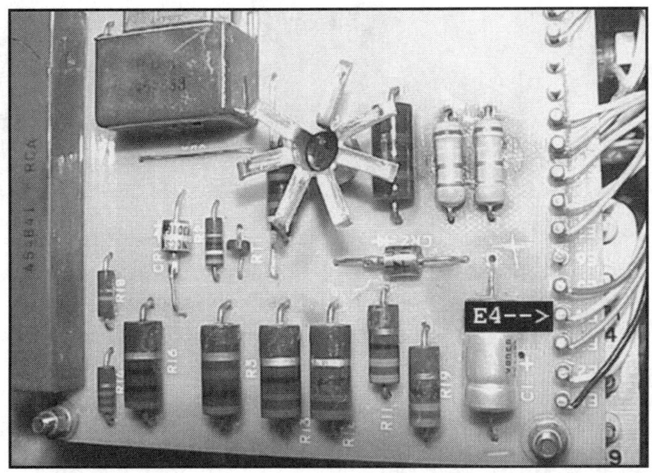

AM-2060 circuit card, top, showing connection point E4

The regulated 13VDC may be adjusted by moving R10 in small increments while monitoring the voltage at connection point E4. E4 is visible at lower right (yellow wire), and is reachable without pulling the board.

The author noticed discoloration on the board beneath the emitter resistors for the audio amp (center-left, near the transistor) -- they run really hot. One was burned in half and needed replacement. Also, the 60uF tantalum filter cap (lower right) on the output of the regulator looked like it'd been inflated with a bike pump. These tantalum caps are a real problem. They're certainly small packages for their capacitance, but their failure rate in this era of equipment is just atrocious. It's always a good idea to do a visual inspection of the board in this unit before powering it up for the first time.

Accessories:

Virtually all the accessories that work with the RT-524/-246 vehicular setup, such as whip antennas and tuner bases, work with the GRC-125 and GRC-160 setups. See the RT-524 section for more info.

AM-598

Application:

This unit is a 28VDC-input power supply that provides necessary voltages for operation of the PRC-8/-9/-10 family of radios in a vehicular setup. Shown above with a PRC-9A installed.

Tech Specs:

Size: 10" x 11" x 11.5"

Weight: 25 pounds

Active devices: vacuum tubes 0C3, 26E6(2), 12AT7(2), 26Z5, plug-in vibrator unit.

Input Power requirements: 28VDC, 2A (radio in receive), 2.5A (radio in transmit)

Output Voltages: -6, 1.5, 67.5 and 135VDC

Other outputs: Internal audio amplifier boosts the radio set output to drive a 600 ohm speaker

Audio input/output devices: H-33, LS-166, other audio accessories with U-77 plugs.

Tech Manuals: TM 11-5055

Construction Style: Heavy-duty cast panel and chassis with slip-on metal case retained by Dzus fasteners. Components easily accessible for maintenance.

Description:

The AM-598 is the 24VDC supply for the PRC-8/-9/-10 family of radios. It's built in the typical style of the era—massive die-cast front panel connected to a heavy chassis and metal cage carrying the components.

Operation:

Important: Before connecting power to the unit, remove the panel/chassis from the case by operating the Dzus fasteners, and verify that the screwdriver switch on the right side of the chassis is set correctly. There's enough of a difference between the unlettered and "A" model radios that a different setting is required for each group. The author believes it's probably due to a different filament load between the two radio types.

Power hookup at the 24V BATT connector is as follows:

Pins A,B: +24VDC
Pins C,D: NEGATIVE

Connect the radio itself to the supply via the octal connector on the radio cable. Clamp the radio to the supply with the side latches as with the regular battery box.

Connect the desired audio accessories such as handsets, speakers, etc., to either or both of the AUDIO connectors. Turn on the whole setup by moving the transceiver POWER switch to ON. It will take about 15 seconds or so for the tubes in the AM-598 to warm up before you hear any audio. Operation of the radio set itself is covered in the PRC-8/-9/-10 section.

Additional Comments:

Caution: The AM-598 uses selenium rectifiers which, if they fail, can produce hazardous vapor. The best bet if one of those things burn up is to get out of the area until the smoke clears.

What To Do If It's D.O.A.:

Obvious things are fuses and tubes. A not-so-obvious thing may be the vibrator unit. If you don't hear a soft buzzing sound from inside when you turn on the unit, the vibrator armature may be stuck. Sometimes it's possible to get the things to start up by tapping on them or, as a last resort, pulling them apart and cleaning the contacts.

The author got his new-appearing AM-598 years ago because it instantly blew fuses when the radio was turned on, and the owner didn't want to mess with it. It became obvious the +24V pins in the 24V BATT connector were dead shorted to ground somewhere.

AM-598 chassis, bottom

The image shows the underside of the author's unit, with the cover over the vibrator socket shield opened. Just above the vibrator socket are toroids used for filtering out vibrator hash. Three of the four cover screws were the same length, but one of them was considerably longer. The screw in the position indicated by the arrow was long enough to actually scrape through the insulation of the top toroid and make contact with the winding! Of course, that winding was the one carrying the +24V supply voltage.

The unit was virtually new-appearing and had certainly never been put into service. In fact, the author believes the unit either never worked, or the screw worked through the winding immediately after the supply was manufactured. The customer probably hooked it up, blew a fuse, and just subbed a different AM-598 and threw this one back into the warehouse. After removing the screw, the supply has worked perfectly. Just goes to show that many problems can have very simple explanations. Be persistent, and you'll find 'em!

AT-1082

Application:

Direction finding antenna used with the PRC-77 and other VHF radios for DF ops in the field.

Tech Specs:

Operational range: 30-80 MHz

Size: 7.5" x 4.25" x 2.75" (loop and whip antennas stowed)

Weight: 5 pounds, entire kit

Active devices: none

Power requirements: none

Antennas: Loop with integral whip

Tech Manuals: unknown, but complete operating directions are included here and on the unit.

Construction Style: Heavy-duty diecast case with recessed controls. Internals repairable, all passive components.

Description:

The AT-1082 is an interesting and useful direction finding antenna that operates through the range of current VHF tactical and vehicular FM gear (30-76 MHz). It uses a loop antenna along with a whip to indicate the true direction of a signal with pretty fair accuracy.

The antenna comes with two lengths of coax—a five foot cable with male BNCs on each end (CG-2840/PRC), and a twelve foot extension with a male BNC on one end and a female on the other (CG-3344A/U). The short coax length has an odd metal cover hanging on it that's obviously meant to fit over the whip antenna connector on a PRC-25 or PRC-77.

It appears the cover is retained by the AB-591 base part of the whip antenna. When this is done, the serrated fingers inside the cover contact the antenna attachment block screws, shorting the whip antenna connection to ground. This is probably an effort to reduce any influence (signal pickup) the whip antenna connector might have on the DF reading.

Operation:

Warning! This is a receive-only antenna. Do not transmit.

The directions as they appear on the back of the unit are shown on the next page.

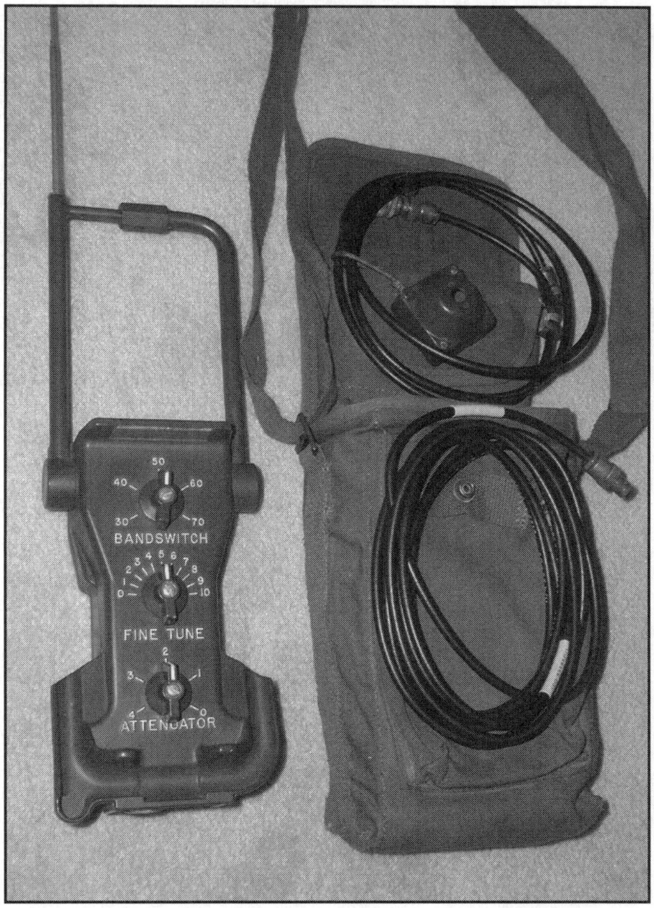

Antenna mount shorting cover

Additional Comments:

The author has had considerable success using the AT-1082 to track down power line noise problems. While the antenna works as specified with a PRC-25 or PRC-77, he found using a Sony PRO-80 receiver makes the setup a little easier to work with. The Sony allows changing modes from FM to AM to SSB,

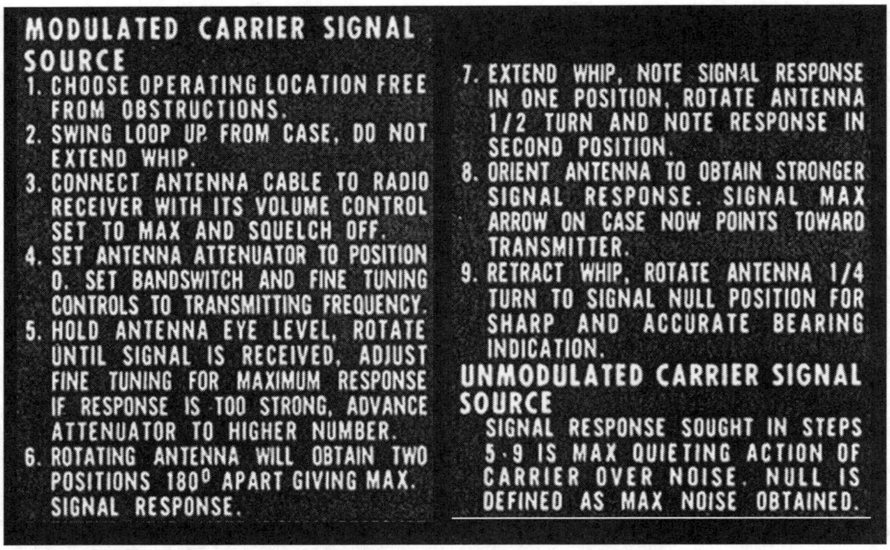

MODULATED CARRIER SIGNAL SOURCE
1. CHOOSE OPERATING LOCATION FREE FROM OBSTRUCTIONS.
2. SWING LOOP UP FROM CASE, DO NOT EXTEND WHIP.
3. CONNECT ANTENNA CABLE TO RADIO RECEIVER WITH ITS VOLUME CONTROL SET TO MAX AND SQUELCH OFF.
4. SET ANTENNA ATTENUATOR TO POSITION 0. SET BANDSWITCH AND FINE TUNING CONTROLS TO TRANSMITTING FREQUENCY.
5. HOLD ANTENNA EYE LEVEL, ROTATE UNTIL SIGNAL IS RECEIVED, ADJUST FINE TUNING FOR MAXIMUM RESPONSE IF RESPONSE IS TOO STRONG, ADVANCE ATTENUATOR TO HIGHER NUMBER.
6. ROTATING ANTENNA WILL OBTAIN TWO POSITIONS 180° APART GIVING MAX. SIGNAL RESPONSE.
7. EXTEND WHIP, NOTE SIGNAL RESPONSE IN ONE POSITION, ROTATE ANTENNA 1/2 TURN AND NOTE RESPONSE IN SECOND POSITION.
8. ORIENT ANTENNA TO OBTAIN STRONGER SIGNAL RESPONSE. SIGNAL MAX ARROW ON CASE NOW POINTS TOWARD TRANSMITTER.
9. RETRACT WHIP, ROTATE ANTENNA 1/4 TURN TO SIGNAL NULL POSITION FOR SHARP AND ACCURATE BEARING INDICATION.

UNMODULATED CARRIER SIGNAL SOURCE
SIGNAL RESPONSE SOUGHT IN STEPS 5-9 IS MAX QUIETING ACTION OF CARRIER OVER NOISE. NULL IS DEFINED AS MAX NOISE OBTAINED.

AT-1082 DF Antenna directions

which sometimes helps in locating line noise problems. Do an operational check on 55.25 MHz, which is one of the carriers for TV channel 2. If you can put channel 2 at roughly the right azimuth, you know you're DFing correctly.

What To Do If It's D.O.A.:

There are no active components inside the AT-1082. It's highly unlikely anything will fail. If operation is erratic, one thing to always check is switch contacts. Remember, do *not* transmit into this antenna!

Accessories:

Two cables mentioned above, and Bag CW-445/PRC.

RT-66, RT-67, RT-68

Application:

Ruggedized 1950s-vintage FM transceivers designed for vehicular, semi-fixed, or portable use in a wide variety of installations. For interoperability data with other gear, see the GRC Gear Spectrum Chart in the Appendix. The RT-66, -67, and -68 differ in minor circuit details and frequency range, with about a one MHz overlap between adjacent radios so they can talk to each other. Photo above is of an RT-67 (left) and its accompanying PP-112/GR 24VDC power supply (right) and CX-1211/U "dogbone" interconnect cable.

Tech Specs:

Frequency Range:
RT-66: 20 - 27.9 MHz
RT-67: 27 - 38.9 MHz
RT-68: 38 - 54.9 MHz

Type of modulation and power out: FM voice and 1600 Hz ringing tone, 2 watts low power, 16 watts high power

Transmit bandwidth: 20 kHz maximum deviation, can be reduced by adding a resistor in-line with the carbon mike audio. (See RT-70 section).

Size: 11" x 9" x 13"

Weight: 34 pounds (R/T unit only, not counting power supply mounting base, etc.)

Active devices: low voltage filament battery type vacuum tubes: 1A3 (2), 1AE4 (2), 1L4 (2), 1R5 (4), 1S5, 1U4 (4), 2E24, 3A4, 3A5 (3), 3B4 (2), 3Q4 (4), 6AK5

Power requirements, assuming capability to transmit on high power is desired:
Filaments: 6.3V 1.5 A
Relay control circuits: 5.6V 500mA
B+: 85V 130mA, 105V 20mA, 150V 40mA, 250V 15mA, 450V 80mA
Transmit bias: -27V 2mA
Applicable power supplies: PP-109/GR, PP-112/GR, and PP-1175A/SR (covered in their own sections)

Audio input/output devices: H-33 handset, LS-166 speaker, H-63 Headset/Boom mike and GSA-6 Chest Set, or pretty much any audio accessory with a U-77 plug.

Antenna connection: 50-ohm BNC jack on front panel

Antennas: Any 50-ohm antenna suitable for operation on the desired frequency.

Antenna tuner: screwdriver adjusts on the front panel for REC ANT TUNE, TRANSMIT ANT COUPLING, and TR ANT TUNE. (See "Operation")

Tech Manuals: There are a wide variety of TMs covering these radios because they were used in so many types of installations. The TM covering the three radios themselves is:

TM 11-289: Receiver-Transmitters RT-66/GRC, RT-67/GRC, and RT-68/GRC; descriptions, theory, troubleshooting and repair, alignment, and schematics.

Construction Style: The radios are integral diecast metal panel and chassis assemblies with heavy-duty ribbed or mottled outer cases. Internal construction is point-to-point between tube sockets and stand-offs. Definitely repairable.

Description:

It's humorous in the era of comm gear stuffed with chips to see a 34 pound beast described in TM 11-289 as a "compact, lightweight radio set". When it was designed around 1950, it *was* "compact and lightweight" by virtue of having the receiver and transmitter in the same case with common circuitry. It was just about as good a design as could be had at the time. This gear was still in use in reserve units into at least the late 60s.

The radios are double-conversion superhet type FM transceivers built in a panel-and-chassis construction with an outer case attached by spring-loaded Dzus

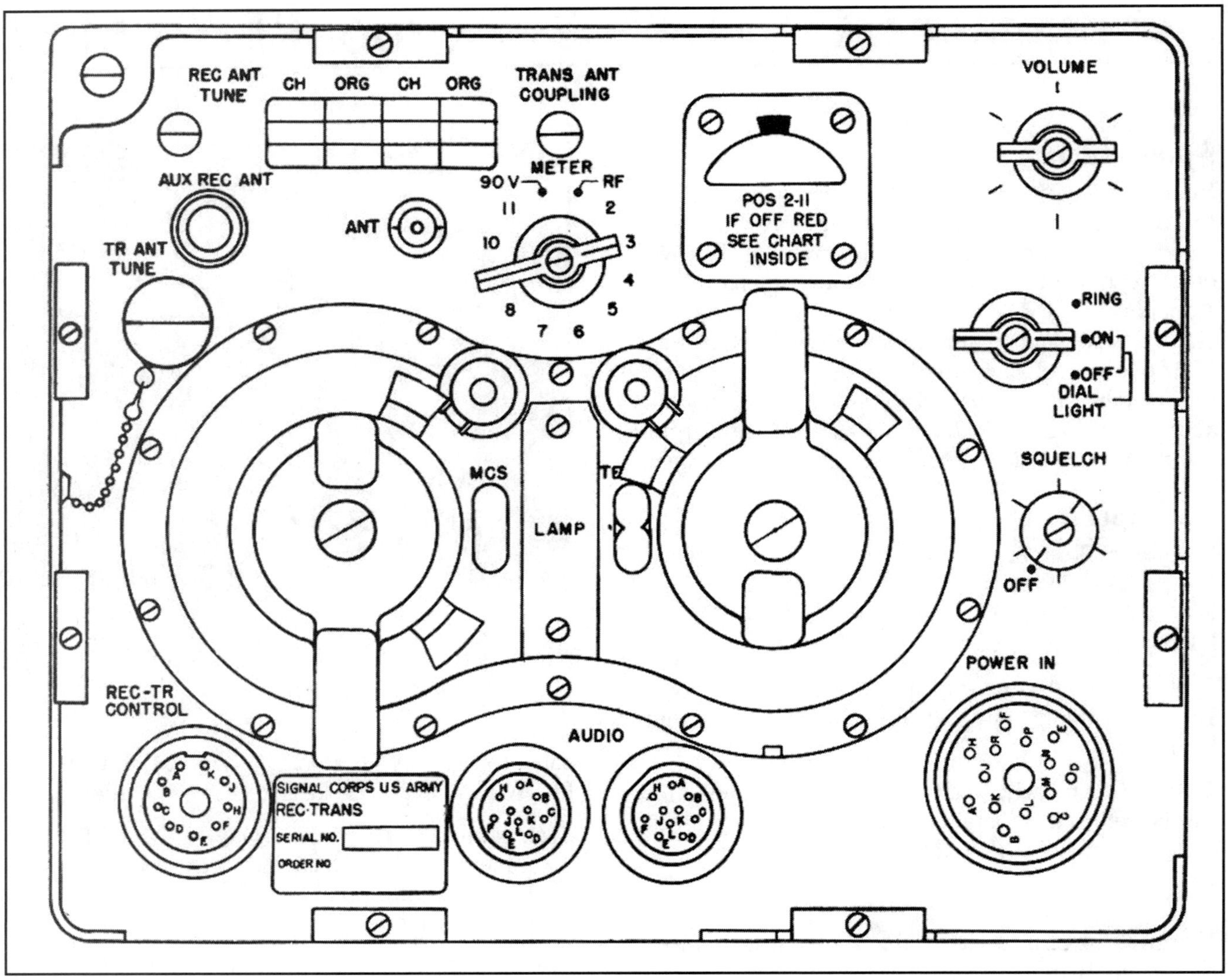

RT-66/-67/-68 front panel

fasteners. When you find one of these units, you'll want to operate these fasteners and pull the radio out of the cover to make sure everything's there. If the cover has been on a long time, sometimes it's necessary to pry gently between the panel and cover to pop it loose.

Tip: It's really not practical (unless you like punishment, both monetarily and mentally) to build a power supply from scratch -- it's easier and cheaper just to buy a PP-112 from Fair Radio Sales.

Operation:

Connect an H-33 handset to either AUDIO connector. Assuming you have a power supply connected to the unit via the CX-1211 "dogbone" cable, apply either 12V (PP-109) or 24V (PP-112) to the power supply. It's much easier and cheaper to buy one of these supplies from Fair Radio Sales and build a 12- or 24volt DC source than it is to build a complex multi-voltage AC supply from scratch.

If you didn't get a DC power input cable with your power supply, the pin-outs on the POWER IN connector are Pin B: POSITIVE, Pin C: NEGATIVE. Your DC source will need to deliver 8 amps at 24VDC to the PP-112 or 12 amps at 12VDC to the PP-109. If you don't have the CX-1211 interconnect cable, it is a straight pin to pin correspondence between the POWER OUT jack on the power supply and the POWER IN jack on the radios using these pins: A, B, C, D, E, F, H, J, K, L, M, N, P, R. The cable is reversible.

Note: If you are not using any sort of mounting base like an MT-298, which supplies a connection to the REC-TR CONTROL jack at the lower left, you must have a jumper from Pin J to Pin H on that jack. These pins normally route the audio through the mounting base circuitry and act as a return audio path. As on the RT-70 POWER jack, if you forget this jumper, you won't get any audio.

Turn the SQUELCH to OFF on the radio and the VOLUME about halfway up. Flip the TRANS POWER switch on the power supply to LOW and the OPERATE switch to TRANS & RECEIVE. You should hear a buzzing inside the supply as the vibrators kick in. You should also get noise in the handset after a short warm-up. The SQUELCH should kick in 1/4 to 1/2 way around.

Now you can perform some simple checks using the METER switch at top center on the panel. During receive, positions 7 through 11 should put the meter within the red zone. This indicates the filaments are good in the receiver section. Also, the "90V" position should also give a reading within the red area, indicating the 90V B+ supply is present.

Hook up an antenna. For receive, ten feet of wire hooked to the center of the BNC jack will do. Tune the MCS and TENTH MCS knobs around. On the RT-68, try 52.525 or 55.25 (Channel 2 TV carrier). On the RT-67, try 10 meters. On the RT-66, try CB. If the TENTH MCS knob is in detent mode -- a detent every 100 kHz -- move it to extreme CCW past "0" until it clicks again. Similarly, you can get back to detent mode by rotating to extreme CW past "9" until it clicks.

To check transmit, hook up a wattmeter and a 50-ohm dummy load. If you have something that will receive in the range, set that up too. The meter pointer should be somewhere near the red in positions 2-6 during transmit. If any of the readings are significantly out of the red zone, you may have some bad filaments. Pull the radio out of the case and see the chart inside for data on which tubes may be bad. Otherwise, set the METER switch to RF and key the handset. You should see some upward deflection on the meter and 1 to 2 watts on your power meter.

Don't despair if you don't see any indication -- the antenna circuits may just need adjustment. The following antenna tuning procedure is adapted from TM 11-611, "Radio Sets AN/VRC-16, AN/VRC-17, and AN/VRC-18", and should be done when the eventual antenna to be used is connected. Read the entire procedure and the following notes before attempting this adjustment. Author's comments in [].

The receiver-transmitter is shipped completely aligned but the receiving and transmitting antenna circuits must be tuned when the set is connected to its particular antenna system. Once the antenna circuits are tuned for a particular system antenna, it is not necessary to retune during normal operation.

a. RECEIVER TUNING. Leave the receiver operating with the SQUELCH control at OFF, and adjust as follows:

(1) Tune for a signal in the top 900 kc of the tuning range. If practicable, arrange to have this signal supplied by another radio set. If another set is not available, radiate the output of an f-m signal generator to the receiver. The signal must be weak so as not to overload the receiver.

[***Tip:*** Generally, atmospheric noise will do for (1) and (2). The REC ANT TUNE behaves like an antenna trim, in that readjusting it after relatively large frequency changes (say, 500 kHz) may result in better receive performance. A gimmick the author uses is to set the squelch with an antenna hooked up so it just barely closes on clear channel noise. If you can get the squelch to open up by tweaking the REC ANT TUNE control, you know you're going in the right direction.]

(2) Using a screwdriver, adjust the REC ANT TUNE control on the receiver-transmitter panel for maximum noise quieting and maximum signal strength in the headset. Reduce the r-f input signal strength as the strength of the a-f signal increases. Repeat the adjustment.

b. TRANSMITTER TUNING. To tune the transmitter antenna, proceed as follows:

(1) Tune the receiver-transmitter to the extreme low end of its tuning range (MCS tuning control at lowest mc calibration; TENTH MCS tuning control at zero). Leave the TENTHS MCS tuning control at zero throughout the tuning procedure. Set the power supply to LOW and TRANS & RECEIVE.

Caution: Do not turn the TENTH MCS tuning control while the TR ANT TUNE control is being adjusted. Serious damage to the tuning mechanism can result.

(2) Remove the cap which covers the TR ANT TUNE control on the receiver-transmitter panel.

Caution: The transmitter should not be operated for more than five minutes continuously. Allow a 15-minute stand-by period between each 5-minute period of continuous operation.

[The author has used an RT-68 on high power for at least two hours in a normal QSO with no problems whatever.]

(3) Key the radio and adjust the TRANS ANT COUPLING adjust for minimum meter reading. This is the point of minimum coupling. [See notes below].

(4) With the radio still keyed, engage the TR ANT TUNE control by pressing in on it with a screwdriver and adjust for a maximum r-f reading on the panel meter. If it is impossible to obtain a reading, advance the setting of the TRANS ANT COUPLING control slightly and repeat the procedure. Keep the coupling as low as possible while adjusting the TR ANT TUNE control for a maximum r-f reading. When the maximum reading is obtained, unkey the radio.

(5) Repeat the adjustments given in (3) and (4) above for each integral mc calibration on the MCS dial and for the highest detented frequency of each receiver-transmitter (27.9, 38.9, and 54.9 mc for Receiver-Transmitters RT-66/GRC, RT-67/GRC, RT-68/GRC, respectively).

(6) When adjustments are complete for each setting of the MCS tuning control, turn the MCS tuning control to the approximate center of the tuning range (ie: 46 MHz for the RT-68), and adjust the TRANS ANT COUPLING control for a maximum reading.

(7) Repeat the adjustments given in (4) through (6) above.

[This procedure is a compromise, and peaks a fixed antenna system (usually a vehicular whip) for receive and transmit operation. If your desire is to operate on only one frequency or group of frequencies -- say 50-54 MHz on the RT-68, with a priority on 51.0 MHz -- you can optimize the tuning procedure around those. You may choose to pay particular attention to peaking the radio right at 51.0 MHz for receive, and the 50, 51, 52, and 53 MHz bands per the procedure for transmit.

In practice the author has found with a 50-ohm load that tuning the TR ANT TUNE in conjunction with the TRANS ANT COUPLING for maximum meter reading on one band, will result in the TRANS ANT COUPLING control at about the right position for all bands. It won't change much across the frequency range. Tuning into oddball impedances like whips may change this situation. The author has seen a wide range of power out figures. Generally, if you can get 1-2 watts on low and 10-16 watts on high, you're doing okay.]

Additional Comments:

The RT-68 will get you on six meters, and the RT-67 will get you on ten meters -- both on a budget. Be sure to do the resistor-in-the-audio-line trick to reduce the transmit bandwidth first (see the RT-70 section). It's pretty tough to come up with anything the RT-66 might be good for. A CB jammer?

What To Do If It's D.O.A.:

If the meter checks outlined above all fall within nominal range, make sure you have J jumped to H on the REC-TR CONTROL jack.

A good deal of basic troubleshooting can be done from the front panel, and most faults can be traced to a bad tube.

Accessories:

Just about any of the GRC-type accessories with U-77 plugs will function with the RT-66/67/68 family of radios: H-33 handsets, H-63 Headset/Boom-mikes, M-29 mikes, etc. There are also several vehicular mounting bases that allow lashing together various combinations of receiver-transmitters, auxiliary receivers (R-108/-9/-10), and remote and local control boxes (C-434 and C-435) into complete systems.

The illustrated set includes an MT-297 mounting base and (left to right) an R-110 auxiliary receiver (38-54.9mHz), RT-68 (38-54.9mHz), PP-112 power supply, and RT-70 transceiver (47-58mHz) perched atop its AM-65 power supply/audio amp. The unit below the RT-68 is a C-435/GRC local control box. The control box allows setting up the set as a repeater,

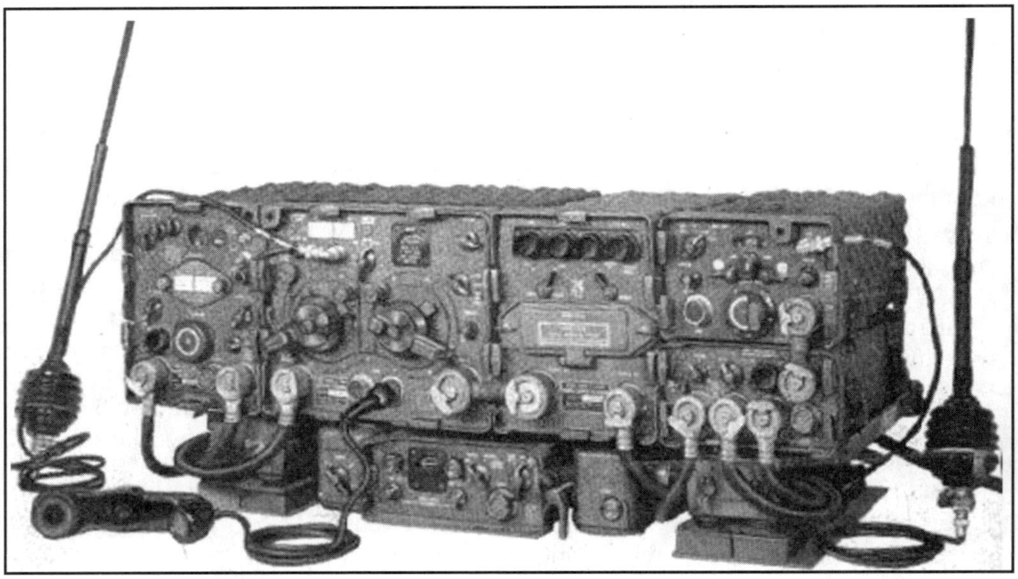

The AN/GRC-7 vehicular radio set

with the ability to receive on the RT-68, route the audio to the RT-70, and retransmit it on a different frequency. When a signal is received on the RT-70, the inverse will occur, with the RT-70 receive audio being retransmitted from the RT-68. The R-110 allows monitoring on a third frequency.

Tip: A nifty setup along these lines is the author's AN/GRC-"8.5", which consists of an RT-68 and an RT-67 on an MT-298 mounting base along with a C-435 control box and two power supplies. This makes a complete six to ten meter FM repeater.

Once the author accumulated a number of GRC-type radio sets, he found he could mix and match to create just about any configuration he wanted. One piece of memorable craziness he did was talk to a friend in Indiana from California by going into an RT-67 on 51.0 MHz using a PRT-4/PRR-9 squad radio set and getting retransmitted out the RT-68 on 29.6 MHz! This setup is covered in its own (GRC-"8.5") section.

RT-70

Application:

VHF low-power wideband FM transceiver originally used in portable, ground, or (more usually) vehicular applications as a secondary transceiver in many different installations.

Tech Specs:

Frequency Range: 47 to 58.4 MHz continuous tuning

Type of modulation and power out: FM voice at about 500mW

Transmit bandwidth: 20kHz maximum, can be easily reduced (see below)

Receive sensitivity: Less than 0.5 uV minimum discernable signal

Size: 7.5" x 5" x 13"

Weight: 15 pounds

Active devices: vacuum tubes, (1) 3B4, (4) 3Q4, (1) 3A5, (2) 1L4, (5) 1U4, (2) 1R5, (1) 1S5, (1) 1AE4

Power requirements: 90VDC 100ma (plate and screen), 6.3VDC 360ma regulated (filaments), 6.3VDC 160ma (transmit/receive relay). Unit may also be powered by the AM-65 audio amplifier/power supply, covered in its own section.

Audio devices: Many of the GRC-type accessories with U-77 plugs like the H-33 Handset, Headset-Microphone HS-63/U with Chest Set Group AN/GSA-6, Loudspeaker LS-166, etc.

Antenna connection: 50ohm BNC jack on the front panel

Tech Manual: TM11-290 (operation, theory, troubleshooting, schematics)

Construction Style: Heavy. The radio is an integral diecast metal panel and chassis assembly with a heavy-duty ribbed or mottled outer case. Internal construction is point-to-point between tube sockets and stand-offs. Definitely repairable.

Description:

The unit illustrated was the author's very first surplus set, and occupies a special place in his green stuff pile. He bought it from Fair Radio Sales in 1976 after much discussion with his more experienced friend about the

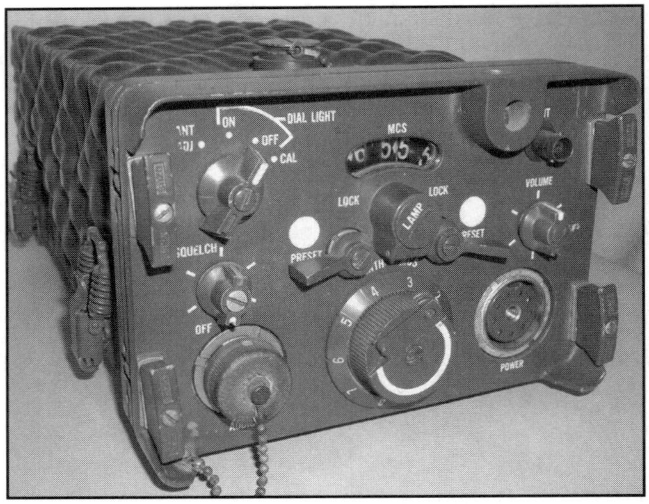

relative merits of various sets. That was when Fair was on Eureka St in Lima, OH -- that amazing old building jammed full of goodies dating back to WWII. Phil Sellati let us wander down into its dark depths to hand pick a unit. We found one in a sealed box that appeared unused, packed with a manual. A power supply using a gaseous regulator circuit was the author's first electronic project of any complexity.

The RT-70 is a great first surplus radio. It takes relatively simple input voltages and tunes continuously through six meters, a useful ham band. If you have someone local to talk to like the author did, it's ideal. The other station had an RT-68 and yagi antenna set up and the author used a simple ground plane.

Operation:

RT-70 POWER connector pin-outs:
Pin J: 90VDC @ 80 ma
Pin F: 6.3VDC regulated @ 360 ma
Pin B: 6.3VDC @ 160 ma for TX/RX relay
Pin D: Ground
NOTE: Pin J must also be jumpered to Pin H

It's easiest to design the regulated 6.3VDC with enough current capability to handle both the filament and relay current. Fair Radio may still have some power connectors, but you can always just poke in wires to check out the radio!

RT-70 AUDIO connector pin-outs:
C: Mike audio in
F: Transmit control (push to talk)
A: Headphone audio
B,E,H: Ground

Attempting to clip-lead to the spring-loaded contacts on the AUDIO jack is an exercise in futility. You really need a suitable handset (H-33). Pin C expects a carbon mike input. In a pinch, you can haywire an old telephone handset carbon element. The male connector (mike end) is a U-77.

Assuming you've connected suitable power, a 50ohm antenna or dummy load, and a handset, turn the SQUELCH to OFF and the VOLUME up about half way and turn on the supply. You should hear a rushing noise in the handset. The SQUELCH should cut off the noise anywhere from 1/4 to 1/2 way clockwise past OFF. Try tuning across the range -- there are sometimes utility stations or businesses on lowband FM. If you have a channel 2 TV signal in your area you should hear a carrier at 55.25 MHz. Of course, if you have anything that transmits in the range, or a sig-gen, etc., you can test the receiver that way too.

Transmit by depressing the mike button. You should get about a half a watt of wideband FM. The transmitter output can be peaked by adjusting the screw accessible beneath the screw cap on top of the radio. Just adjust for max power out into a 50ohm load. This setting shouldn't change much across the range.

To turn on the dial light, flip the function switch at upper left on the panel to ON. This is a very weak bulb. There's a calibrator that gives a zero beat at each MHz point on the dial. To use this, rock the function switch to CAL. If the MHz point on the dial is off substantially, you can slip the dial ring while holding the tuning knob.

Two frequencies may be pre-set by rotating the PRESET levers to LOCK when tuned to the desired frequency to be pre-set.

Additional Comments:

You can easily remove the radio from the case by rotating the Dzus fasteners on the front. In fact, you should do this to check that the "TANK-VEH-FIELD" switch on the left side is set to either "TANK" or "VEH" (vehicle). The "TANK" and "VEH" positions equalize the filament drain between transmit and receive, and the "TANK" position raises the sidetone level for noisy tank environments. The "FIELD" position does not equalize filament drain for use with batteries in the field. This is not critical, and the radio will work on any of the three positions, but depending on how you built your power supply it may be happier with equal drain on the filaments.

Tip: The standard transmit deviation is too wide for use with modern FM ham gear, BUT there is a quick and dirty fix. Just put a resistor in series with the mike input. You can do this inside the handset switch, or inside the radio itself. Start with a haywired 5K pot to determine the value you want, then replace it with a corresponding fixed resistor. Make up for the distant station's lower deviation by simply turning up the RT-70 volume. This works just as well with the other wideband gear like the RT-66, RT-67, RT-68, PRC-10, VRC-12, etc.

What To Do If It's D.O.A.:

The author has owned numerous RT-70s, including a sizable stack purchased through a government surplus bid back in the late 1970s. Incredibly, the only defective one was his original unit purchased new-in-the-box from Fair Radio! The problem turned out to be a physically broken tube -- the little nub had snapped off the envelope, destroying the vacuum. Most problems will be traced to a bad tube.

Reminder: If you are using your own power supply with the radio, remember the J to H jumper on the POWER connector! Without that, you won't get any audio and it'll seem like the set is dead.

Accessories:

Technical manual, tubes, handsets, and connectors are still available from Fair Radio Sales. A useful accessory is the AM-65 audio amplifier/power supply that allows operation from 6, 12, or 24VDC (see the AM-65 section).

C-435, MT-298 and the "GRC-8.5"

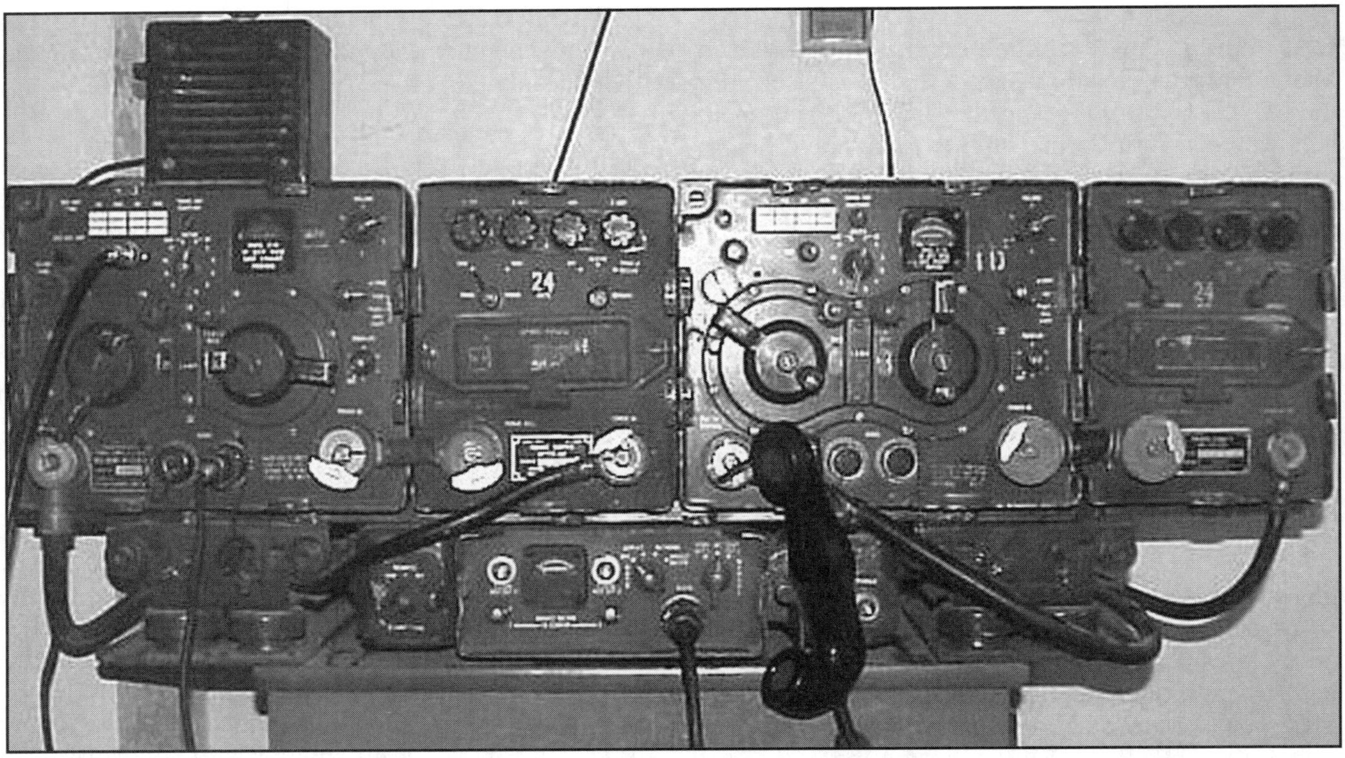

Application:

Example of how several pieces of GRC gear can be integrated to produce a useful suite -- in this case a six- to ten-meter crossband FM retransmission setup. The set consists of an RT-67 and an RT-68 transceiver, two PP-112 24VDC power supplies, a C-435 local control box, an MT-298 mounting base, and two CX-1211 "dogbone" interconnect cables (power supply to radio). The MT-298 supplies the remaining interconnect cables to complete the set.

Tech Specs:

Specs for the individual units will be found in their respective sections, except for the C-435 and MT-298 which are covered here.

Size: 37" x 14" x 13"

Weight: Somewhere in the region of 200+ pounds. With one of these over the rear axle in your M-151, you won't have to worry about traction!

Power requirements: With PP-112 power supplies, 24VDC at 9 amps or so. With PP-109 supplies, 12VDC at a little less than double that. These figures are for one radio receiving and the other transmitting on HIGH power setting.

Tech Manuals: The closest TM the author has is TM 11-284 "Radio Sets AN/GRC-3, -4, -5, -6, -7, and -8" which covers sets similar to the one shown at the end of the RT-66, -67, and -68 section in this book. The interconnections are very similar, however, and the C-435 local control is used for the retrans capability in exactly the same way.

Construction Style: The radios are integral diecast metal panel and chassis assemblies with heavy-duty ribbed or mottled outer cases. Internal construction is point-to-point between tube sockets and stand-offs. Definitely repairable.

Description:

If you study the GRC FM Frequency Spectrum Chart in the appendix, you'll see there is no designation for a set consisting of two RT-66/67/68 units, of the same frequency range or otherwise. Hence the author's GRC-"8.5" designation. It's hard to believe this idea of using two of the higher-power sets didn't occur to anyone in the Army. The author's setup consists of an RT-67 and RT-68 with their associated PP-112 power supplies on an MT-298 vehicular mounting base with a C-435 control box. The radios are described in their own sections. The rest of the components will be covered here.

The C-435/GRC local control box plugs into the mounting base and extends the set function to include duplex operation and retransmission. The MT-298 is a heavy-duty (and just plain heavy) steel mounting base that accepts various pieces of GRC FM gear and supplies cables for the POWER jacks on the power supplies, and the REC-TR CONTROL (receive-transmit control) jacks on the radios. The REC-TR CONTROL cables carry push-to-talk (PTT) keying and audio between the sets for retransmit purposes. The MT-298 has a main power switch OFF-REMOTE-ON shown just to the left of the installed C-435 above. This switch must be ON to supply power to the system. The pilot light visible just to the right of the control box will illuminate to show you have power.

The control box rails will accept any of the following controls depending on the application: C-433, C-434, or C-435. (The C-433 and C-434 are covered in separate sections).

Operation:

If you are unfamiliar with the operation of the basic radio and power supply, see the "RT-66, RT-67, RT-68" section.

Connect the two left-hand MT-298 cables to the left-hand radio set, and the right-hand cables to the right-hand set. The POWER connector on the power supply and REC-TR CONTROL connector on the radios have different pin configurations and can't be interchanged. Each set will also have a CX-1211 "dogbone" cable between power supply and radio.

Connect a handset to the AUDIO connector on the C-435, and set the SQUELCH control on both radios fully CW (fully squelched). Set the AUTO switch on the C-435 to OFF. Turn the main mounting base switch to ON, and the switches on the two power supplies to LOW and RECEIVE.

You should be able to back off the SQUELCH on each radio in turn and hear receiver noise. Un-squelch SET 1 (left-hand radio). While holding the AUTO switch to the ADJUST METER position, adjust SIG REC SET 1 for center on the meter. Re-squelch SET 1. Un-squelch SET 2 (right-hand radio). While holding the AUTO switch to the ADJUST METER position, adjust SIG REC SET 2 for center on the meter. This procedure equalizes the audio levels for retransmit or duplex operation. Notice that when squelch breaks on either radio, the corresponding SET 1 or SET 2 light illuminates.

Assuming SET 1 is the RT-67, tune it to 29.6 MHz. Tune the RT-68 (SET 2) to 51.0 MHz. Connect suitable antennas or dummy loads to the two radios and readjust the SQUELCH so it's positively closed. Set the OFF-RECEIVE-TRANS & RECEIVE switch on both supplies to TRANS & RECEIVE. Verify the AUTO switch on the C-435 is set to RETRANS. You should now be able to transmit into the RT-68 on 51.0 MHz and have it retransmitted by the RT-67 on 29.0 MHz. The process should also work in the other direction.

If the AUTO switch is set to the DUPLEX position, SET 1 (in this case the RT-67) will transmit and SET 2 will receive. This order can't be swapped without physically switching the radios. The MANUAL switch allows for straight PTT operation of SET 1 and SET 2. Note that while using a handset plugged into the C-435, you need to depress the PTT button on the handset while keying SET 1 or SET 2 with the MANUAL switch. The GSA-6 chest group used with the HS-63 headset can be rigged to hold the PTT down constantly.

Tip: Audio accessories plugged into the C-435 will hear all audio channels mixed together. Accessories plugged into a particular radio will hear only audio from that radio.

Additional Comments:

The author once used this setup to retransmit a PRT-4A/PRR-9 squad radio combo on 51 MHz through to 10 meters. Getting the power supply vibrators buzzing and the lash-up retransmitting again brought back old memories.

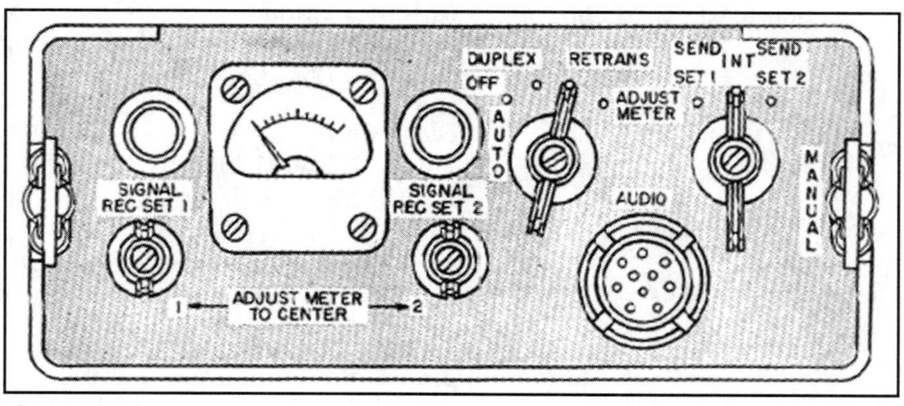

C-435 Control, front panel

RT-524, RT-246

Application:

Tactical vehicular FM VHF transceiver used by virtually every branch of the U.S. military (yes, the Navy too—more on that below) from the early 1960s into the 1980s.

Tech Specs:

Frequency range: 30.00 to 75.95 MHz in 50kHz steps, 920 channels

Type of modulation and power out: FM
Low power -- .5 to 8W with 25VDC input; .5 to 10W with 30VDC input
High power -- 35W minimum with 25VDC input

Transmit bandwidth: audio input of 2.8 mV produces 8 +/- 2kHz deviation

RF sensitivity: .1uV, minimum

Squelch: 150Hz tone activation or standard carrier squelch

Transmission distance: 5mi on low power, or 25mi on high power (see text)

Transmit duty cycle:
Continuous in high power in 120degF ambient, input power of 22VDC
One hour continuous, same conditions, except input power of 30VDC

Size: 15.5" x 6.5" x 12"

Weight: 52 pounds

Active devices: Solid state, except for tubes 6205(2) first and second RF amp; 5686 transmitter driver; 7843 power amp

Power requirements: 22 to 30VDC

Current draw:
Receive: 750ma at 25VDC
Transmit: 3A at 25VDC(low power); 10A at 25VDC (high power)

Audio input/output devices: Any of the new family audio accessories with U-229 plugs, such as M-80 mike, H-138, H-189, H-250, H-161, etc.

Antenna connection: BNC 50-ohm

Whip antenna: AS-1729/VRC whip

Other antennas: RC-292 beam antenna for extended range

Antenna tuner: MX-2797/VRC base containing remotely operated antenna matching circuits, used with AS-1729 whip.

Tech Manuals:
TM11-5820-401-10-1 "OPERATOR'S MANUAL, RADIO SETS: AN/VRC-12, AN/VRC-43, AN/VRC-44, AN/VRC-45, AN/VRC-46, AN/VRC-47, AN/VRC-48, AN/VRC-49 (USED WITHOUT INTERCOM SYSTEMS)"; operating instructions only, no theory or schematics.

TM 11-5820-401-34-2 "DIRECT SUPPORT AND GENERAL SUPPORT MAINTENANCE MANUAL

RADIO SETS AN/VRC-12, AN/VRC-43, AN/VRC-44, AN/VRC-45, AN/VRC-46, AN/VRC-47, AN/VRC-48, AN/VRC-49, (RECEIVER-TRANSMITTERS RADIO RT-246/VRC AND RT-246A/VRC AND RT-524/VRC AND RT-524A/VRC"; theory, troubleshooting, repair, alignment, schematics
Equivalent Navy TM number for above: NAVELEX 0967-LP-432-3030

TB 11-5820-401-34-1 "ALIGNMENT PROCEDURES FOR RECEIVER-TRANSMITTERS RADIO RT-246/VRC, RT-246A/VRC AND RT-524/VRC, RT-524A/VRC)"; complete electronic alignment procedures (see text)
Equivalent Navy TM number for above: EE 150-JA-MMI-010/E154 RT 246

Construction Style: Heavy-duty ruggedized cast metal outer case. Elaborate, modularized construction, all modules are repairable/replaceable, although extender cards are necessary in many cases.

Description:

The RT-524 and RT-246 are complex electromechanical hybrid radios of typical design for the period (early to mid-1960s). They were the standard U.S. short range vehicular FM radio from the early 1960s into the 80s until the SINCGARS sets began taking their place. The author believes they may still be found in reserve units. The type has also been manufactured by other countries (TADIRAN in Israel comes to mind). The author's unit has proven to be very reliable, with only one fault (a problem in the I.F.) popping up since he acquired it in the early 80s.

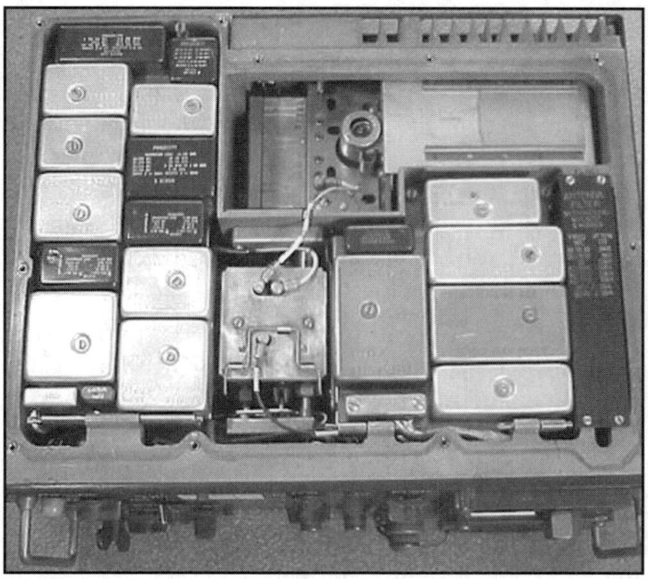

RT-524, top cover removed

RT-524, top cover removed, module trays hinged up

The illustrations show the amazing detail and compact construction of this radio. Note the profusion of Oldham couplers. This kind of mechanical construction is seldom seen any more due to the advent of microprocessor control (see the RF-5020 section).

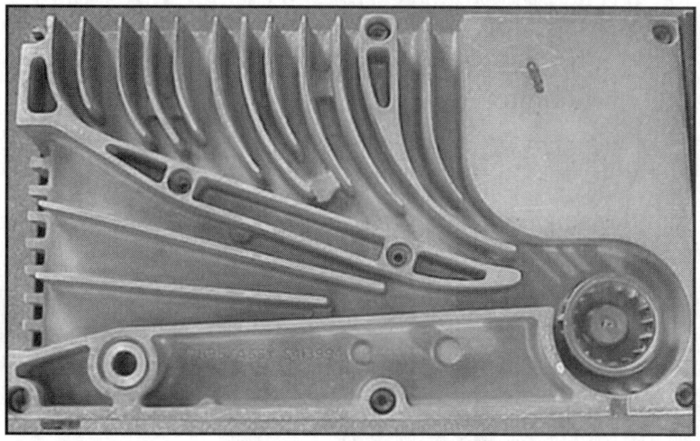

RT-524 blower and convective heat sink, cover removed

Cooling air is directed around the side of the R/T and over the DC-DC converter transistors. The final tube, an external anode 7843 type, is clamped to the heat sink inside and cooled by conductive means.

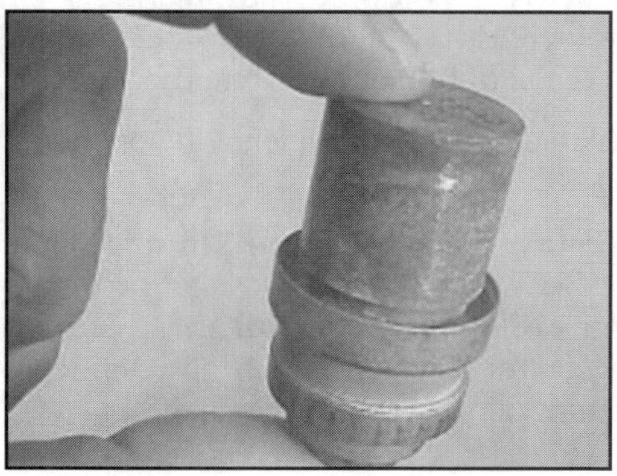

7843 power amp tube

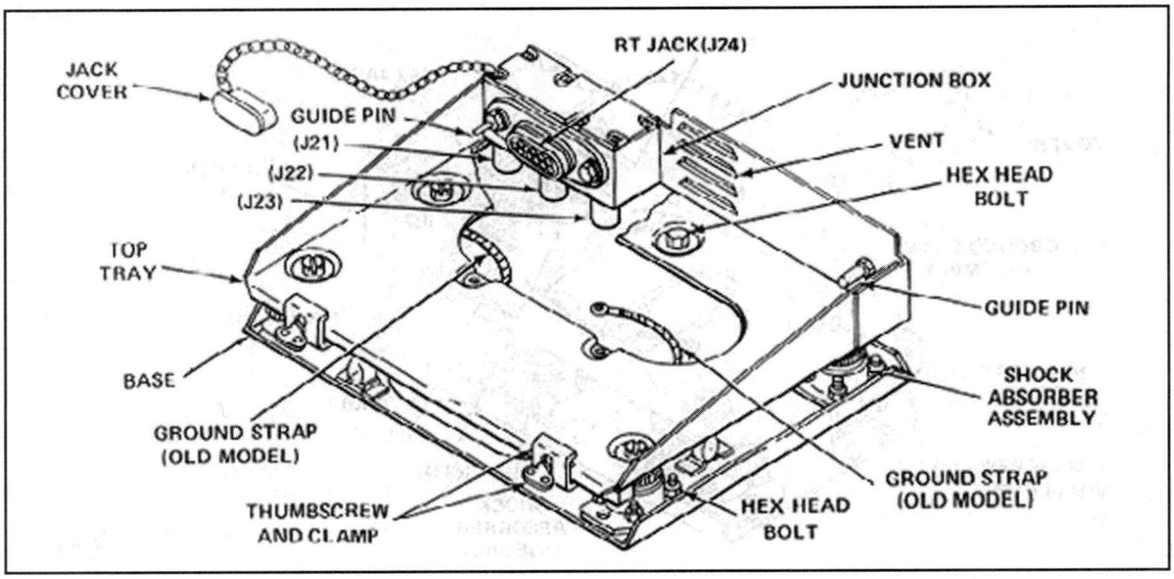

Mount MT-1029/VRC, for one RT-524 or RT-246

Operation:

If you have Mounting MT-1029/VRC the power cord has red and black leads with spade lugs corresponding to +26VDC and -26VDC respectively.

For those without a mounting base, the pin-outs at the rear connector are: Pin A GROUND; Pins B,J +26VDC. B and J are paralleled together to handle the ten amp current on the high-power transmit setting.

Condensed operating instructions, from the author's experience:

1) Connect power per above

2) Connect a handset to either audio connector, and a 50-ohm load to the ANT connector

3) SPEAKER switch to ON

4) SQUELCH to OLD-ON

5) VOLUME to 1/4 or so

6) POWER switch to LOW

A correctly operating RT-524 will come up in an unsquelched state. After a few seconds, the I.F. noise will gradually increase as the tubes in the receiver front end warm up until the squelch abruptly closes.

7) Enter the desired operating frequency by operating BAND, MC and KC dials

CAUTION: (From operator's manual) "In some units of the RT-524/VRC, the KC control cannot be turned beyond the limits of 00 and 95 positions. To prevent damage to the control mechanisms of these units, do not attempt to turn the KC control beyond 00 and 95 positions."

8) If operation with radios using 150Hz tone-operated squelch is desired, set the SQUELCH control to NEW-ON.

9) To transmit, hit the PTT button and talk. From the blower noise you will surmise this is not a radio that should be used while sneaking up on the enemy! If necessary, set the POWER switch to HIGH for approx 35W out.

Note: The blower runs during any transmitting periods, and also whenever the internal temperature goes above a certain value. So don't be surprised if, in extremely warm conditions, you unkey and the blower keeps running. The author has never had this happen under usual conditions indoors.

Additional Comments:

The caution concerning the KC control is true only for some sets, possibly a very small minority of sets. It's worth (carefully) investigating on your set because the author's group has taken advantage of the fact for many years now that many RT-524s will operate well below 30mHz.

Tip: Set your operating frequency to 30.00 MHz. CAREFULLY attempt to move the KC control below "00". It should move with the same feel as usual from "00" to "95" and the left half (megahertz) window should display a red flag. This is something to check as you're contemplating buying that RT-524.

Caution: If you experience any undo resistance, or the KC control seems jammed, DO NOT attempt to force it—you may have one of the transceivers mentioned in the caution above.

If the KC knob moves freely, try transmitting below 30.00 MHz. The author's group has had varying degrees of success with this gimmick. ALL of our radios function down to 29.60, and several exhibit good performance well below that. The author's unit transceives with nearly full output down to 29.30 MHz. Your mileage may (and will) vary—these radios were never guaranteed to specs below 30.00 MHz.

Tip: Remember, these radios have a nominal 8 kHz transmit bandwidth. You will disturb people on the ham bands with a signal that wide! A simple fix is to mount a resistor in a spare handset to limit the deviation, in series with the element. For receive, just turn up the volume. Start with a haywired 5k ohm pot and figure out the correct resistance with on-air tests, then substitute a fixed resistor. Label that handset as your "ham" unit. The RT-524 or RT-246 are great ways to get on ten meters FM!

What To Do If It's D.O.A.:

This is an outrageously complicated radio, both electronically and mechanically. I'm sure some who have worked on them may disagree, but the author sure wouldn't want to service 'em for a living. Something else that might be illustrative. The third manual listed above is actually a Technical Bulletin, published in 1983. There's a very interesting paragraph on the first page:

"Many people throughout the Army have developed quick-fix alignment procedures. These procedures have helped to overcome problem areas in specific cases; however, unless a complete alignment was performed on the radio, the overall efficiency of the radio set may not have been upgraded. As a result, the majority of FM radio sets being used in the field will not meet final performance test standards. To ease this problem, all the AN/VRC-12 series radios should be subjected to a complete alignment on a periodic basis (approximately annually). The performance guides in this technical bulletin will cover those procedures that must be used to do the job properly."

This is a bit of a hoot, because it amounts to both the Army and Navy scolding their techs for sloppy aligning jobs! Actually, it's sort of easy to understand why the techs developed the quick-fixes; alignment of these radios is complicated. You'll need the TM with complete theory and schematic for any serious work on these radios. See Appendix C for manual sources.

Other interesting warnings and informational messages in the Operator's Manual:

Caution: Do not use the following frequencies

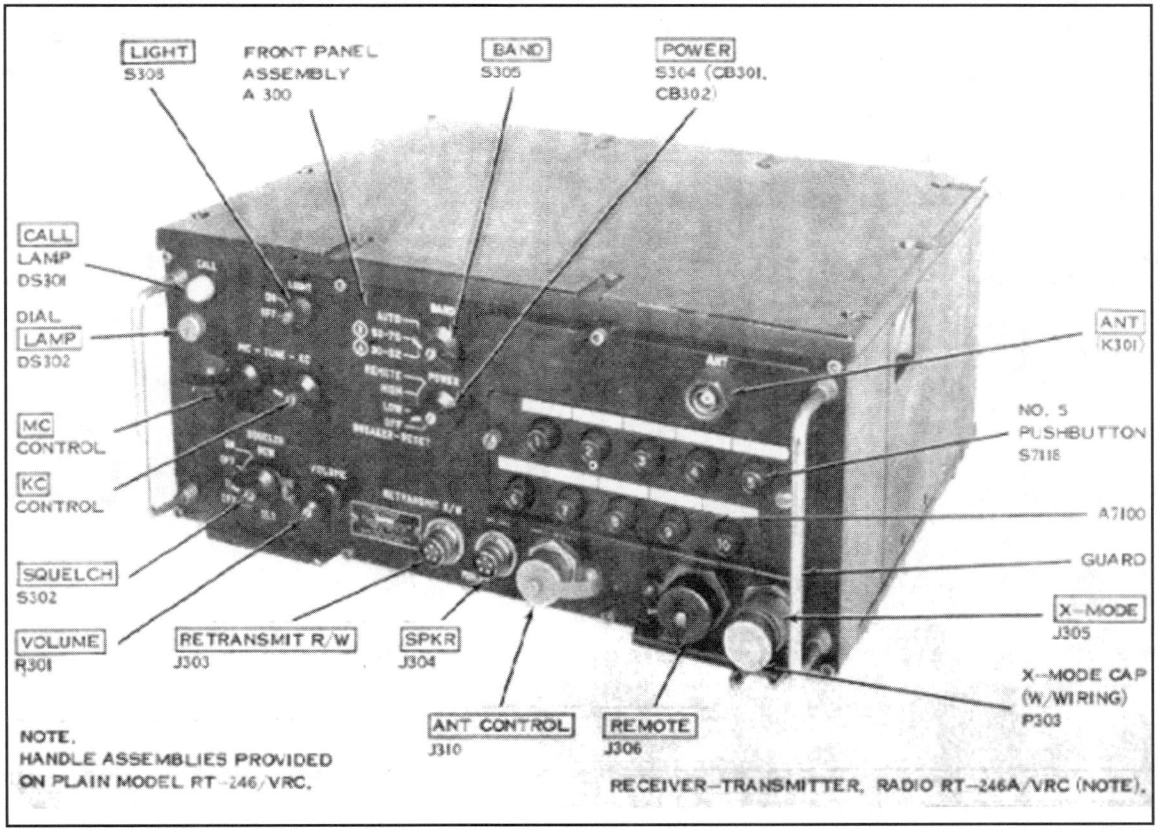

RT-246/VRC, for comparison

because they cause improper operation of the set: 33.90, 45.20, 56.50, 67.8 MHz.

Caution: Do not use frequencies which are separated exactly by 5.75 or 23.00 MHz.
[This must apply when two radios are set up for retransmission duty].

Caution: Do not start vehicle while radio is ON.

The RT-246 is the autotune version of the VRC-12 R/T unit. The speaker is eliminated and an autotune mechanism is built into the space thus freed up.

Power hookups are the same. The RT-246 has an extra AUTO position on the BAND switch. In the AUTO position, the front panel pushbuttons allow tuning to one of ten preset frequencies. The quick version of how to preset the buttons follows.

1) Loosen the captive screws on the hinged pushbutton cover and drop the cover.

2) For the channel button you wish to preset, set the A-B switch to the correct range; A: 30.00-52.95mHz, B: 53.00-75.95mHz

3) Push TUNE button (upper left between buttons 1 and 2), and tune preset control of the corresponding channel (center row of screws) to the correct frequency as shown on the dial window. Release TUNE button to lock in frequency.

4) Repeat for other desired presets

In 1985, the author was working for Rockwell International, Autonetics Marine Division on the systems engineering for the Comm Center for the new DDG-51 Arleigh Burke class AEGIS guided missile destroyer. While reviewing the prints, he noticed they had specified an FM VHF radio to be used for ship-to-ship and ship-to-shore short-range comm. A requirement existed for the captain or radio room personnel to select the operating frequency remotely. The prints had the correct control box, the C-2742/VRC. But the radio specified was the RT-524. The author went to his boss and pointed out the goof, and showed how they needed to change the transceiver to an RT-246. The supervisor's eyebrows went up and he asked how a young upstart engineer was so sure, whereupon he was informed said upstart had an RT-524 sitting on his bench at home! That got his boss's attention, and it's how yours-truly saved Rockwell several thousand bucks in prototype costs.

Accessories:

The MT-1029, pictured earlier, is the mounting base that accepts a single RT-524 or RT-246.

The autotuned whip antenna is the AS-1729/VRC, which consists of the AS-1730 and AT-1095 whip sections and the Antenna Base, Matching Unit MX-6707.

There's a square base version called the AT-912/VRC.

An interesting note given for both antennas: "Disconnect the cables before manually tuning the

Antenna AS-1729/VRC (Round Base)

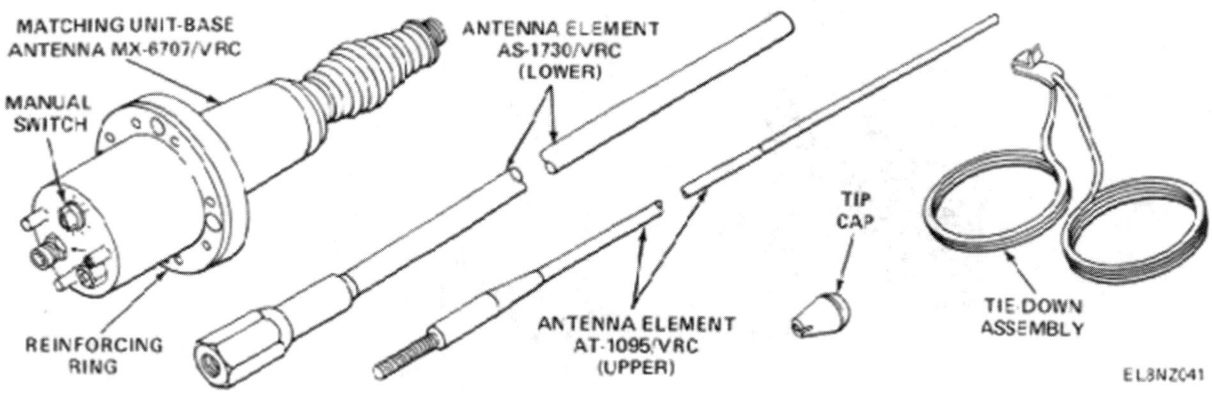

The AS-1729/VRC is used with vehicles which have the radio sets AN/VRC-12 and AN/VRC-43 through AN/VRC-49. The Antenna AS-1729 consists of two whip antenna elements (AT-1095/VRC and AS-1730/VRC), a tip cap, the Matching Unit MX-6707/VRC and miscellaneous hardware.

The MX-6707/VRC tunes the antenna elements either by automatic or manual selection. The manual tuning switch is on the bottom of the matching unit.

Complete mobile whip antenna setup

antenna matching unit." Not having messed with either of these antennas, the author isn't sure if that means the RF cable or the control cable, or both. He's guessing it means the control cable, since by manually tuning the matching circuits in the base, you may end up on a setting different than is specified by the internal R/T circuits. Since the circuits are just the open- (or ground-) seeking reentrant type, if the control cable were hooked up and you moved the tuner off the setting specified by in the R/T, the internal switch in the matching base would just return to that setting. The manual setting appears to be for when you do not have the CX-4722 control cable.

For those who have the AS-1729 or AT-912 but are missing the control cable, here's the info to make your own: Pins A, B, C, D, E, F, H, J, K, L, M and N on the multipin MX-6707 (or CX-4722) connector go to corresponding pins on the ANT CONT jack on either the RT-524 or RT-246. Once these connections are made, changing frequencies on the R/T will automatically tune the whip matching bases.

NOTE: The author remembers hearing about numerous instances of burned-up matching bases. The tuned circuits in the bases are apparently somewhat prone to failure if mis-tuned.

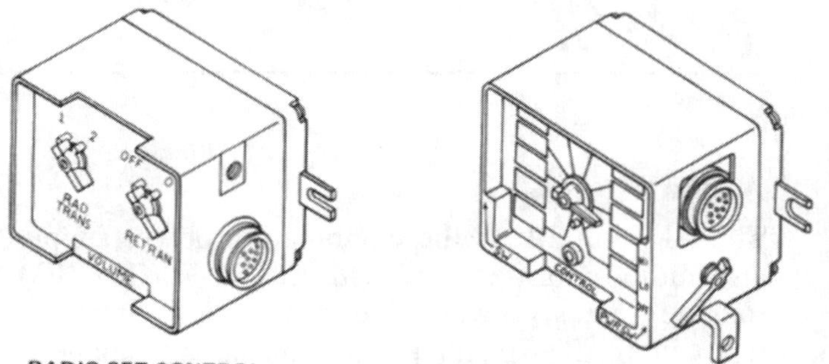

RADIO SET CONTROL BOX C-2299/VRC

FREQUENCY SELECTOR CONTROL BOX C-2742/VRC

VRC-12 control boxes

The C-2299 is the retransmit control, and the C-2742 is used only with the RT-246 to control it remotely. (This is the box that ended up on the bridge of the Arleigh Burke class destroyers!)

AM-65

Application:

The AM-65/GRC is the combination power supply and audio amplifier used with the RT-70/GRC FM transceiver. It's shown above in the usual configuration with the RT-70 on top. The units are interconnected with the CX-1213/U "dogbone" cable.

Tech Specs:

Size: 4" x 7.5" x 13"

Weight: 15 pounds

Active devices: Vacuum tubes, 12AU7(3), 6AK6(2), 0B2(2)

Power requirements: 6VDC 10A, 12VDC 10A, 24V 10A, depending on which plug-in power supply (PP-448, PP-281, PP-282) is installed. See text.

Power output: All operating voltages necessary for operation of the RT-70 transceiver.

Audio input/output devices: Any of the GRC gear audio accessories—H-33 handset, M-29 mike, HS-63 headset/boom mike with GSA-6 chest group, LS-166 speaker, etc.

Tech Manuals: TM 11-5039 (detailed theory), and also TM 11-284 "Radio Sets AN/GRC-3, -4, -5, -6, -7, and -8" (info on larger radio sets that include the AM-65, with schematics)

Construction Style: Integral chassis and diecast panel with heavy-duty outer case. Point-to-point wiring inside. Repairable unit.

Description:

The AM-65 is referred to as an "interphone amplifier" in the TMs covering larger systems of which it is a part. But the unit is probably more important to the surplus collector because it also functions as the power supply for the RT-70 transceiver. The fact it amplifies the RT-70 output so it will drive a speaker is a plus. The unit will supply operating voltages for the RT-70 from 6, 12, or 24VDC input depending on which plug-in power supply module is installed.

Tip: Determine which module is installed at purchase time. The little voltage "flag" at center left on the AM-65 panel is supposed to reflect the required voltage input, but this was not always kept up to date! Turn the Dzus fasteners to release the unit from the outer case. You may have to pry gently between the panel and case to pop it loose. The plug-in is clearly marked on the back side of the chassis.

Caution: The screwdriver switch on top of the chassis behind the front panel must be set to the same voltage as on the plug-in.

Power connections to the POWER IN jack on the AM-65 are as follows:

POSITIVE (+) 6, 12, or 24VDC: pins A, B (paralleled for current handling)
NEGATIVE (-): pin C

These connections are automatically taken care of if you use the power connector supplied on any of the various mounting bases, MT-297, MT-298, MT-300, etc.

Important: You must jump pin J to pin H on the AM-65 REC-TR CONTROL jack in order to get any audio at either the RT-70 or AM-65 AUDIO jacks. This connection is taken care of by the corresponding connector on the mounting base. If a mounting base is not used, simply use a wire jumper pushed into J and H.

For those of you who cannot find a CX-1213/U interconnect "dogbone" cable, here are the cable connections for wiring an expedient solution between the RT-70 POWER jack on the AM-65 and the POWER jack on the RT-70: A to A, B to B, C to C, D to D&E, E to D&E, F to F, H to H, J to J, K to K. Note the cable is reversible.

Operation:

Assuming power is hooked up, the two units are interconnected, and there's a J to H jumper in the AM-65 REC-TR CONT jack, apply the correct input voltage and move the function switch from OFF to INT. You'll hear the vibrator start. The interphone amp is now running, but not the RT-70. To start the RT-70, move the switch to "RT-70". The RT-70 can now be operated per the instructions in the RT-70 section.

Caution: Don't move the switch to "RT-70" unless there's an RT-70 connected. Without a load on the supply, excessive filament voltage will be applied to the AM-65 audio amplifier tubes, shortening their life.

Additional Comments:

An AM-65, PP-281 (12VDC) plug-in, and RT-70 make an interesting mobile 6m setup that'll be sure to get recognition from your incredulous (and uninitiated) ham friends...

Q: "What the &%^$ is that?"
A: "Oh, that's my new thirty pound, half-watt six-meter handheld. Cool huh?"

What To Do If It's D.O.A.:

As always, check the fuse. It's under the screw-in cap above the REC-TR CONT jack. There should be a spare inside the case.

If the fuse is good, and you still don't hear a buzzing in the INT or RT-70 switch position, the vibrator may have taken a powder. The vibrator is located in the plug-in power supply unit. Sometimes opening the plug-in supply then removing the vibrator and rapping it on a hard surface a few times will break loose contacts that have become stuck over the years. If you have another power supply plug-in, substitution is the next thing to try. If the RT-70 works okay, but you have no audio amplifying action at the AM-65 AUDIO connector, check the tubes in the AM-65. If the vibrator is obviously running and you have no RT-70 function, more advanced troubleshooting is necessary.

Accessories:

Any of the usual GRC accessories will work with the AM-65... H-33 handset, H-63 headset/boom mike, M-29 mike, LS-166 speaker, etc.

For a picture of the AM-65/RT-70 integrated into a larger radio set, see the AN/GRC-7 shown in the "RT-66, RT-67, RT-68" section. The AM-65, RT-70, and MT-300 mounting base comprise the AN/VRC-7 set.

A VRC-7 radio set

The author built up this VRC-7 for a friend back in the late 70s. The MT-300 mounting base supplies the interconnect cables only, with no further control circuits. The set is shown with an AB-15 mast base, antenna sections, H-33 handset, and LS-166 speaker. The power cable can be seen just to the right. All the user had to do was hook up 24V.

C-433

Application:

Remote Control C-433/GRC allows push-to-talk transmission control of a suitable radio set over phone lines from a remote position using a handset or chest set (combination headset and mike). It also provides telephone comm between the remote control station and the local operator. The radio sets used are generally the older GRC FM gear. The C-433 is used with a wide variety of radio sets, including the VRC-18 (RT-68, R-110, on an MT-327 base).

When used in conjunction with Local Control C-435/GRC at the local station, the resulting system extends monitoring and push-to-talk functions up to two miles over phone lines to the remote unit. The C-433 is also intended for portable use as a remote control box carried by the operator with a shoulder strap.

Tech Specs:

Size: 3.5" x 8.25" x 7"

Weight: 7 pounds, including battery

Active devices: None

Power requirements:
2 1.5V D-cells
1 45V Battery BA-414/U (not necessary in all applications—see text)

Terminal impedances:
Line: 600 ohms (balanced)
Mike: 150 ohms
Receiving: 600 ohms (balanced or unbalanced, depending upon circuit selections)

Remote control line voltage: 24 to 45VDC

Audio input/output devices: Any of the audio accessories terminated by the U-77 plug: H-33, H-63 and GSA-6 Chest set, M-29 mike, etc.

Tech Manuals:
TM 11-5038, "CONTROL GROUP AN/GRA-6", covers both the C-434 and C-433, description, operation, theory, troubleshooting, schematics.

Also, any TM covering an installation containing the C-433 as a component will generally have the above detailed info. Examples are:

TM 11-611, "RADIO SETS AN/VRC-16, AN/VRC-17, AND AN/VRC-18", (RT-66 through RT-68 with receivers R-18 through R-110 respectively).

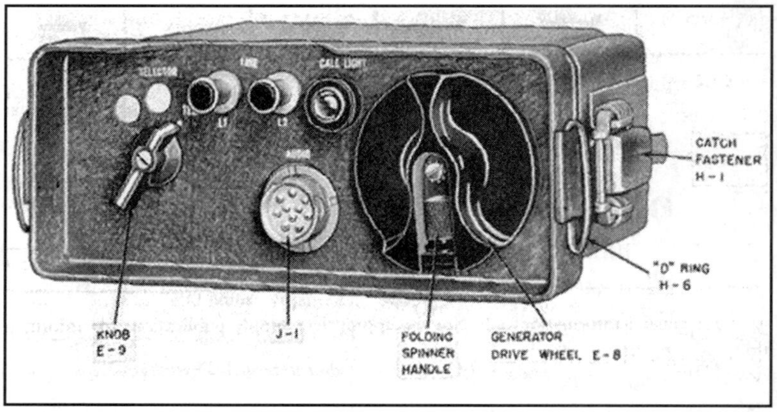

TM 11-284, " RADIO SETS AN/GRC-3, -4, -5, -6, -7 AND -8", (one example: RT-66, R-108, PP-112, AM-65, RT-70, C-434)

Construction Style: Heavy-duty die-cast panel and chassis fits into a sheet metal outer case. Components should be easy to reach and work on, except for the hermetically sealed relay, which could be a problem if it fails.

Description:

The C-433 was generally used with older GRC FM gear, but there's no reason it couldn't be integrated with newer systems, given the correct interconnection.

Construction is typical panel-and-chassis for the time, with internal battery compartments for the two standard D-cells and the 45V BA-414/U. The BA-414 may be the single best reason to use the more modern GRA-39 control box, which uses six D-cells.

The illustration on the next page shows the C-433 in a typical application.

Operation:

Release the two snap catches and pull the panel/chassis out of the case. Install two D-cell batteries. The BA-414 is basically extinct. If you wish to remotely control a set, you can stuff the compartment with five 9V alkaline batteries.

Note: 45V is not needed if you only intend to monitor the distant set, or converse with the distant operator.

CAUTION: The keyline has 45V or so on it when remotely keying the distant set. The older GRC gear is designed for this. This may not be true with more modern equipment. Be advised.
Set the internal LAMP-BELL switch to the type of call notification desired.

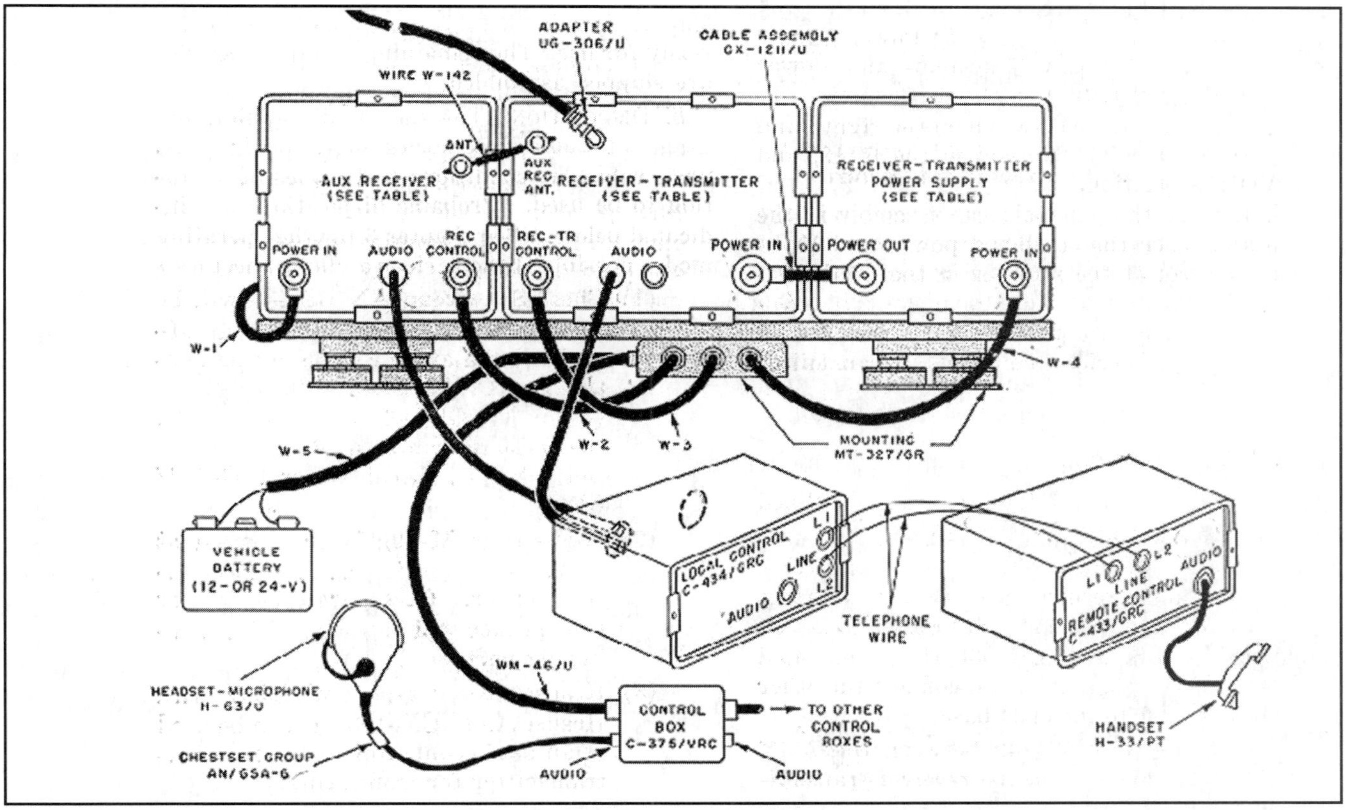

VRC-16/-17/ or -18 setup, showing application of C-434 and C-433

Hint: Use BELL; the lamp is an NE-51 neon type and is pretty dim!

Re-install the unit in its case and you're ready to go.

Assuming you have the unit connected to a distant radio set which includes a C-434 Local Control Box, here are the basic operating instructions directly from TM 11-5038, See author's comments in the []:

27. Push-to-talk Operation from Remote Control C433/GRC

The procedures for implementing the push-to-talk operating procedures possible at the remote control unit require that the desired mode of operation be established in advance by telephone conversation between the local and remote operators. The operating procedures are described in the following subparagraphs.

a. REMOTE PUSH-TO-TALK OPERATION OF SET 1 WITH LOCAL CONTROL OF RADIO SET POWER.

(1) At the local control unit [C-434], set the REMOTE switch to the SET 1 position and turn the equipment power on.

(2) At the remote control unit [C-433], set the SELECTOR switch to the left-hand write-in position, and momentarily operate the microphone push-to-talk switch.

(3) Operate the microphone push-to-talk switch in order to talk.

(4) Release the microphone push-to-talk switch in order to listen.

(5) At the end of the conversation, release the microphone push-to-talk switch.

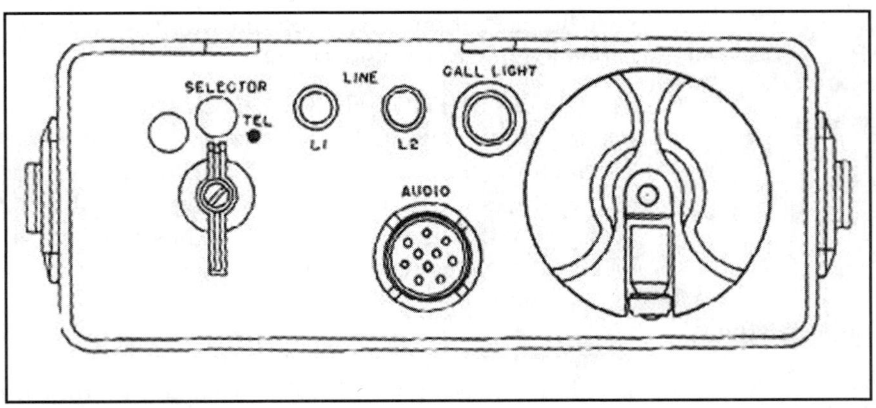

C-433, front panel

(6) Set the SELECTOR switch in the TEL position. At the local control, set the REMOTE switch to the TEL ONLY position; this returns the equipment to the stand-by condition.

b. REMOTE PUSH-TO-TALK OPERATION OF SET 2 WITH LOCAL CONTROL OF RADIO SET POWER.

(1) At the local control unit, set the REMOTE switch to the SET 2 position and turn the equipment power on.

(2) At the remote control unit, set the SELECTOR switch in the left-hand write-in position, and momentarily operate the microphone push-to-talk switch.

(3) Operate the microphone push-to-talk switch in order to talk.

(4) Release the microphone push-to-talk switch to listen.

(5) At the end of the conversation, release the microphone push-to-talk switch.

(6) Set the SELECTOR switch to the TEL position. At the local control, set the REMOTE switch to the TEL ONLY position.

c. REMOTE PUSH-TO-TALK OPERATION OF EITHER SET 1 OR SET 2 WITH LOCAL CONTROL OF RADIO SET POWER.

(1) At the local control unit, set the REMOTE switch to the SET 1 & 2 position and turn the equipment power on.

(2) At the remote position, turn the SELECTOR switch to the left-hand write-in position for Set 1 operation. Operate the microphone push-to-talk switch in order to transmit over Set 1; release it to listen to either or both Set 1 and Set 2.

(3) Turn the SELECTOR switch to the right-hand write-in position for Set 2 operation. Operate the push-to-talk switch in order to transmit over Set 2; release it to listen to either or both Set 1 and Set 2.

d. REMOTE PUSH-TO-TALK OPERATION OF SET 1 OR SET 2 WITH REMOTE CONTROL OF RADIO SET POWER. This mode of operation is possible only if provisions are made for it in the radio set whose power is to be controlled. The local control unit may be left unattended after the switches have been set as described below.

(1) At the local control unit, place the REMOTE switch in the SET 1 or the SET 2 position, depending on which set is to be used.

(2) Turn the equipment power switch to the remote position.

(3) At the remote control unit, turn the SELECTOR switch to the left-hand write-in position.

(4) Momentarily press the push-to-talk switch on the handset at the remote control unit. The circuit is now ready for operation over the set selected by the REMOTE switch of the local operator.

(5) To talk, operate the push-to-talk switch.

(6) To listen, release the push-to-talk switch.

(7) To turn off the power at the end of an operation, release the push-to-talk switch and turn the SELECTOR switch to the right-hand write-in position. Momentarily operate the push-to-talk switch.

Note: The procedure of step (7) above turns off the power for the radio set, but leaves the control of turning the power on or off at the remote control. The use of the left-hand write-in position as an on position and the right hand write-in position as an off position is based on the assumption that the line wires have been connected in the manner described in paragraphs 19 and 20. If the lines are reversed, the functions of the two write-in positions will be reversed.

[See "SETTING UP THE PHONE LINES CORRECTLY" section below for more details on this].

(8) To restore the telephone circuit to stand-by, set the REMOTE switch in the TEL ONLY position and set the SELECTOR switch in the TEL position.

Note: Power control of radio set. Control Group AN/GRA-6 may be made to operate with either one of two types of radio sets.

(1) In one type, power is turned on exclusively by means of switches on the radio set itself.

(2) In another type, switching is provided to give power control to either Control Group AN/GRA-6 (remote) or to the power on-off switch of the radio set. In this latter case, if power is to be controlled by Control Group AN/GRA-6, be sure that the radio set switches are set properly, according to the instructions contained in the radio set manual.

SETTING UP THE PHONE LINES CORRECTLY. Polarity of the lines makes a difference when it comes to which remote set is controlled for a given position of the SELECTOR switch. In order for the following behavior to occur, corresponding binding posts on the local and remote control units must be connected: L-1 to L-1, and L-2 to L-2.

From the T.M.:
19. Filling in Write-in Blanks on Remote Control Unit

The control circuits are so arranged that, with the

telephone line properly connected [see immediately above], the coordination of the write-in positions of the remote control unit SELECTOR switch with the local control unit REMOTE switch positions is as follows:

a. When the SELECTOR switch is in the left-hand write--in position and the REMOTE switch is in the SET 1 position, operation of the microphone push-to-talk switch of the remote operator will turn on the Set 1 transmitter. If the radio set is arranged for remote power control, the radio set power will be turned on. In this case, the right-hand write-in position of the SELECTOR switch will turn the radio set power off when the microphone push-to-talk switch is operated and then released.

b. When the SELECTOR switch is in the left write-in position and the REMOTE switch is in the SET 2 position, the remote operator's push-to-talk switch will turn on the Set 2 transmitter. If remote power control is provided for, then the radio set power will be turned on. The right-hand write-in position is again the power off and Set 2 off position.

c. When the REMOTE switch is in the SET 1&2 position, power to the radio set is controlled locally, and the left- and right-hand write-in positions of the SELECTOR switch turn on the Set 1 or Set 2 transmitters respectively.

Note: The above set-up is the intended arrangement. If the line connections are reversed, the coordination between the SELECTOR and REMOTE switches also will be reversed.

Tip: The C-433 can be used as a stand-alone field phone. Simply connect the two-wire phone line from the distant installation to the LINE L1 and LINE L2 binding posts, and connect an H-33 handset. Rotating the crank generator on the C-433 will ring the distant field phone (and vice versa). Push the H-33 button to talk.

Note: No power source other than the internal D-cells is necessary when using the C-433 as a field phone.

Additional Comments:

See the C-434 and C-435 sections for additional info about how these units interact.

What To Do If It's D.O.A.

There's virtually nothing to go wrong in these things, since they contain little more than switching circuits.

C-434

Application:

Local Control C-434/GRC allows push-to-talk transmission over a suitable radio set from a handset or chest set (which includes headset and mike). It also provides telephone comm between the local and remote control station. The radio sets are generally the older GRC FM gear. The C-434 is used with a wide variety of radio sets, including the VRC-18 (RT-68, R-110, on an MT-327 base).

When used in conjunction with Remote Control C-433/GRC at the distant station, the resulting system allows extension of monitoring and push-to-talk functions up to two miles over phone lines to the remote unit.

The C-434 can be used standalone as a field phone, as part of radio systems without provision for mounting it, or as part of radio system with mounting bases such as the MT-298 which allow for mounting of the control.

Tech Specs:

Size: 3.5" x 10.5" x 8.75"

Weight: 10.5 pounds

Active devices: None

Power requirements: 2 1.5V D-cells

Terminal impedances:
Line: 600 ohms (balanced)
Mike: 150 ohms
Receiving: 600 ohms (balanced or unbalanced, depending upon circuit selections)

Remote control line voltage: 24 to 45VDC (see C-433 section)

Audio input/output devices: Any of the audio accessories terminated by U-77 plugs: H-33, H-63 and GSA-6 Chest set, M-29 mike, etc.

Tech Manuals:
TM 11-5038, "CONTROL GROUP AN/GRA-6", covers both the C-434 and C-433, description, operation, theory, troubleshooting, schematics.

Also, any TM covering an installation containing the C-434 as a component will generally have the above detailed info. Examples are:

TM 11-611, "RADIO SETS AN/VRC-16, AN/VRC-17, AND AN/VRC-18", (RT-66 through RT-68 with receivers R-18 through R-110 respectively).

TM 11-284, " RADIO SETS AN/GRC-3, -4, -5, -6, -7 AND -8", (one example: RT-66, R-108, PP-112, AM-65, RT-70, C-434)

Construction Style: Heavy-duty die-cast panel and chassis fits into a sheet metal outer case. Components are easy to reach and work on, except for the three hermetically sealed relays, which could be a problem if one of them fails.

Description:

The C-435 was generally used with older GRC FM gear, but there's no reason it couldn't be integrated with the newer systems, given the correct interconnection. The GRA-39 group of control units is the newer equivalent made for the later gear with U-229 ("new family") connectors.

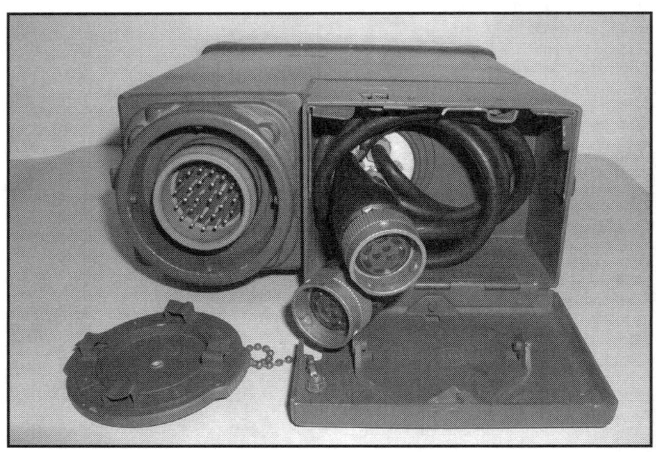

C-434 rear, covers removed/opened

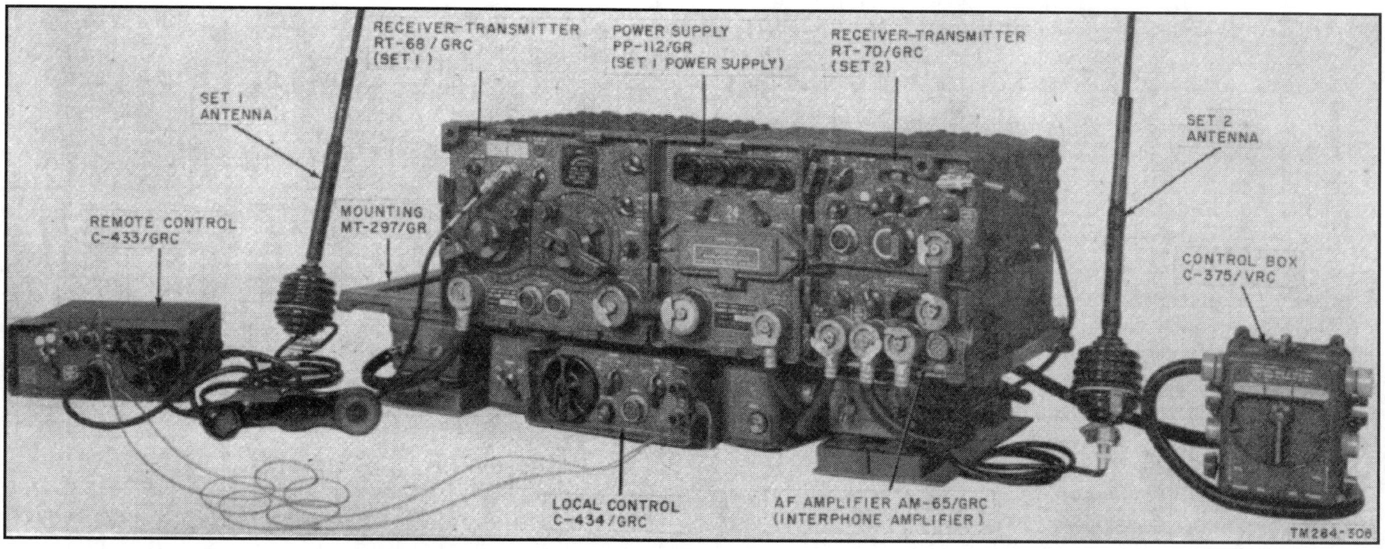

C-434 installed in a system with MT-297 mounting base

The rear compartment houses a pair of cables with U-77 plugs, which are typically connected to the audio jacks on the transceiver and the auxiliary receiver in installations like the VRC-16/-17/-18 where slip-in provisions on the mounting base are not provided. See the C-433 section for an illustration of the C-434 in use in this set.

The large connector on the back of the C-434 is engaged when the unit is installed in a mounting base with such provision, like the MT-297 or MT-298.

In these systems, the audio and control interconnects are made via the rear connector on the C-434 and the circuitry in the mounting base itself. The rear lid has a provision for stowing the connector cover when the unit is installed in a mounting base.

Operation:

To install batteries and select lamp/bell operation, loosen the two captive wing nuts on the front panel and work the panel/chassis out of the case.

Note: Be sure to open the rear door and stretch out the cables -- they must be pulled through a hole in the rear of the case. Also, you will almost certainly have to pry the panel/chassis out of the case for the first half-inch or so to overcome the O-rings on the large connector.

Install two D-cells in the battery compartment. Also, set the internal LAMP-BELL switch to the type of call notification desired.

Hint: Use BELL; the lamp is an NE-51 neon type and is pretty dim!

Re-install the unit in its case and you're ready to go.

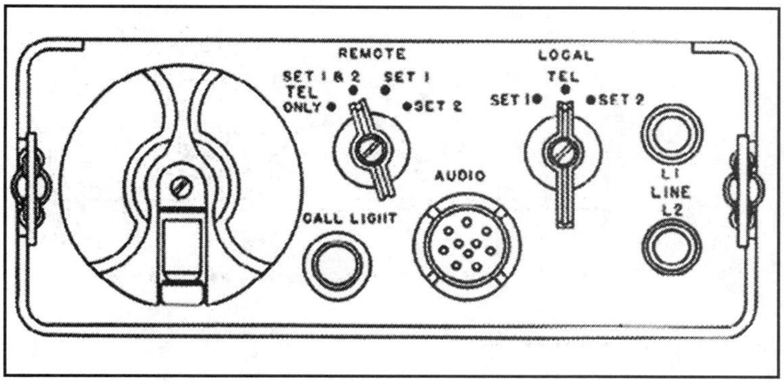

C-434, front panel

Basic operating instructions adapted from TM 11-5038, assuming you have a suitable radio installation set up:

Local push-to-talk operation of Set 1. The LOCAL switch is used as the push-to-talk switch for all PTT operation.

1) Turn on power for set with the main radio set power switch.

2) Turn the REMOTE switch to the SET 1 & 2 position.

3) Hold the LOCAL switch in the SET 1 (spring return) position.

4) Push the mike PTT switch to talk over SET 1. Release the LOCAL switch to listen.

5) At the end of the exchange restore the circuit to standby by by turning the LOCAL switch to TEL and the REMOTE switch to TEL ONLY.

Follow a similar procedure for PTT operation of SET 2 (assuming there is a second set in the system).

Monitoring operation at Local Control C-434.

a) With the REMOTE switch in a position except the TEL ONLY position the local operator can hear the audio from all receivers in the installation simultaneously.

b) The local operator cannot monitor directly the transmission of a remote operator (using a C-433 and phone lines) through the C-434. However, since most of the transceivers used in these installations have sidetone, the operator will hear the remote operator's transmission via that route. This is because the sidetone is routed back through the C-434 as part of the audio mentioned in part a) above.

c) The local operator can monitor simultaneously the output of all receivers in the set and speech from the remote operator, provided that the remote (C-433) SELECTOR switch is in the TEL position and the (C-434) LOCAL switch is in the the TEL position.

Tip: The C-434 can be used as a stand-alone field phone. The author has used his with his EE-8 field phone. Simply connect the two-wire phone line from the distant installation to the LINE L1 and LINE L2 binding posts, and connect an H-33 handset. Set the REMOTE switch to TEL ONLY. Rotating the crank generator on the C-434 will ring the distant field phone (and vice versa). Push the H-33 button to talk. No power source other than the internal batteries is necessary when using the C-434 as a field phone.

Additional Comments:

See the C-433 section for additional info about how these units interact.

What To Do If It's D.O.A.:

There's virtually nothing to go wrong in these things, since they are nothing but switching circuits. The single exception might by a failure of one of the hermetically sealed relays. They are plug-in units, and you could replace one with a spare out of a junker C-434. Or if you're really ambitious you could try cutting the recalcitrant thing open. The author actually fixed a stuck T/R relay in a T-195 that way, but it wasn't easy.

PP-109, PP-112

Application:

These DC supplies provide operating power for Receiver-Transmitters RT-66, -67, and -68 in a wide variety of vehicular and semi-fixed applications. The 24VDC input PP-112 is shown above. The PP-109 12VDC input supply is identical in appearance except for "12 VOLTS" above the SPARE FUSE door.

Tech Specs:

Size: 7.25" x 9" x 13"

Weight: 34 pounds

Active devices: Vacuum tubes 1007(2), 0A2, 0B2, three vibrator units.

Power requirements:
PP-109 25.2 VDC nominal at 7.0 A
PP-112 12.6 VDC nominal at 11.5 A

Output voltages:

	All voltages are DC	
	High Power	Low Power
Receiver filaments	6.3, 595ma	6.3, 595ma
Transmitter filaments	6.3, 1.4A	6.3, 1.4A
Relay control circuits	5.6, 431ma	5.6, 431ma
Receiver plates	85, 55ma	85, 55ma
	105, 20ma	105, 20ma
Transmitter plates	85, 71ma	90, 75ma
	100, 19ma	90, 17ma
	150, 37.5ma	90, 22.5ma
	250, 11.5ma	105, 4.8ma
	450, 75ma	200, 33.5ma
Transmitter bias	-27, 2ma	0

Tech Manuals: TM 11-5036 (schematics are also found in all manuals covering systems in which the supplies are a part—see the sections on the other related FM gear).

Construction Style: heavy die-cast front panel and chassis with slip-on reinforced metal outer case. Point to point wiring, easy access to replaceable components.

Description:

These are vibrator type power supplies. Construction, like all the related GRC gear, is very rugged -- they were made to be bounced around in the back of military vehicles over rough terrain.

Operation:

Input power connections to the POWER IN jack for both supplies are as follows:

Pin B: POSITIVE
Pin C: NEGATIVE

Here are the pin-outs at the POWER OUT jack:

A: +6V relay
B: +85V
C: +150 (HIGH power), +85V (LOW power)
D: CHASSIS GROUND
E: GROUND RETURN
F: +250
K: POWER CONTROL (PTT)
L: +6.3 receive filaments
M: +105
N: +6.3 transmit filaments
P: +450
R: -27 bias
All other pins unused

Note: Grounding Pin K, energizes voltages for Pins R, P, F, C, and N. Also, Pin E is grounded on the radio side. (Pins E and Pin D are shown connected together on the RT-68 radio schematic).

PP-112 chassis, showing construction

Assuming the correct voltage is hooked up for the supply in use, and an RT-66/-67/-68 is connected to the POWER OUT jack, proceed as follows. Move the OPERATE switch to RECEIVE and you should hear a vibrator start up in the supply. Select the desired power setting with the TRANS POWER switch. It's always best to use LOW power setting until you're sure the associated radio is working correctly. Radio audio should come up fairly rapidly if the SQUELCH is not closed. To enable transmit, move the OPERATE switch to TRANSMIT & RECEIVE. You should hear a second vibrator kick in. Operation of the transceiver itself is covered in the "RT-66, RT-67, RT-68" section.

What To Do If It's D.O.A.:

If no vibrator noise is heard at all, check the fuses. There should be some spares behind the SPARE FUSES door. If the fuses are good but the unit still doesn't operate, the armatures in one or more vibrators might be stuck. Try tapping on the cans with a screwdriver handle. A last resort may be to open up the vibrators and check the armatures, or actually buff the contacts with 400- then 600-grit emery cloth.

For troubleshooting, a schematic printed on fabric has been thoughtfully provided. It's rolled up in an inside compartment on the upper side of the chassis.

Additional Comments:

The interconnections between the supply and the radio are usually handled with a CX-1211 "dogbone" cable. If you don't have the CX-1211, the cable is a straight pin-to-pin correspondence between the POWER OUT jack on the power supply and the POWER IN jack on the radios, using pins A, B, C, D, E, F, H, J, K, L, M, N, P, and R.

PP-1175

Application:

Power supply that produces all necessary DC output voltages for the RT-66, -67, and –68 from 117VAC input. It will also power an R-108, -109, or -110 auxiliary receiver if present on the mounting base.

Tech Specs:

Size: 7.25" x 9" x 13"

Weight: 38 pounds

Active devices: 60s-era semis 2N1412, 2N1229, 2N697, 1N248, 1N540M, 1N750AM; gaseous regulator tubes 0A2, 0B2

Power requirements: 117VAC 60 Hz

Tech Manuals: Unknown. See appendix for possible source.

Construction Style: Integral chassis and die-cast metal panel with a heavy-duty slip-on metal outer case. Several circuit cards mounted on the chassis. Repairable.

Description:

The PP-1175/SR was manufactured for the Navy by Chatham Electronics Division of Tung Sol Electric, most likely for shipboard installations of the FM GRC gear used for comm with shore units. It is something of a rare item, and the author has not seen them generally since Fair Radio sold out quite a few years ago.

Operation:

Operation is exactly the same as for the PP-109 or PP-112, except the power input is 117VAC 60 Hz. Also, a very bright red pilot light is illuminated when the supply is on—hardly anything you'd want while deployed tactically on land!

Important: A simple mod must be done to the supply before using it either stand-alone with a radio, or on a mounting base such as the MT-298. Remove the supply chassis from the outer case, cut the wire leading to Pin H of the POWER OUT jack, and stub it off. If this isn't done and the supply is connected to a radio and turned on, you won't be able to turn it off and the radio won't work correctly. This is because a remote control relay inside the supply ends up connected across the 90V supply. This not only holds the supply on, but also drags down the 90V and kills off the audio. The PP-1175 was originally built for a special Navy setup that didn't use the same control circuits as the regular mobile GRC mounting bases like the MT-298.

AC power input connections at the POWER IN jack are pins B and C. Radio operation details are covered in the "RT-66, RT-67, RT-68" section.

Additional Comments:

Tip: A complete schematic is located inside the case on the chassis.

What To Do If It's D.O.A.:

A fuse check is in order right off. Use the schematic inside to troubleshoot. The components are well marked, but some look tough to reach.

PRC-41

Application:

UHF comm, widely used in the forward air control role in Southeast Asia from the mid 1960s on.

Tech Specs:

Frequency range: 225.0 to 399.9 MHz in 100kHz intervals, 1750 total channels with an additional guard receiver, fixed-tuned to 243.0mc

Type of modulation and power out: AM, 3W average unmodulated into 50 ohms

Sensitivity: 3uV modulated 30% at 1000Hz produces at least 7mW at a signal-plus-noise to noise (S+N)/N ratio of 10dB or greater

Size: 11.5" x 4.75" x 21" with BB-451/U battery or PP-3700/PRC-41 AC supply

Weight: 42 (!!) pounds (radio, AS-1404 24" antenna, BB-451/U battery)

Active devices: Solid state, and vacuum tubes 6442, 7077(4), 7544(4).

Power requirements: 24VDC 3A

Audio input/output devices: audio accessories with U-77 plug, such as H-33, LS-166

Antenna connection: Type N, 50 ohms

Whip antenna: AS-1404/PRC41, 24" cylindrical whip

Other antennas: AS-1405/PRC-41 collapsible Yagi

Antenna tuner: none accessible by user, internal circuits compensate during whip operation. When external yagi is used, the elements can be adjusted.

Tech Manuals: NAVSHIPS 0967-872-5020 "REPAIR AND MAINTENANCE TECHNICAL MANUAL FOR RADIO SET AN/PRC-41A"; description, specs, theory, repair, alignment. (No operation instructions).
Equivalent Army TM: TM 11-5820-510-35-1

Construction Style: Elaborate front panel and chassis arrangement with repairable replaceable modules.

Description:

While the so-called "manpack" radios designed by Collins in the 1960s -- the PRC-47 and the PRC-41 -- seem big and clunky by today's standards, they performed their tasks quite effectively. The PRC-41 is an amazingly complicated electromechanical contrivance

(some would say nightmare) that's a real challenge to work on without extender cables/cards. The mechanical gear train will drive you bananas if gets out of sync.

In designing the PRC-47 and PRC-41 the Collins engineers pushed the state of the synthesizer art to the limit with then current technology. So they naturally ended up with hybrids (part solid-state, part hollow state) stuffed with DC to DC converters, analog phase locked loops and spectrum generators, and all sorts of other hideous stuff.

When's the last time you saw a packset with a differential in it?

More seriously, the PRC-41 was used widely in forward (sometimes very forward) air control missions. It certainly did the job. Even across all these years, the author feels for the radiomen who had to hump a 50-

(Part of) the PRC-41 tuning mechanism

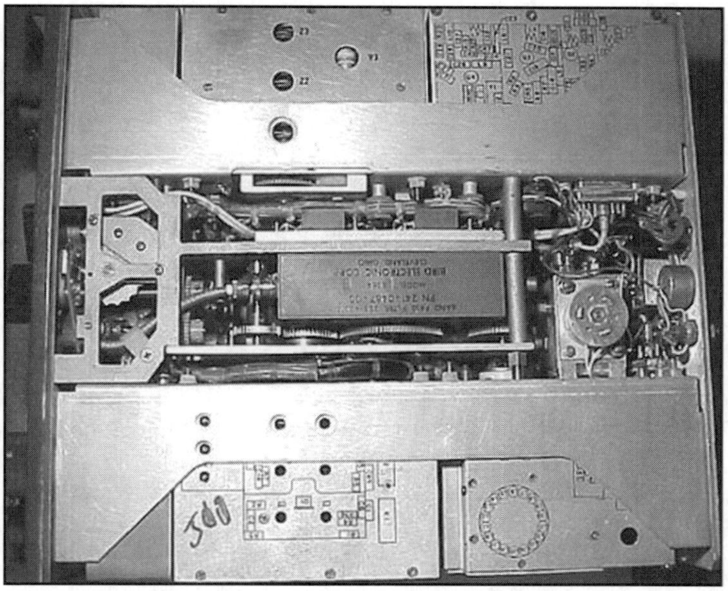

PRC-41 innards, opposite side, showing modular construction

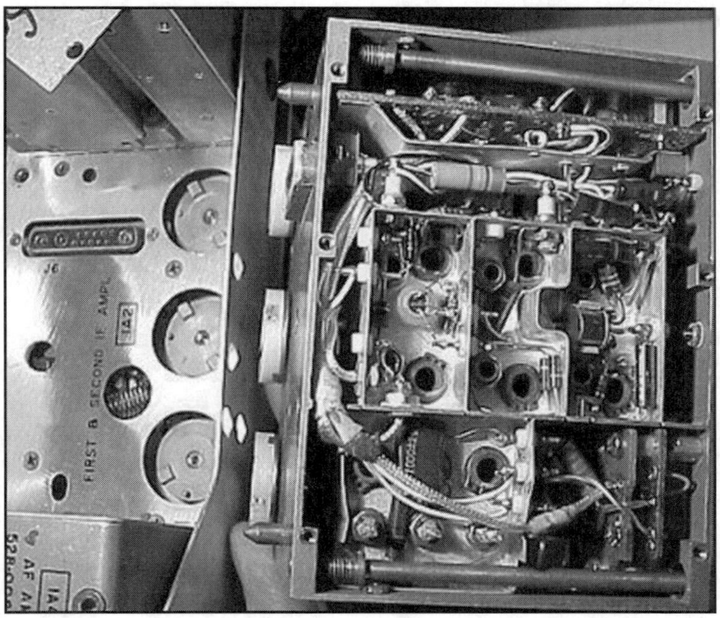

First and Second I.F. module, cover removed

pound radio… and their weapon and their ammo and their food and their water and, and, and…

This is the kind of Oldham-coupler, mechanically sandwiched, squished-in construction that'll give you nightmares in the daytime. These modules were made to be yanked and replaced wholesale until the radio started working. Then the bad modules were sent back to a depot for repair.

Rhetorical question: Could that be the reason why Fair Radio still has a big pile of PRC-41s in their warehouse?

PRC-41 case, bottom, showing power connections

Operation:

The original Battery BB-451/U is a silver-zinc alkaline unit of about 20Ah capacity. It is seldom encountered because it contained about $200 worth of recoverable silver. Junk dealers often bought entire PRC-41 sets in the transit cases to get the batteries, then scrapped the remainder.

PRC-41 with PP-3700 AC supply attached

The PP-3700 and BB-451 have the same form factor, and weigh almost exactly the same. Since a quarter-wave at these frequencies is about a foot, and the antenna is 24 inches long, the AS-1404 shown is probably an omnidirectional vertical dipole with some gain over a plain whip.

Once the antenna and handset are attached, operation is as follows:

1) Turn the Function Switch from OFF to T/R.

2) Set desired operation frequency by rotating frequency knobs: LEFT knob - tens of MHz; CENTER knob - 1 MHz; RIGHT knob - 100 kHz

3) Adjust SQUELCH until noise just disappears.

4) To transmit, depress PTT button on handset.

5) If desired, 243mHz fix-tuned guard receiver is enabled by moving Function Switch to T/R/G.

Note: if the guard receiver is enabled and a signal comes in on 243mHz, you'll hear it, no matter what frequency the main display shows. Interestingly, the manual states the guard receiver performance will be degraded if the main display is set exactly to the guard frequency (usually 243mHz).

6) Activate dial light by momentarily advancing the Function Switch to DIAL LIGHT.

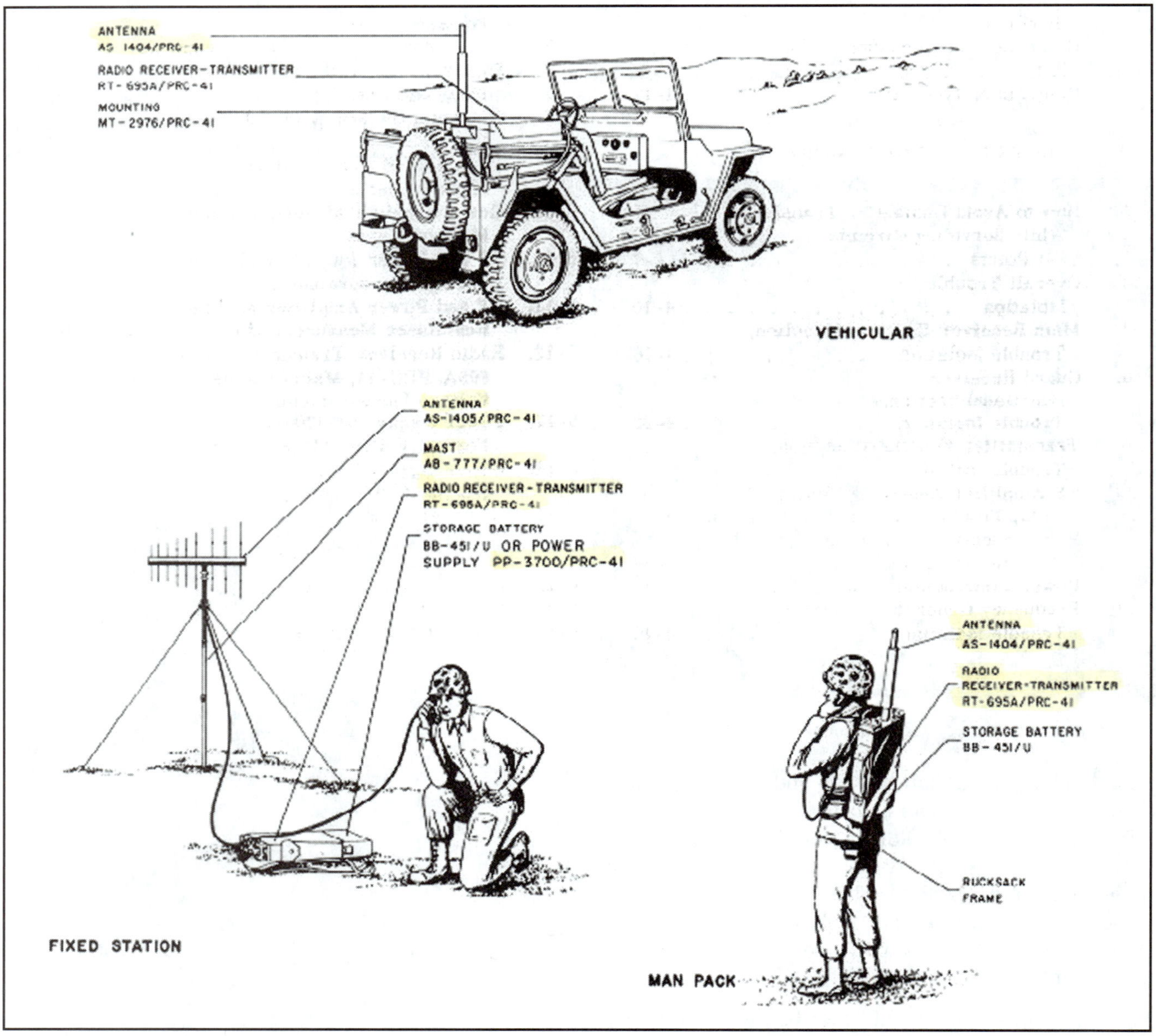

Typical PRC-41 installations

Additional Comments:

The PRC-41 is *not* a browsing receiver. Try rotating those knobs for twenty minutes!

What To Do If It's D.O.A.:

If the radio is totally dead (not even dial light function) pull the radio out of the case and check the 4A fuse F1. If the dial light comes on in the DIAL LIGHT position, put your ear to the case and check for the high hum of the internal DC-DC converter. If that's missing, the converter has a problem and the tubes in the receiver front end aren't getting any B+. Anything beyond that, and you're going to need the TM, and a heckuva lotta patience.

Accessories:

The original transit case comes with a number of accessories including power cables, coax, vehicular mounting brackets for the R/T and the AS-1404 antenna, module extender cables, a tool kit, and the interesting AS-1405 yagi antenna.

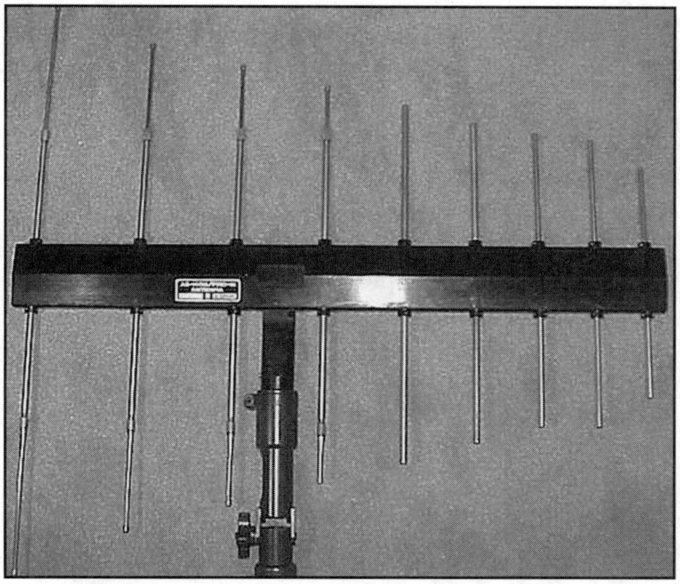

AS-1405/PRC-41 antenna and AB-777 mast

H-33/PT

Application:

The H-33 is the commonly used handset for the GRC gear with "old family connectors" that accept the U-77 male plug. Examples are the RT-70, RT-66, RT-67, RT-68, GRC-106, PRC-47, and others.

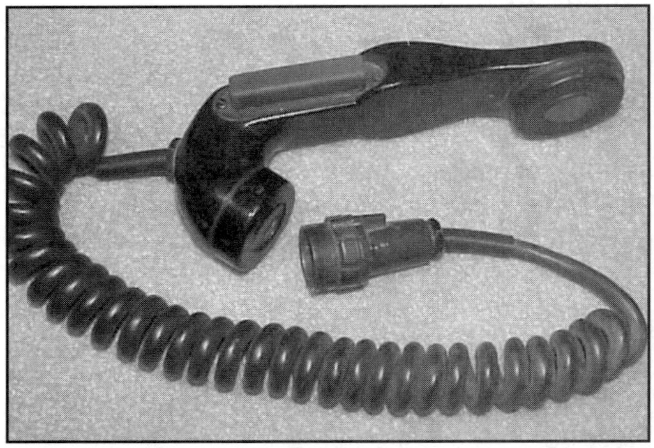

Tech Specs:

Size: Handset itself about about 8" long

Receiver impedance: 300 ohms

Mike impedance: 40 ohms (carbon element)

Construction Style: Repairable. Both the earpiece and mike element can be changed out.

Description and Tips:

The H-33 is a conventional push to talk handset with a dynamic receiver and carbon mike. Versions were made with either coiled or straight cords. Connections at the U-77 plug are as follows:

C: Mike
F: Transmit control (push to talk)
A: Headphone audio
B,E,H: Ground

What To Do If It's D.O.A.:

The author has never experienced a failure of an earpiece, but the carbon mike elements sometimes get cranky. Sometimes removing the element by unscrewing the mike cap and tapping the element lightly on a hard surface a few times will help.

Tip: When you find an H-33 at a swap meet, be sure to make sure the mike element is in there!

The other possible failure mode is corrosion in the PTT switch contacts. These can be reached by removing the metal plate and rubber switch boot. There's enough room in the switch compartment to install a fixed resistor in the mike audio line to lower the transmit deviation on FM sets if so desired. See the "RT-70" section for a description of this mod.

LS-166

Application:

This loudspeaker is used with any of the GRC gear that accepts U-77 plugs. These sets include the RT-70, RT-66/-67/-68, PRC-41, GRC-106, PRC-47, and others.

Tech Specs:

Type: 4" dynamic permanent magnet speaker, 5' cord and U-77 connector.

Size: 5" x 5" x 3", with 2" mounting bracket post

Weight: About 2 pounds

Tech Manuals: Covered in almost any TM that deals with complete GRC-type radio sets.
One example is "TM11-284 Radio Sets AN/GRC-3, -4, -5, -6, -7, and -8"

Construction Style: Repairable. The speaker and internal transformer can be changed out if necessary.

Description:

This is a heavy-duty splash-resistant speaker in a die-cast metal case.

Tip: The impedance of the speaker voice coil itself is 8 ohms, but there is an internal 8 ohm to 600 ohm transformer for matching the speaker with the 600 ohm audio of the associated gear. The LS-166 can be converted to 8 ohms rapidly by removing the back cover and jumping the transformer out of the circuit.

Operation:

Use the FIELD OR PACK SET USE switch position with low audio output packsets and the VEHICULAR SET USE position with higher output sets.

Additional Comments:

Tip: These speakers can sound tinny or rattle at times.

If this happens, remove the back cover and stuff the compartment with rags. This will often stop the rattling and improve the audio.

What To Do If It's D.O.A.:

Sometimes the U-77 plug contacts can get dirty or oxidized; try cleaning them with an eraser. Try a different position of the PACK SET-VEHICULAR SET switch.

H-138/U

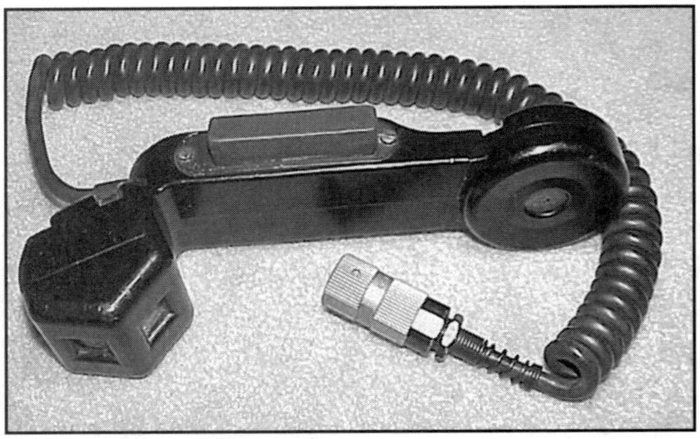

Application:

Handset for later model mil radio gear like the PRC-74, RT-524/VRC-12, PRC-77, and others that accept "new family" accessories with U-229 connectors.

Tech Specs:

Size: Handset itself about 8.5" long

Receiver impedance: 120 ohms (measured with a VOM)

Mike impedance: 150 ohms (dynamic element)

Construction Style: PTT switch, mike and audio elements are retained by screws and are repairable/ replaceable.

Description:

The H-138 is a noise-canceling push-to-talk handset with a dynamic receiver and mike. This handset is easily recognizable by its distinctive shape.

Connections at the U-229 plug are as follows:

A: Ground
B: Earpiece
C: Push to Talk
D: Mike
E: No connection

Additional Comments:

The H-138 is a heavier-weight handset for the new family radios that accept U-229 plugs. This handset has become a favorite of the author because of its excellent crisp mike audio—almost all mids and highs. An older type handset, they have become relatively rare. There also seem to have been fewer made than the H-189 and H-250.

Avoid dropping or rough handling these handsets because the plastic tends to crack, especially the areas around the mike and earphone elements.

What To Do If It's D.O.A.:

Problems can often be traced to dirty PTT switch or element contacts. The author's two H-138s have been reliable for many years.

H-189/GR

Application:

Handset for later model mil radio gear like the PRC-74, RT-524/VRC-12, PRC-77, and others that accept "new family" accessories with U-229 connectors.

Tech Specs:

Size: Handset itself about 8" long

Receiver impedance: 1000 ohms (measured with a VOM)

Mike impedance: 150 ohms (dynamic element)

Construction Style: PTT switch is retained by screws; mike and audio elements are retained by screw-on caps. Both are repairable/replaceable.

Description:

The H-189 is a push-to-talk handset with a dynamic receiver and mike. It's midway in weight between the H-250 and H-189, and does not have a noise-canceling mike.

Connections at the U-229 plug are as follows:

A: Ground
B: Earpiece

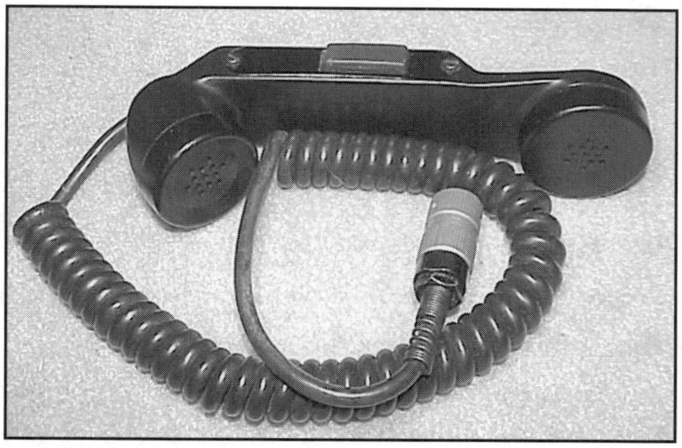

C: Push to Talk
D: Mike
E: No connection

Additional Comments:

The H-189 generally sounds better than most H-250s, but not as good as the H-138.

What To Do If It's D.O.A.:

Problems can often be traced to dirty PTT switch or element contacts.

H-250/U

Application:

Handset for later model mil radio gear like the PRC-74, RT-524/VRC-12, PRC-77, and others that accept "new family" accessories with U-229 connectors.

Tech Specs:

Size: Handset itself about 8" long

Receiver impedance: 1000 ohms

Mike impedance: 150 ohms (dynamic element)

Construction Style: PTT switch is reachable/repairable but elements are generally not -- handset is glued together.

Description:

The H-33 is a lightweight noise-canceling push-to-talk handset with a dynamic receiver and mike.

Connections at the U-229 plug are as follows:

A: Ground
B: Earpiece
C: Push to Talk
D: Mike
E: No connection

Additional Comments:

The H-250 is the lightest-weight handset commonly available for the new family radios that accept U-229 plugs. H-250s come in two basic varieties shown in the illustration, with large and small diameter cords. The author has learned to avoid the variety with the small diameter cords. This H-250 type has proven overly prone to RF pickup when used with HF gear

due to what appears to be poor shielding on the cord. The large-cord H-250s seem to be somewhat more immune to RF pickup, but still often suffer from poor mike audio quality.

H-250 overall quality varies all over the map. Some of the mike elements sound good and crisp, and some have a muddy, mushy sound to them. These are probably good handsets to avoid if you have a choice. (See the H-138 and H-189 sections).

What To Do If It's D.O.A.:

If the fault is not in the switch contacts, throw it out. This is literally true -- H-250 elements cannot be reached without cutting open the handset.

Tip: If you are specifically looking for an H-250 at a hamfest, take along your radio and some sort of receiver to test any examples you find.

LS-454

Application:

This loudspeaker is used with any of the later model gear that accepts U-229 plugs. These sets include the RT-524 (VRC-12), PRC-74, PRC-70, and others.

Note: LS-454s must be modified before they will function correctly with many of the radios listed, especially the HF packsets. See text.

Tech Specs:

Type: 4" dynamic permanent magnet speaker, 5' cord and U-229 connector.

Size: 5" x 5" x 3", with 2" mounting bracket post

Weight: About 2 pounds

Tech Manuals: Covered in almost any TM that deals with maintenance or installation of later model radio sets with the five- and six-pin new family connectors.

Construction Style: Repairable. The speaker and internal transformer can be changed out if necessary.

Description:

This is a heavy-duty splash-resistant speaker in a die-cast metal case.

Tip: The impedance of the speaker voice coil itself is 8 ohms, but there is an internal 8 ohm to 600 ohm transformer for matching the speaker with the 600 ohm audio of the associated gear. The speaker can be converted to 8 ohms rapidly by removing the back cover and jumping the transformer out of the circuit. The author has a stock LS-454 for use with the various 600 ohm equipment, and one labeled "8 Ohms" with the transformer bypassed for use with his PRC-1099. The 1099 has an op-amp audio output stage and is much more efficient into a low impedance.

Operation:

Important: A box-stock LS-454 will not function with many of the packsets like the PRC-74 or PRC-77. The speaker is connected to pins A and E on the U-229 plug, which is incorrect for most of the HF/VHF packsets and will key the radio or cause other erratic operation. In virtually every case, the correct wiring is to pins A (ground) and B (headset or speaker audio). You'll need to take the U-229 apart and move the connection on pin E to pin B.

Additional Comments:

Tip: These speakers can sound tinny or rattle at times. If this happens, remove the back cover and stuff the compartment with rags. This will often stop the rattling and improve the audio.

What To Do If It's D.O.A.:

Sometimes the U-229 plug contacts can get dirty or oxidized; try cleaning them with an eraser.

TRC-3

Application:

Remote sensor/transmitter (three are pictured) from a seismic intrusion set. The author believes the original set was supplied with a receiver that had a small window for displaying a number corresponding to which sensor/transmitter had been activated.

The original receivers are seldom encountered, and the transmitters can be found on the web sold in "kits" that include a consumer electronics-type of receiver. The prices of these "kits" are often preposterously high. Be advised.

Tech Specs:

Type of modulation and power out: AM; pulsed tones, the number of which depends on the particular transmitter/sensor unit (see text). The output power is probably in the under-1W class, possibly even 100mW.

Frequency of operation: The author's units are nominally on 129.9 MHz.

Size: 6" x 2" x 2.5 (each sensor/transmitter, stowed)

Weight: about 1.5 pounds per unit

Active devices: Solid state

Power requirements: One 9V battery, or two 9V batteries in parallel

Battery life: with two 9V batteries, 30 days typical

Whip antenna: 23" metal tape-type whip antenna

Tech Manuals: Unknown

Construction Style: Sensor/transmitters are built into sealed, hard O.D. plastic cases

Description:

The TRC-3 is a simple VHF transmitter hooked to a very sensitive sensor that sends one, two, three, or four amplitude-modulated beeps whenever it detects vibration in the ground. The transmitter cases have bumps molded into the plastic just above the function switch indicating how many beeps that particular unit transmits.

The illustrated unit transmits three beeps each time the ground sensor is agitated.

There is a single rotary control with six settings: OFF---1---2---3---4---5. The sensitivity increases with increasing numbers. The author did some rudimentary tests:

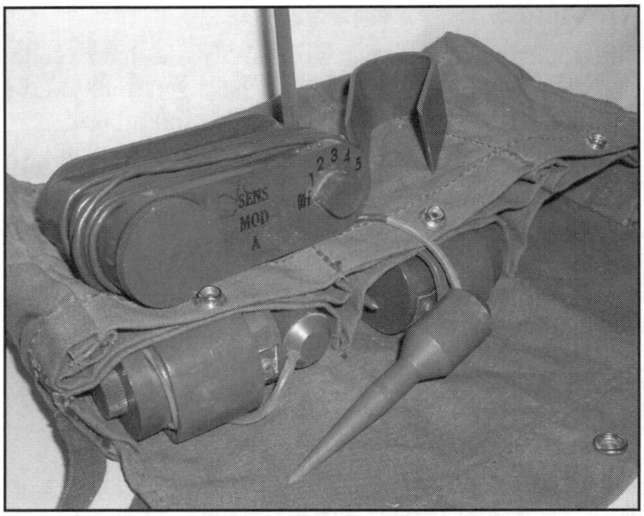

TRC-3 sensor/transmitter, showing battery holder, sensitivity control, and "beep-indicator bumps"

TRC-3 Sensor/Transmitter Sensitivity:

Setting	Performance
1	Soft footsteps—10 feet
	Foot stomp—20 feet
5	Soft footsteps—20 feet
	Foot stomp—200 feet+

These things will work for foot traffic but are best for vehicular traffic (see story below).

The complete sets had small specialized receivers that displayed which sensor was reacting to movement, but the receivers seem to be much more scarce than the sensors. It doesn't matter, because any VHF AM receiver will work. The author has used both his Sony Pro-80 and AN/ARC-73 in the past. In a pinch, you can even discern the beeps with an FM receiver.

Operation:

Per the illustration above, unscrew the battery cover and install one or two 9V alkaline batteries and replace the cover securely. The author's units seem to exhibit higher sensitivity to ground vibrations with two batteries.

Tip: If you're using two batteries, put a piece of paper or cardboard between them.

The location will obviously vary depending on your application. With careful placement, the flat, O.D. whip antenna will be mistaken for a blade of local vegetation. If the unit or units are expected to be outside for significant periods, put them in zip-lock plastic bags with just the antenna and the lead to the sensor head poking through. Once you have the unit located, push the sensor head into the ground. Check your concealment efforts.

You'll need to experiment with the sensitivity setting to find out what's best for your situation. Extremely dry rocky soil will conduct vibration differently than moist humid soil. The author carries his Sony Pro-80 scanner and stomps around (literally), adjusting the sensitivity until it's set right. That's really all there is to it. The sensors will return one, two, three, or four quick beeps, telling you which location is taking the hits.

Tip: You may have to tune quite a few kHz off the frequency on the TRC-3 label to get a readable signal.

Additional Comments:

Several years ago, the author was surprised to see a crude lean-to go up on some unoccupied property adjoining his property. More activity followed. The owner of record was contacted— he said he'd sold the property and he didn't know nuthin' about nuthin'. Then one day the author was alarmed to discover signs of heavy automobile traffic about 150 feet onto his land, in an area not visible from his house. He knew the dingbats were driving over his land to get back to their nearly-inaccessible parcel, but he couldn't catch them in the act. The property lines were clearly staked out. (Survey stakes? Survey stakes? We don' see no steeeeenking survey stakes!)

So he placed two TRC-3s, one about halfway down his neighbor's long drive (which he figured the insurgents had to be using to get back there) and the other near the mashed down area across his property. An ARC-73 aircraft transceiver did the monitoring duties, 24-7. Sure enough, the warning came and he caught 'em in the act. They turned out to be members of a "church" (a cult, actually) based in Denver, who thought they could come into the boonies, buy land, and do anything they wanted—including dance around in robes, chant gibberish all night long, light illegal campfires, and sacrifice chickens on stone altars (I am not making this up). They have since lost interest and resold the property.

What To Do If It's D.O.A.:

These things usually work -- the author hasn't encountered a bad one yet. Fixing them is problematic since they're sealed up tightly in a hard plastic case.

Accessories:

There's a lightweight O.D. canvas bag for holding six TRC-3s and one receiver. The plastic U-clip shown holds the antenna blade and sensor cable once they've been reeled onto the unit.

Appendix A—GRC Gear Frequency Chart

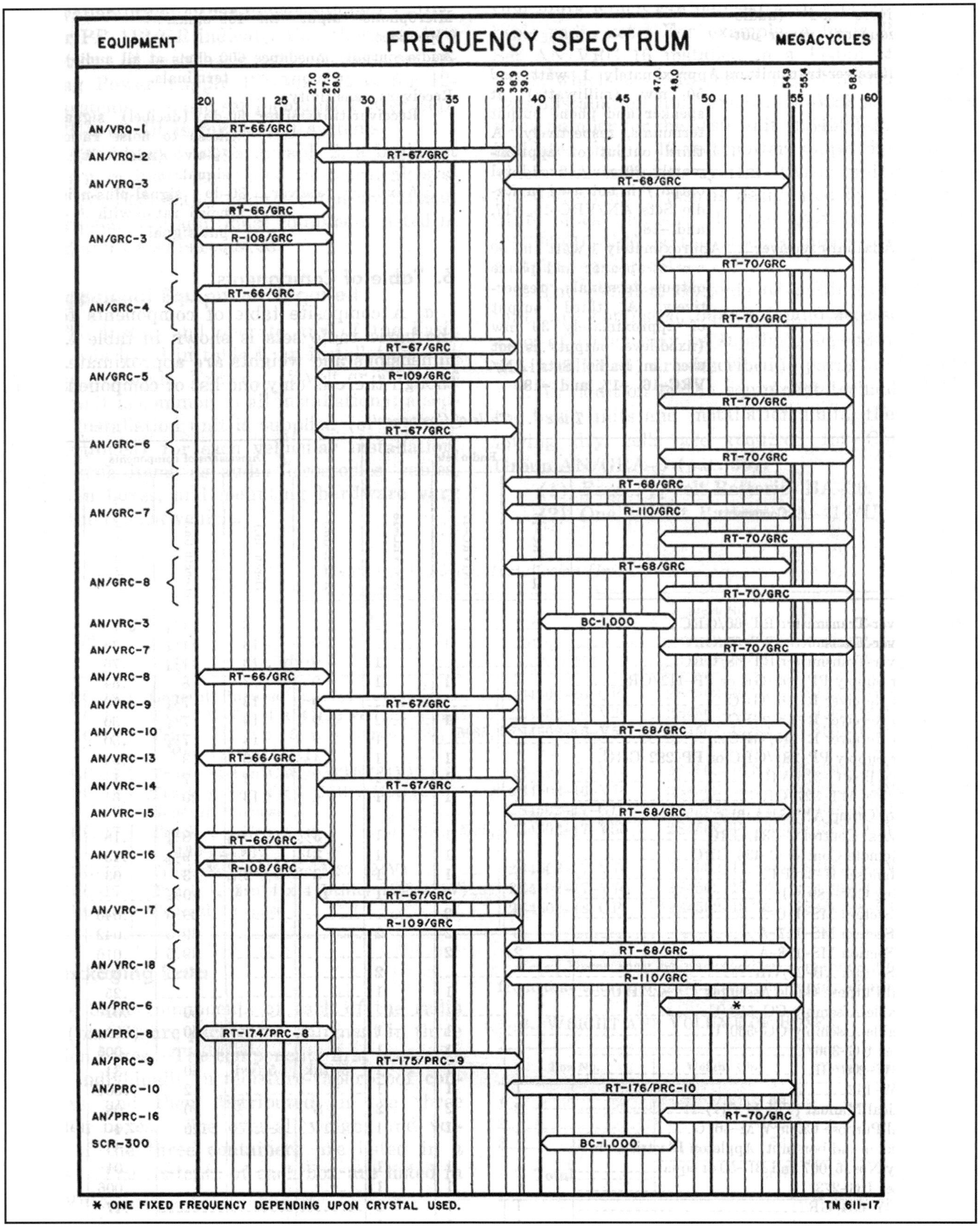

Appendix B—A Simple Solar Panel Regulator

The open-circuit voltage on the author's solar panels is almost 18 volts, and he was a little nervous about lighting off the PRC-74B receiver at a voltage so close to its absolute maximum specified capability. Even if the voltage drops very rapidly once current is actually drawn, this is still a concern.

Here's a simple regulator made from an LM340K three-terminal device. Note the large margin added to the voltage rating on the tantalum capacitors! These caps ensure stability of the regulator.

Solar panel regulator circuit

The parts are mounted on a metal lid from an I.F. can pulled out of something long-forgotten. This little doodad does the trick when running the PRC-74B receiver directly off solar panels, and can certainly be used for lots of other purposes. Total cost: $1.63

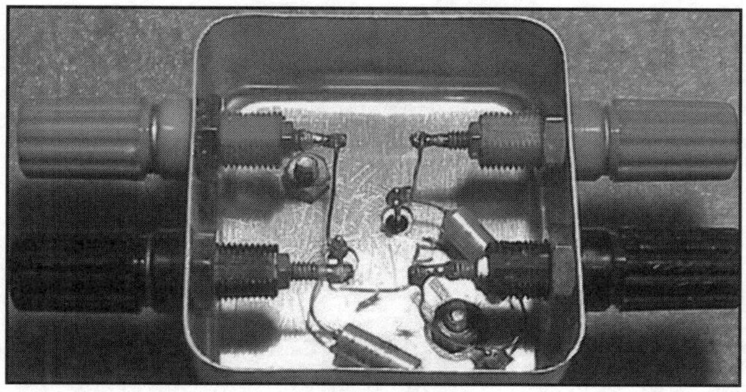

Regulator, bottom view

Finished regulator

Appendix C—Surplus Sources

Here are a few surplus dealers the author has done business with and has found to be straight-shooters. Links were current as of Winter 2005.

William Perry Company Inc.
92 Beechwood Rd. (Rear)
Louisville, KY 40207
(502) 893-8724
(502) 893-9220 (fax)
Bill has been an excellent source of military connectors for years. If you need an oddball MS-type mil connector, or other mil connectors, try him.

ESRC
1599 SW 30th Ave #4
Boynton Beach, FL 33426
(561) 735 3397
(561) 737 3541 (fax)
<http://home.att.net/~esrc/esrcmain.html>
Good source for 7 and 9-pin vacuum tubes.

The Battery Station, Inc.
215 S. Union St
Ottumwa, IA 52501
<www.batterystation.com>
(515) 683-7621/fax 7631
Good source for batteries of all types for making battery packs.

Fair Radio Sales
P.O. Box 1015
2395 St. Johns Rd
Lima, OH 45802
(419) 223-2196/227-6573
(419) 227-1313 (fax)
<www.fairradio.com>
Around since 1947, Fair is the classic surplus dealer. Their catalog is free, and a yearly must-have. A great source for complete radios, parts, and especially technical manuals.

Toronto Surplus and Scientific (Canadian company)
<http://www.torontosurplus.com/>
Source for high-end premium HF receivers and other surplus.

Murphy's Surplus Warehouse
401 N. Johnson Ave.
El Cajon, CA 92020
(619) 444 7717
FAX 444 6750
<http://www.murphyjunk.bizland.com/>
Mike Murphy's prices are often pretty high, but he has many items you won't find otherwise. He also has a 30 day right of refusal, rare in the surplus biz.

Military Surplus Repair Source.
The author's friend Jim Karlow, KA8TUR, has kindly given permission to be listed as a surplus repair source. Jim is not an electronic tech by profession, but he might as well be! Like many of us, surplus is his hobby. If you have a problem and get totally stuck, contact him to see what he can do. He handles each situation a little differently. With dealers who may acquire a number of broken sets of the same type, he'll get as many working as possible by swapping parts and troubleshooting, then retain a couple of sets as payment. For individuals, he may trade for other equipment or work something else out. Contact him by email at <jkarlow@comcast.net>

Companies the author has not dealt with, but are of interest:

Army Radio Sales Company (British company)
<http://www.armyradio.co.uk/>

Combat Radio (British Company)
<http://www.combatradio.org.uk/>

A.G. Tannenbaum (source for manuals)
<http://www.agtannenbaum.com/m_cat.html>

Suggestions for Further Research

An extensive history of U.S. military portable radios by Alan Tasker and Dennis Starks originally published in the "Military Collector Group Post," an international email magazine dedicated to the preservation of history and the equipment that made it.
http://hereford.ampr.org/history/portable.html

A compendium of military surplus home pages
http://www.pacificsites.com/~brooke/MSPHP.shtml